Lecture Notes in Mathematics

Edited by A. Dold, Heidelberg and B. Eckmann, Zürich

348

A. M. Mathai
R. K. Saxena
McGill University, Montreal/Canada

Generalized Hypergeometric Functions with Applications in Statistics and Physical Sciences

Springer-Verlag
Berlin · Heidelberg · New York 1973

AMS Subject Classifications (1970): Primary: 33-02, 33 A 30, 33 A 35
Secondary: 62 E 15, 62 H 10, 80 A 20, 94 A 05

ISBN 3-540-06482-6 Springer-Verlag Berlin · Heidelberg · New York
ISBN 0-387-06482-6 Springer-Verlag New York · Heidelberg · Berlin

Offsetdruck: Julius Beltz, Hemsbach/Bergstr.

PREFACE

This book deals with generalized hypergeometric functions. The main topic is Meijer's G-function and its applications to various practical problems in Statistical Distributions, Prior and Posterior Distributions, Characterizations of Probability Laws. Hard Limiting of Several Sinusoidal Signals in Communication Theory, Conduction of Heat and Cooling and Dual integral Equations. This book gives a balanced treatment of theory and applications. Out of the seven chapters the last three are devoted to applications. This book can be used as a text or as a reference book.

SPECIAL FEATURES:

(1) Most of the materials in Chapter 3 and all the materials in the Chapters 4 to 7 are based on recent publications and thus they are not available in any of the books in this field.

(2) Each chapter contains a list of exercises and most of these problems are from recent research papers.

(3) This is the first book in which the logarithmic cases and computable representations of Meijer's G-function are discussed in detail. Chapter 5 deals with computable representations of hypergeometric and G-functions in the most general logarithmic cases.

(4) This is the first book in which applications of Meijer's G-function in a number of topics in Statistics, Communication Theory and Theoretical Physics, are given. These results are from papers published in 1970,1971,1972 and papers to appear in 1972-73.

(5) This is the first book in which hypergeometric functions with matrix arguments, together with their application to statistical problems are discussed. These generalized functions are defined in terms of Zonal Polynomials of real symmetric matrices.

(6) On each topic the key formulas, together with the outlines of the proofs, are given in the text. From these key formulas all the results available on elementary Special Functions, which are given in various books, are available. Thus this book supplements other books and tables in this field.

(7) About 360 papers appear in the list of references and all these papers are directly or indirectly used in the text or in the exercises in the various chapters.

Chapter 1 deals with the definitions, various elementary properties and asymptotic expansions of Meijer's G-function. This chapter is mainly based on the work of Meijer, Saxena and others.

Chapter 2 gives the various elementary Special Functions which are particular cases of a G-function, representations of these elementary functions in terms of G-functions and G-functions expressed in terms of elementary functions. Results in this chapter supplement the results available in other books and tables.

In Chapter 3 the various type of integrals associated with a G-function, product of G-functions and product of G and other functions, are given. For example, we give an infinite integral involving products of two G-functions in which the argument of one of the G-functions contains rational exponents. Integrals of G-functions in which the arguments contain certain special factors of the variable are among the interesting and useful results given in this chapter.

Chapter 4 deals with the various finite and infinite summation formulas for the G-function and for the products of G and other functions. Infinite series of G-functions are summed in terms of products of gamma functions which are some interesting examples of summation formulas of this section. Such expansions have motivated the various workers to obtain the sum of infinite series of products of two G-functions in terms of a single G-function. Results of Verma, Sharma, B.L. and Abiodun are presented in this direction.

The main topic of disucssion in Chapter 5 is the computable representation of G-function in the logarithmic cases. Due to their individual importance in statistical problems three cases of G-functions, namely, $G_{0,p}^{p,0}(\cdot)$, $G_{p,p}^{p,0}(\cdot)$ and $G_{p,q}^{m,n}(\cdot)$, in their most general forms of the logarithmic cases, are discussed separately. Various statistical problems, where these representations are applicable, are also pointed out.

Chapter 6 is devoted to the applications of G-functions in the various topics in Statistics. Exact null and non-null distributions of multivariate test criteria and their representations in computable forms are discussed in some detail. Other problems considered are Prior and Posterior distributions useful in Bayesian Inference, Zonal Polynomials, hypergeometric functions of matrix arguments, product and ratio distributions, structural setup of probability laws and generalized statistical distributions.

Chapter 7 is concerned with the applications of G-functions in the conduction of heat, signal suppression in Communication Theory and dual integral equations.

Throughout this book the notations are made consistent, back references and references to later sections are kept a minimum. Most of the research workers in the field of generalized hypergeometric functions such as Meijer's G-function and H-function may not have seen the applications of these functions in Statistics and other fields because most of the papers dealing with applications are published in journals devoted to these applied fields. Therefore the authors feel that this book will generate a new awakening among the research workers in Special Functions and it will open up new realms of applications of the vast resources of theoretical results available on Special Functions. Some open problems are also pointed out in each chapter.

The authors would like to express sincere thanks to Miss Hildegard Schroeder for taking up the hard job of typing this manuscript and the National Research Council of Canada for financial assistance.

A.M. Mathai
R.K. Saxena

July 1972

TABLE OF CONTENTS

Chapter IV

Chapter V

CHAPTER I

MEIJER'S G-FUNCTION

During the last one decade Meijer's G-function has found various applications in a number of fields such as Statistical Distributions, Functional equations, characterizations, Theoretical Physics, Hydrodynamics, Hard Limiting of Sinusoidal Signals and Optimization problems. A detailed account of some of these applications are given in Chapters V, VI and VII. On account of its various interesting and important properties a number of problems of Physical Sciences are capable of being represented in terms of this function in an elegant form. Besides this the problem becomes quite general and due to greater freedom in its parameters m,n,p and q in comparison with other functions, as will be seen later, the analysis is much simplified.

The importance of this function further lies in the fact that a great many of the Special Functions occurring in Applied Mathematics. Mathematical Physics and Statistics follow as its particular cases. A study of this function will, therefore, give rise to several new results not hitherto available in the literature on Special Functions.

Throughout this chapter we have followed the original papers of Meijer. The appropriate references are cited in each section. Some of the theorems appearing in this chapter are in the same sequence as the original theorems given by Meijer. Notations are made consistent, the details of lengthy proofs are omitted and outlines of some of the proofs are given. The sequence and notations are kept so that the readers will be able to consult the original papers without spending much time in translating our notations to those of Meijer's.

Some of the extensions of the results of Meijer, various representation theorems and related topics, due to Saxena, Bhise, Sundararajan, Sharma, K.C. and others are given in the text and in the exercises at the end of this chapter as well.

In this chapter we have used gamma function and hypergeometric function. No other elementary function is used. Hence the definitions of these functions are given in Chapter II along with the definitions of other Special Functions and particular cases of G-functions expressed in terms of elementary Special Functions.

1.1 DEFINITION OF THE G-FUNCTION

In an attempt to give a meaning to the symbol ${}_pF_q$, when $p > q + 1$, Meijer introduced the G-function into Mathematical Analysis. Firstly the G-function was defined by Meijer in 1936 by means of a finite series of the generalized hypergeometric functions resembling (1.1.9). Later on, this definition was replaced by a general definition in terms of Mellin-Barnes type integral (Meijer, [205], p.83; [208]), as follows:

$$G^{m,n}_{p,q}\left(z \,\middle|\, \begin{matrix} a_1,\dots,a_p \\ b_1,\dots,b_q \end{matrix}\right) = \frac{1}{2\pi i} \int_C \chi(s)\, z^{-s} ds , \qquad (1.1.1)$$

where $i = (-1)^{1/2}$, $z \neq 0$ and

$$z^s = \exp\{s(\mathrm{Log}\,|z| + i \arg z)\} \qquad (1.1.2)$$

in which $\log|z|$ denotes the natural logarithm of $|z|$ and arg z is not necessarily the principal value,

$$\chi(s) = \frac{\prod_{j=1}^{m} \Gamma(b_j+s) \prod_{j=1}^{n} \Gamma(1-a_j-s)}{\prod_{j=m+1}^{q} \Gamma(1-b_j-s) \prod_{j=n+1}^{p} \Gamma(a_j+s)} \qquad (1.1.3)$$

where m,n,p and q are integers with

$$0 \leq n \leq p, \quad 0 \leq m \leq q \qquad (1.1.4)$$

and

$$a_j (j = 1,\dots,p), \quad b_j (j = 1,\dots,q) \qquad (1.1.5)$$

are complex numbers such that

$$a_j - b_h \neq 0,1,2,3,\dots (j = 1,\dots,n;\ h = 1,\dots,m). \qquad (1.1.6)$$

The parameters are such that the points

$$-s = (b_j + \nu)\ (j = 1,\dots,m;\ \nu = 0,1,\dots) \qquad (1.1.7)$$

and

$$-s = (a_j - \nu - 1)\ (j = 1,\dots,n;\ \nu = 0,1,\dots) \qquad (1.1.8)$$

are separated. This is possible on account of (1.1.6). An empty product is interpreted as unity. These assumptions will be retained throughout.

The existence of three different contours C is discussed by Erdelyi, A. et. al [85] which we enumerate below with slight modifications. (i) C runs from $k - i\infty$ to $k + i\infty$ such that all the points (1.1.7) lie to the right and all the points (1.1.8) to the left of it. The integral converges absolutely if $m + n - \frac{p}{2} - \frac{q}{2} > 0$ and $|\arg z| < (m + n - \frac{p}{2} - \frac{q}{2})\,\pi$. If $|\arg z| = (m+n- \frac{p}{2} - \frac{q}{2})\pi \geq 0$, the integral converges absolutely when $p = q$ if $R(\mu) < -1$; and when $p \neq q$, if with $s' = \sigma + ik$ and σ and k being real, σ is chosen so that for $k \to \pm\infty$, $(q-p)\sigma > R(\mu) + 1 - (\frac{q-p}{2})$, where

$$\mu = \sum_{j=1}^{q} b_j - \sum_{j=1}^{p} a_j .$$

(ii) C is a loop starting and ending at $+\infty$ and encircling all the points given by (1.1.7) once in the negative direction, but none of the points given by (1.1.8). The integral converges if $q \geq 1$ and either $p < q$ or $p = q$ and $|z| < 1$.

(iii) C is a loop starting and ending at $-\infty$ and encircling all the points given by (1.1.8) once in the positive direction, but none of the points given by (1.1.7). The integral converges if $p \geq 1$ and either $p > q$ or $p = q$ and $|z| > 1$.

It is further assumed that the parameters and the variable z are such that at least one of the above definitions (i), (ii) and (iii) make sense. When more than one of these definitions have a meaning, they lead to the same results and therefore no ambiguity arises.

From the definition (1.1.1) it is evident that the G-function is a symmetric function of the parameters $a_1,\ldots,a_n$; likewise of $a_{n+1},\ldots,a_p$; of $b_1,\ldots,b_m$, and of $b_{m+1},\ldots,b_q$. It is an analytic function of z.

Whenever there is no confusion we will use any one of the following notations for a G-function:

$$G(z),\ G^{m,n}_{p,q}(z),\ G^{m,n}_{p,q}\left(z\middle|{a_p \atop b_q}\right),\ G^{m,n}_{p,q}\left(z\middle|{a_1,\ldots,a_p \atop b_1,\ldots,b_q}\right).$$

Also whenever there is no confusion the generalized Hypergeometric Function will be written in one of the following forms:

$${}_pF_q(a_1,\ldots,a_p;\ b_1,\ldots,b_q;\ z),\quad {}_pF_q\left({a_1,\ldots,a_p \atop b_1,\ldots,b_q};\ z\right),$$

$${}_pF_q\left({a_p \atop b_q};\ z\right),\quad {}_pF_q(a_p;b_q;z),\quad {}_pF_q\left[{(a_p) \atop (b_q)}\ ;\ z\right].$$

For the definition of a generalized hypergeometric function, see Chapter II. The notation ${}_pF_q(+\,1)$ will mean that the parameters are $a_1,\ldots,a_p$; $b_1,\ldots,b_q$ but the variable $z = 1$.

If we assume that no two $b_j (j = 1,\ldots,m)$ differ by an integer or zero, all the poles are simple and the integral can be evaluated as a sum of residues by using (ii) as,

$$G^{m,n}_{p,q}\left(z\middle|{a_p \atop b_q}\right) = \sum_{h=1}^{m} \frac{\prod_{j=1}^{m}{}' \Gamma(b_j-b_h) \prod_{j=1}^{n} \Gamma(1+b_h-a_j)}{\prod_{j=m+1}^{q} \Gamma(1+b_h-b_j) \prod_{j=n+1}^{p} \Gamma(a_j-b_h)}\ z^{b_h}$$

$$\times\ {}_pF_{q-1}\left[\begin{matrix}1+b_h-a_1,\ldots,\ 1+b_h-a_p\\ 1+b_h-b_1,\ldots*,1+b_h-b_q\end{matrix};\ (-1)^{p-m-n}z\right], \tag{1.1.9}$$

$$p < q \quad \text{or} \quad p = q \quad \text{and} \quad |z| < 1 .$$

Here the prime in Π' denotes the omission of the factor $\Gamma(b_h-b_h)$ and the asterisk in ${}_pF_q$ indicates the omission of the parameter $1+b_h-b_h$.

Similarly, if no two of the $a_k(k = 1,\ldots,n)$ differ by an integer, then by means of the result (iii), we obtain

$$G^{m,n}_{p,q}\left(z\middle|\begin{matrix}a_p\\ b_q\end{matrix}\right)= \sum_{h=1}^{n} \frac{\prod_{j=1}^{n}\Gamma(a_h-a_j)\ \prod_{j=1}^{m}\Gamma(b_j-a_h+1)}{\prod_{j=n+1}^{p}\Gamma(1+a_j-a_h)\ \prod_{j=m+1}^{q}\Gamma(a_h-b_j)}\ z^{a_h-1}$$

$$\times\ {}_qF_{p-1}\left(\begin{matrix}1+b_1-a_h,\ldots,\ 1+b_q-a_h\\ 1+a_1-a_h,\ldots*,1+a_p-a_h\end{matrix};\ \frac{(-1)^{q-m-n}}{z}\right) \tag{1.1.10}$$

$$p > q \quad \text{or} \quad p = q \quad \text{and} \quad |z| > 1.$$

The preceding results (1.1.9) and (1.1.10) as well as the other expansions as given by Meijer ([208],[209]) available in the literature are mostly under the conditions that the poles of the integrand in (1.1.1) are simple. But in various statistical problems usually the parameters, in the G-function appearing therein, differ by integers, consequently the poles are not simple and hence the expansions (1.1.9) and (1.1.10) can be used only in very special cases. A general series expansion of computable form, when the poles are not restricted to be simple, is given in Chapter V. Various statistical and other applications of such computable representations are given in Chapters V, VI and VII.

It is worthwhile to consider separately the cases with $m = 1$ and $n = 1$, since many of the results then take simple and elegant forms. We, therefore, have

$$G^{1,n}_{p,q}\left(z\middle|\begin{matrix}a_p\\ b_q\end{matrix}\right)=\frac{\prod_{j=1}^{n}\Gamma(1+b_1-a_j)z^{b_1}}{\prod_{j=2}^{q}\Gamma(1+b_1-b_j)\ \prod_{j=n+1}^{p}\Gamma(a_j-b_1)}$$

$$\times\ {}_pF_{q-1}\left(\begin{matrix}1+b_1-a_1,\ldots,\ 1+b_1-a_p\\ 1+b_1-b_2,\ldots,1+b_1-b_q\end{matrix};\ (-1)^{p-n-1}z\right) \tag{1.1.11}$$

$$p < q \quad \text{or} \quad p = q \quad \text{and} \quad |z| < 1 .$$

$$G^{m,1}_{p,q}\left(z\,\middle|\,{a_p \atop b_q}\right) = \frac{\prod_{j=1}^{m}\Gamma(b_j-a_1+1)\; z^{a_1-1}}{\prod_{j=2}^{p}\Gamma(1+a_j-a_1)\;\prod_{j=m+1}^{q}\Gamma(a_1-b_j)}$$

$$\times\; {}_qF_{p-1}\left({1+b_1-a_1,\ldots,\, 1+b_q-a_1 \atop 1+a_2-a_1,\ldots,1+a_p-a_1}\;;\;\frac{(-1)^{q-m-1}}{z}\right), \qquad (1.1.12)$$

$p > q$ or $p = q$ and $|z| > 1$.

A particular but interesting case of the G-function has been studied by Boersma [46].

It is interesting to observe that the E-function introduced by MacRobert [157] and the G-function are connected by the relation

$$E(\alpha_1,\ldots,\alpha_p;\beta_1,\ldots,\beta_q;z) = G^{p,1}_{q+1,p}\left(z\,\middle|\,{1,\beta_1,\ldots,\beta_q \atop \alpha_1,\ldots,\alpha_p}\right). \qquad (1.1.13)$$

For convenience we will also denote $E(\alpha_1,\ldots,\alpha_p;\beta_1,\ldots,\beta_q;z)$

by $E(p;\alpha_r;q;\beta_s:z)$.

The G-function has been expressed by Sharma, K.C. [290] in a finite series of E-functions in the following forms.

$$G^{m,n}_{p,q}\left(z\,\middle|\,{a_p \atop b_q}\right) = \sum_{h=1}^{n} \frac{\pi^{m+n-q-1}\prod_{j=m+1}^{q}\sin(a_h-b_j)\pi}{\prod_{\substack{j=1\\ j\neq h}}^{p}\sin(a_h-a_j)\pi}\; z^{a_h-1}$$

$$\times\; E\left[{1-a_h+b_1,\ldots,1-a_h+b_q \atop 1-a_h+a_1,\ldots,1-a_h+a_{h-1},1-a_h+a_{h+1},\ldots,\, 1-a_h+a_p}\;;\;\frac{(-1)^{q-m-n-1}}{z}\right]$$

$$= \sum_{h=1}^{m} \frac{\pi^{m+n-p-1}\prod_{j=n+1}^{p}\sin(a_j-b_h)\pi\;\; z^{b_h}}{\prod_{\substack{j=1\\ j\neq h}}^{q}\sin(b_j-b_h)\pi}$$

$$\times\; E\left[{1+b_h-a_1,\ldots,\, 1+b_h-a_p \atop 1+b_h-b_1,\ldots,\, 1+b_h-b_{h-1},1+b_h-b_{h+1},\ldots,1+b_h-b_q}\;;(-1)^{m+n+p-1}\,\frac{1}{z}\right].$$

$$(1.1.15)$$

1.2. ELEMENTARY PROPERTIES OF THE G-FUNCTION

The properties given in this section are immediate consequences of the definition (1.1.1). These properties are used in later sections of this chapter as well as in the following chapters and hence they will be discussed here. The properties (1.2.5), (1.2.7) and (1.2.8) are important properties which are not enjoyed by many Special Functions. The results recorded in this section are due to Meijer ([205],[207]) and Saxena [258]. Asymptotic expansion of the algebraic order for the G-function is also obtained.

Property 1.2.1. Since

$$z^{\mu} = \exp(\mu \log z) = \exp(\mu \log |z| + i\mu \arg z), \tag{1.2.1}$$

it follows that the G-function is, in general, a many valued function of z with a branch point at the origin.

Property 1.2.2.

$$G^{m+1,n}_{p+1,q+1}\left(z\middle|{a_p, 1-r \atop 0, b_q}\right) = (-1)^r \; G^{m,n+1}_{p+1,q+1}\left(z\middle|{1-r, a_p \atop b_q, 1}\right), \tag{1.2.2}$$

$r = 0,1,2,\ldots$.

Property 1.2.3.

$$G^{m,n}_{p,q}\left(z\middle|{a_p \atop b_q}\right) = \frac{1}{2\pi i}\{\exp(i\pi b_{m+1})\; G^{m+1,n}_{p,q}\left(ze^{-i\pi}\middle|{a_p \atop b_q}\right)$$

$$- \exp(-i\pi b_{m+1})\; G^{m+1,n}_{p,q}\left(ze^{i\pi}\middle|{a_p \atop b_q}\right)\}, \quad m \leq q-1. \tag{1.2.3}$$

$$= \frac{1}{2\pi i}\{\exp(i\pi a_{n+1})\; G^{m,n+1}_{p,q}\left(ze^{-i\pi}\middle|{a_p \atop b_q}\right)$$

$$- \exp(-i\pi a_{n+1})\; G^{m,n+1}_{p,q}\left(ze^{i\pi}\middle|{a_p \atop b_q}\right)\}, \quad n \leq p-1. \tag{1.2.4}$$

Property 1.2.4. If r is a positive integer, then

$$G^{m,n}_{p,q}\left(z\middle|{a_p \atop b_q}\right) = (2\pi)^{(1-r)c^*}\, r^{U}\; G^{rm,rn}_{rp,rq}\left[z^r r^{r(p-q)}\middle|{\Delta(r,a_1),\ldots,\Delta(r,a_p) \atop \Delta(r,b_1),\ldots,\Delta(r,b_q)}\right], \tag{1.2.5}$$

where $c^* = m+n-\frac{p}{2}-\frac{q}{2}$, $U = \sum_{j=1}^{p} b_j - \sum_{j=1}^{q} a_j + \frac{p}{2} - \frac{q}{2} + 1$ and the symbol $\Delta(r,a)$ represents the set of r parameters

$$\frac{a}{r}, \frac{a+1}{r}, \ldots, \frac{a+r-1}{r}.$$

This formula is due to Saxena [258] and it can be established by the application of the multiplication formula for the gamma functions due to Gauss and Legendre, namely

$$\Gamma(kz) = (2\pi)^{\frac{1}{2}-\frac{k}{2}} k^{kz-\frac{1}{2}} \prod_{j=0}^{k-1} \Gamma(z+\frac{j}{k}), \tag{1.2.6}$$

$k = 2,3,4,\ldots$.

For $r = 2$, (1.2.5) reduces to a known result (Erdelyi, A. et.al [85], Vol. I, p. 209 (10)).

Property 1.2.4. An interesting and useful result which readily follows from (1.1.1) is

$$z^{\sigma}\, G^{m,n}_{p,q}(z|{}^{a_p}_{b_q}) = G^{m,n}_{p,q}(z|{}^{a_p+\sigma}_{b_q+\sigma}). \tag{1.2.7}$$

Property 1.2.5.

$$G^{m,n}_{p,q}(z|{}^{a_p}_{b_q}) = G^{n,m}_{q,p}(\frac{1}{z}|{}^{1-b_q}_{1-a_p}), \quad \arg(\frac{1}{z}) = -\arg z. \tag{1.2.8}$$

This is an important property of the G-function because it enables us to transform a G-function with $p < q$ to one with $p > q$ and vice versa. Thus in the discussion of the G-function, $p \leq q$ may be assumed without loss of generality.

Property 1.2.6. If one of the b_j's, $j = m+1,\ldots,q$ is equal to one of the a_k's, $k = 1,2,\ldots,n$, the G-function reduces to one of lower order and the parameters p,q and n are each decreased by unity. Thus

$$G^{m,n}_{p,q}(z|{}^{a_1,\ldots,a_p}_{b_1,\ldots,b_{q-1},a_1}) = G^{m,n-1}_{p-1,q-1}(z|{}^{a_2,\ldots,a_p}_{b_1,\ldots,b_{q-1}}), \tag{1.2.9}$$

$p,q,\ n \geq 1$.

An analogous formula is

$$G^{m,n}_{p,q}(z|{}^{a_1,\ldots,a_{p-1},b_1}_{b_1,\ldots,b_q}) = G^{m-1,n}_{p-1,q-1}(z|{}^{a_1,\ldots,a_{p-1}}_{b_2,\ldots,b_q}) \tag{1.2.10}$$

$p,q,\ m \geq 1$.

1.2.1. Asymptotic Expansion of Algebraic Order.

From (1.1.9) it is seen that if $p \leq q$, then

$$G^{m,n}_{p,q}(z) = O(|z|^{\alpha})$$

for small z, where $\alpha = \min R(b_h)$, $h = 1,2,\ldots,m$.

Similarly, in view of the expansion (1.1.10) we find that if $p < q, n \geq 1$, $m+n-\frac{p}{2}-\frac{q}{2} > 0$, then

$$G^{m,n}_{p,q}(z) = O(|z|^{\beta}), \tag{1.2.12}$$

for large z, with $|\arg z| < (m+n-\frac{p}{2}-\frac{q}{2})\pi$, where $\beta = \max R(a_j)-1$, $j = 1,2,\ldots,n$.

In (1.2.11) and (1.2.12) the usual convention is used. That is, a function $f(z)$ is said to be of order $g(z)$ and is written as $f(z) = O(g(z))$ as $z \to z_o$ in R if there exists a number A independent of z, such that $\left|\frac{f(z)}{g(z)}\right| \leq A$ for all z in R, where z and z_o are two points in a region R of complex plane with $g(z) \neq 0$.

1.3. DIFFERENTIAL PROPERTIES AND RECURRENCE RELATIONS

Meijer [207] has shown that

$$z^k \frac{d^k}{dz^k}\{G^{m,n}_{p,q}(z|{}^{a_p}_{b_q})\} = G^{m,n+1}_{p+1,q+1}(z|{}^{0,\ a_p}_{b_q,\ k}) \tag{1.3.1}$$

and

$$z^k \frac{d^k}{dz^k}\{G^{m,n}_{p,q}(\tfrac{1}{z}|{}^{a_p}_{b_q})\} = (-1)^k\ G^{m,n+1}_{p+1,q+1}(\tfrac{1}{z}|{}^{1-k,a_p}_{b_q,1}), \tag{1.3.2}$$

where k is a non-negative integer and $z \neq 0$. If $m = 1$ and b_1 is a non-negative integer, then the formula (1.3.1) also holds for $z = 0$.

The proof of these results are based on the formulae

$$z^k \frac{d^k}{dz^k}(z^r) = r(r-1)\ \ldots\ (r-k+1)z^r$$

$$= \frac{\Gamma(1+r)}{\Gamma(1-k+r)}\ z^r\ , \tag{1.3.3}$$

and

$$z^k \frac{d^k}{dz^k}(z^{-r}) = (-1)^k\ \frac{\Gamma(k+r)}{\Gamma(r)} z^{-r}\ , \quad \text{where } z \neq 0\ . \tag{1.3.4}$$

The equations (1.3.1) and (1.3.2) readily follow from the definition (1.1.1) in view of the results (1.3.3) and (1.3.4).

The following results due to Bhise [37] can be established in a similar manner.

$$\frac{d^k}{dz^k}\{z^{a_1-1}\ G^{m,n}_{p,q}(\tfrac{1}{z}|{}^{a_p}_{b_q})\} = (-1)^k\ z^{a_1-k-1}\ G^{m,n}_{p,q}(\tfrac{1}{z}|{}^{a_1-k,a_2,\ldots,a_p}_{b_1,\ldots,b_q}),\ n \geq 1; \tag{1.3.5}$$

$$\frac{d^k}{dz^k}\{z^{a_p-1}\,G^{m,n}_{p,q}(\tfrac{1}{z}\,|\,{}^{a_p}_{b_q})\} = z^{a_p-k-1}\,G^{m,n}_{p,q}\left(\tfrac{1}{z}\,\middle|\,\begin{matrix}a_1,\dots,a_{p-1},a_p-k\\ b_1,\dots,b_q\end{matrix}\right),\ n<p; \tag{1.3.6}$$

$$\frac{d^k}{dz^k}\{z^{-b_1}\,G^{m,n}_{p,q}(z\,|\,{}^{a_p}_{b_q})\} = (-1)^k\,z^{-b_1-k}\,G^{m,n}_{p,q}\left(z\,\middle|\,\begin{matrix}a_1,\dots,a_p\\ b_1+k,b_2,\dots,b_q\end{matrix}\right),\ m\geq 1; \tag{1.3.7}$$

$$\frac{d^k}{dz^k}\{z^{-b_q}\,G^{m,n}_{p,q}(z\,|\,{}^{a_p}_{b_q})\} = z^{-b_q-k}\,G^{m,n}_{p,q}\left(z\,\middle|\,\begin{matrix}a_1,\dots,a_p\\ b_1,\dots,b_{q-1},b_q+k\end{matrix}\right),\ m<q. \tag{1.3.8}$$

For k = 1, (1.3.5) to (1.3.8) give rise to the formulae due to Meijer [205].

$$\frac{d}{dz}\{z^{a_1-1}\,G^{m,n}_{p,q}(\tfrac{a}{z}\,|\,{}^{a_p}_{b_q})\} = -z^{a_1-2}\,G^{m,n}_{p,q}\left(\tfrac{a}{z}\,\middle|\,\begin{matrix}a_1-1,a_2,\dots,a_p\\ b_1,\dots,b_q\end{matrix}\right),\ n\geq 1; \tag{1.3.9}$$

$$\frac{d}{dz}\{z^{a_p-1}\,G^{m,n}_{p,q}(\tfrac{a}{z}\,|\,{}^{a_p}_{b_q})\} = z^{a_p-2}\,G^{m,n}_{p,q}\left(\tfrac{a}{z}\,\middle|\,\begin{matrix}a_1,\dots,a_{p-1},a_p-1\\ b_1,\dots,b_q\end{matrix}\right),\ n<p; \tag{1.3.10}$$

$$\frac{d}{dz}\{z^{-b_1}\,G^{m,n}_{p,q}(az\,|\,{}^{a_p}_{b_q})\} = -z^{-1-b_1}\,G^{m,n}_{p,q}\left(az\,\middle|\,\begin{matrix}a_1,\dots,a_p\\ 1+b_1,b_2,\dots,b_q\end{matrix}\right),\ m\geq 1; \tag{1.3.11}$$

$$\frac{d}{dz}\{z^{-b_q}\,G^{m,n}_{p,q}(za\,|\,{}^{a_p}_{b_q})\} = z^{-1-b_q}\,G^{m,n}_{p,q}\left(za\,\middle|\,\begin{matrix}a_1,\dots,a_p\\ b_1,\dots,b_{q-1},1+b_q\end{matrix}\right),\ m<q. \tag{1.3.12}$$

The above results (1.3.9) to (1.3.12) can be put into the following alternative forms:

$$z\frac{d}{dz}G^{m,n}_{p,q}(az\,|\,{}^{a_p}_{b_q}) = G^{m,n}_{p,q}\left(az\,\middle|\,\begin{matrix}a_1-1,a_2,\dots,a_p\\ b_1,\dots,b_q\end{matrix}\right) + (a_1-1)\,G^{m,n}_{p,q}(az\,|\,{}^{a_p}_{b_q}),\ n\geq 1; \tag{1.3.13}$$

$$z\frac{d}{dz}G^{m,n}_{p,q}(az\,|\,{}^{a_p}_{b_q}) = (a_p-1)\,G^{m,n}_{p,q}(az\,|\,{}^{a_p}_{b_q}) - G^{m,n}_{p,q}\left(az\,\middle|\,\begin{matrix}a_1,\dots,a_{p-1},a_p-1\\ b_1,\dots,b_q\end{matrix}\right); \tag{1.3.14}$$

$$z\frac{d}{dz}G^{m,n}_{p,q}(az\,|\,{}^{a_p}_{b_q}) = b_1\,G^{m,n}_{p,q}(az\,|\,{}^{a_p}_{b_q}) - G^{m,n}_{p,q}\left(az\,\middle|\,\begin{matrix}a_1,\dots,a_p\\ b_1+1,b_2,\dots,b_q\end{matrix}\right); \tag{1.3.15}$$

$$z\frac{d}{dz}G^{m,n}_{p,q}(az\,|\,{}^{a_p}_{b_q}) = b_q\,G^{m,n}_{p,q}(az\,|\,{}^{a_p}_{b_q}) + G^{m,n}_{p,q}\left(az\,\middle|\,\begin{matrix}a_1,\dots,a_p\\ b_1,\dots,b_{q-1},b_q+1\end{matrix}\right); \tag{1.3.16}$$

From (1.3.13) and (1.3.14) we get,

$$(a_p-a_1)\, G^{m,n}_{p,q}\left(az\,\middle|\,{a_p \atop b_q}\right) = G^{m,n}_{p,q}\left(az\,\middle|\,{a_1-1,a_2,\ldots,a_p \atop b_1,\ldots,b_q}\right) + G^{m,n}_{p,q}\left(az\,\middle|\,{a_1,\ldots,a_{p-1},a_p-1 \atop b_1,\ldots,b_q}\right)$$

$$1 \leq n \leq p-1. \qquad (1.3.17)$$

Similarly from (1.3.13) and (1.3.15) we obtain

$$(1-a_1+b_1)\, G^{m,n}_{p,q}\left(z\,\middle|\,{a_p \atop b_q}\right) = G^{m,n}_{p,q}\left(z\,\middle|\,{a_1-1,a_2,\ldots,a_p \atop b_1,\ldots,b_q}\right)$$

$$+ \; G^{m,n}_{p,q}\left(z\,\middle|\,{a_1,\ldots,a_p \atop b_1+1,b_2,\ldots,b_q}\right), \quad m,n \geq 1. \qquad (1.3.18)$$

Finally four relations similar to the ones given above, can be easily deduced by subtracting (1.3.15) from each of the results (1.3.12), (1.3.13) and (1.3.14) respectively.

1.4 ASYMPTOTIC EXPANSIONS FOR $G^{m,n}_{p,q}(z)$ IN SPECIAL CASES.

Barnes [34] has given the asymptotic expansion of the functions $G^{q,1}_{p,q}(z)$ and $G^{q,0}_{p,q}(z)$ which will be discussed in this section. The results of this section are applicable in obtaining the asymptotic expansion of the general G-function. It may be observed here that the formulas of Barnes [34] are written in a different form. The various symbols used here do not appear in his papers. The results described in this section are based on the work of Meijer [208].

In the discussion that follows the poles of the integrands of G-functions under consideration are assumed to be simple.

In a contracted form $G^{p,1}_{p,q}(z\|a_t)$ will be used to denote the function,

$$G^{q,1}_{p,q}\left(z\,\middle|\,{a_t,a_1,\ldots,a_{t-1},a_{t+1},\ldots,a_p \atop b_1,\ldots,b_q}\right) = G^{q,1}_{p,q}(z\|a_t). \qquad (1.4.1)$$

The following definitions are needed in order to simply the presentation of the theorems that follow.

Definitions. We formally write

$$E_{p,q}(z\|a_t) = \frac{z^{a_t-1}\prod_{j=1}^{q}\Gamma(1+b_j-a_t)}{\prod_{\substack{j=1\\ j\neq t}}^{p}\Gamma(1+a_j-a_t)}\; {}_qF_{p-1}\left({1+b_1-a_t,\ldots,1+b_q-a_t \atop 1+a_1-a_t,\ldots*\ldots,1+a_p-a_t}; \pm z^{-1}\right), \qquad (1.4.2)$$

where p,q and t are integers with $1 \leq t \leq p < q$ and

$a_t - b_j \neq 1,2,3,\ldots$ $(t = 1,\ldots,n;\ j = 1,\ldots,q)$.

We also need the condition

$$a_t - b_j \neq 1,2,3,\ldots \quad (t = 1,\ldots,n;\ j = 1,\ldots,m). \tag{1.4.3}$$

From (1.4.2) it readily follows that

$$E_{p,q}[z \exp(2i\pi\alpha) \| a_t] = \exp(2i\pi\alpha\, a_t)\, E_{p,q}(z \| a_t). \tag{1.4.4}$$

When $q = p$, we have

$$E_{p,p}(z\|a_t) = G^{1,p}_{p,p}\left(z^{-1}\middle|{1-b_1,\ldots,1-b_p \atop 1-a_t, 1-a_1,\ldots,1-a_{t-1}, 1-a_{t+1},\ldots,1-a_p}\right) \tag{1.4.5}$$

$$= \frac{z^{a_t - 1} \prod_{j=1}^{p} \Gamma(1+b_j - a_t)}{\prod_{\substack{j=1 \\ j\neq t}}^{p} \Gamma(1+a_j - a_t)}$$

$$\times\ {}_pF_{p-1}\left({1+b_1-a_t,\ldots,1+b_p-a_t\ ; \atop 1+a_1-a_t,\ldots*\ldots,1+a_p-a_t;\ -z^{-1}}\right). \tag{1.4.5}$$

The symbol $\Delta^{m,n}_q(t)$ will be used to denote the following quantity.

$$\Delta^{m,n}_q(t) = \pi^{m+n-q-1} \frac{\prod_{j=m+1}^{q} \sin(b_j - a_t)\pi}{\prod_{\substack{j=1 \\ j\neq t}}^{n} \sin(a_j - a_t)\pi} \tag{1.4.6}$$

$$= (-1)^{m+n-q-1} \frac{\prod_{j=1}^{n}{}' \{\Gamma(a_t - a_j)\Gamma(1+a_j - a_t)\}}{\prod_{j=m+1}^{q} \{\Gamma(a_t - b_j)\ \Gamma(1+b_j - a_t)\}} \tag{1.4.7}$$

where m,n,p,q and t are integers with $1 \leq t \leq n$, $0 \leq m \leq q$ and

$$a_j - a_t \neq 0, \pm 1, \pm 2,\ldots,(j,t = 1,\ldots,n;\ j \neq t). \tag{1.4.8}$$

The following condition is also required in the analysis.

$$a_j - a_h \neq 0, \pm 1, \pm 2,\ldots,(j,h = 1,\ldots,p;\ j \neq h). \tag{1.4.9}$$

When no confusion arises, we simply write $\Delta(t)$ for $\Delta_q^{m,n}(t)$.

In order to discuss the nature of some asymptotic expansions we employ a notation $H_{p,q}(z)$ defined as,

$$H_{p,q}(z) = [\exp\{(p-q)z^{\frac{1}{q-p}}\}]\, z^{\rho^*} \left\{ \frac{(2\pi)^{\frac{q-p-1}{2}}}{(q-p)^{1/2}} + \frac{M_1}{z^{\frac{1}{q-p}}} + \frac{M_2}{z^{\frac{2}{q-p}}} + \ldots \right\} \tag{1.4.10}$$

where

$$\rho^* = \frac{1}{q-p}\left[\sum_{j=1}^{q} b_j - \sum_{j=1}^{p} a_j + \frac{p-q+1}{2} \right], \tag{1.4.11}$$

and $M_1, M_2, \ldots$ are independent of z but are complicated functions of the parameters a_j and b_h ($j = 1,\ldots,p$; $h = 1,\ldots,q$). For details, see [34].

Remark: It is interesting to observe, in view of (1.4.10) and (1.4.11), that the function

$H_{p,q}(z)$ satisfies the equation

$$H_{p,q}(z) = (-1)^{p-q+1} \left[\exp\{2\pi i \left(\sum_{j=1}^{q} b_j - \sum_{j=1}^{p} a_j \right)\}\right]$$

$$\times H_{p,q}[z \exp\{i\pi(2p-2q)\}].$$

Theorem 1.4.1. If $1 \leq t \leq p < q$, then the asymptotic expansion

$$G_{p,q}^{q,1}(z\|a_t) \sim E_{p,q}(z\|a_t), \tag{1.4.12}$$

holds for large values of $|z|$ with $|\arg z| < (1 + \frac{q-p}{2})\pi$.

Theorem 1.4.2. If m,n,p and q are integers with $1 \leq m \leq q$, $1 \leq n \leq p < q$, $m+n - \frac{p}{2} - \frac{q}{2} > 0$ and the conditions (1.4.3) and (1.4.8) are satisfied, then

$$G_{p,q}^{m,n}(z) \sim \sum_{t=1}^{n} \exp[i\pi a_t(m+n-q-1)]$$

$$\times \Delta_q^{m,n}(t)\, E_{p,q}[z \exp\{(q-m-n+1)i\pi\}\|a_t], \tag{1.4.13}$$

for large values of $|z|$ with $|\arg z| < (m+n- \frac{p}{2} - \frac{q}{2})\pi$.

Theorem 1.4.3. The function $G_{p,q}^{q,0}(x)$ admits for large values of $|z|$ with $|\arg z| < (q-p+\overset{*}{\epsilon})\pi$, the asymptotic expansion

$$G^{q,0}_{p,q}\left(z\Big|{a_p \atop b_q}\right) \sim H_{p,q}(z) \tag{1.4.14}$$

where $H_{p,q}(z)$ is defined in (1.4.10), $\epsilon^* = \frac{1}{2}$ if $q = p+1$ and $\epsilon^* = 1$ if $q \geq p+2$.

Theorem 1.4.4. If $m+n \geq p+1$, $|\arg z| < (m+n-p)\pi$ and the conditions (1.4.3) and (1.4.8) hold, then the function $G^{m,n}_{p,q}(z)$ can be continued analytically outside the circle $|z| = 1$ by the expression

$$G^{m,n}_{p,p}(z) = \sum_{t=1}^{n} \exp\left[-i\pi a_t(p-m-n+1)\right]\Delta^{m,n}_{p}(t)$$

$$\times E_{p,p}\left[z \exp\{i\pi(p-m-n+1)\}\|a_t\right]. \tag{1.4.15}$$

An interesting particular case of this theorem for $m = p$ and $n = 1$ can be stated as follows.

Theorem 1.4.4a. If $|\arg z| < \pi$ and (1.4.3) is satisfied with $n = 1$ and $m = p$, then $G^{p,1}_{p,p}(z\|a_t)$ can be continued analytically outside the unit circle by the equation

$$G^{p,1}_{p,p}(z\|a_t) = E_{p,p}(z\|a_t). \tag{1.4.16}$$

Remark 1: The asymptotic expansion discussed in Sections 1.4 and 1.8 have applications in Statistical theory of Distributions as well. The literature on asymptotic statistical distributions is vast and they won't be given in this book. In Chapters V and VI we have concentrated on exact distributions. As indicated in these chapters, the exact distributions are representable in terms of G-functions and thus the result on asymptotic expansion of G-functions are directly applicable to these problems in Statistics.

Remark 2: A further discussion of the analytic continuation of the function $G^{m,n}_{p,p}(z)$ can be found in Sections 1.6 and 1.8.

1.5 DIFFERENTIAL EQUATION SATISFIED BY $G^{m,n}_{p,q}(z)$.

By adopting a procedure similar to that used by Barnes [35] to establish the Gauss's hypergeometric function differential equation, it can be shown that the function $u = G^{m,n}_{p,q}(z)$ satisfies the homogeneous linear differential equation

$$\left[(-1)^{p-m-n}\, z \prod_{j=1}^{p} (\theta - a_j+1) - \prod_{j=1}^{q} (\theta - b_j)\right] u = 0. \tag{1.5.1}$$

where the operator $\theta = z\frac{d}{dz}$.

This equation is of degree max (p,q) and as remarked earlier we may assume $p \leq q$. If $p < q$, the only singularities of (1.5.1) are $z = 0$ and $z = \infty$. The former is regular and the latter an irregular one. In case $p = q$, then $z = 0$ and $z = \infty$ are regular singularities, and besides there is a regular singularity at $z = (-1)^{m+n-p}$. The fundamental solutions of (1.5.1) in the neighbourhood of $z = (-1)^{m+n-p}$ are not recorded in the literature. For this case, see the work of Norlund [222]. In the remaining cases, Meijer ([208], pp.354-356) has completely investigated the fundamental solutions of (1.5.1) which are enumerated below.

From (1.1.9) it readily follows that the q-functions,

$$\exp[(m+n-p-1)i\pi b_h]\, G^{1,p}_{p,q}\left[z \exp\{(p-m-n+1)i\pi\} \,\Big|\, \begin{matrix} a_1,\dots,a_p \\ b_h, b_1,\dots,b_{h-1}, b_{h+1},\dots,b_q \end{matrix}\right]$$

$$= \frac{\prod_{j=1}^{p} \Gamma(1+b_h-a_j)}{\prod_{\substack{j=1\\ j\neq h}}^{q} \Gamma(1+b_h-b_j)}\, z^{b_h}\; {}_pF_{q-1}\left[\begin{matrix} 1+b_h-a_1,\dots,1+b_h-a_p; (-1)^{p-m-n} z \\ 1+b_h-b_1,\dots *\dots, 1+b_h-b_q \end{matrix}\right] \tag{1.5.2}$$

where $h = 1,\dots,q$ satisfy the differential equation (1.5.1) provided $b_h - b_j \neq 0, \pm 1, \pm 2, \dots$. ($h = 1,\dots,q$; $j = 1,\dots,q$; $h \neq j$). The functions given by (1.5.2) are linearly independent and, therefore, form a fundamental system of solutions of (1.5.1) valid in the neighbourhood of $z = 0$. When the poles are of higher order, G-functions can be expressed in terms of series associated with ψ-function, generalized Zeta-function and logarithmic functions. In this connection see Chapter V.

In order to determine the fundamental solutions of (1.5.1) in the neighbourhood of the irregular singularity at $z = \infty$, the following cases are discussed.

<u>Case I, $p < q$:</u> Meijer has determined two integers λ and ω for every value of arg z such that

$$\left(\frac{p-q}{2} - 1\right)\pi < \arg z + (q-m-n-2\lambda + 1)\pi < \left(\frac{q-p}{2} + 1\right)\pi, \tag{1.5.3}$$

$$(p-q-\epsilon^*)\pi < \arg z + (q-m-n-2\psi)\pi < (q-p+\epsilon^*)\pi, \tag{1.5.4}$$

for $\psi = \omega, \omega + 1, \dots, \omega + q-p-1$, where ϵ^* is defined in (1.4.14)

If the condition (1.4.9) is satisfied, then the equation (1.5.1) is satisfied by the p-functions

$$G^{q,1}_{p,q}[z \exp\{i\pi(q-m-n-2\lambda+1)\} \| a_t], \quad \text{for } t = 1,\dots,p. \tag{1.5.5}$$

It appears from the results (1.4.2), (1.4.12) and (1.5.3) that these functions tend algebraically to zero or to ∞ for large values of $|z|$. Further, it is also evident in view of (1.4.2) and (1.4.9) that they have a mutually different algebraic behaviour for large values of $|z|$. Thus they form a system of p linearly independent solutions of (1.5.1). If, however, the condition (1.4.9) is not satisfied, that is the poles are of higher orders, we can get independent solutions by employing the techniques of Chapter V and with the help of the method of Frobenius. But these linearly independent solutions have not been investigated so far.

To investigate the remaining q-p solutions, we consider the functions

$$G^{q,0}_{p,q} \; [z \exp\{(q-m-n-2\psi)\pi i\}] \;, \tag{1.5.6}$$

$(\psi = \omega, \omega + 1, \ldots, \omega + q-p-1)$.

Evidently the differential equation (1.5.1) is satisfied by them. From (1.5.4) and (1.4.14) it follows that they tend exponentially to zero or to ∞ as $|z| \to \infty$ and that they are mutually linearly independent. Hence they provide the q-p remaining solutions of (1.5.1).

Thus it is observed that if $p < q$ and the conditions (1.4.9), (1.5.3) and (1.5.4) are fulfilled then the p-functions (1.5.5) and the q-p functions (1.5.6) provide a fundamental system of solutions of (1.5.1) valid in the vicinity of $z=\infty$.

Case II, $p = q$: Let us assume that the condition (1.4.9) and

$$\arg z + (p-m-n)\pi \neq 0, \pm 2\pi, \pm 3\pi, \ldots \tag{1.5.7}$$

are satisfied.

An integer λ can be determined from (1.5.3) such that

$$-\pi < \arg z + (p-m-n-2\lambda + 1)\pi < \pi \,. \tag{1.5.8}$$

Clearly the function

$$G^{p,1}_{p,p}[z \exp\{(p-m-n-2\lambda + 1)\pi i\}\|a_t] \;, \tag{1.5.9}$$

for $t = 1,2,\ldots,p$ satisfy (1.5.1) with $q = p$. In view of (1.4.5),(1.4.9) and (1.4.16) these solutions are also linearly independent.

Remark: In certain results there may occur not the p-functions given by (1.5.5) and (1.5.9) but only the first n of these functions, namely

$$G^{q,1}_{p,q} \; [z \exp \{i\pi(1+q-m-n-2\lambda)\}\|a_t], \; (t = 1,2,\ldots,n) \tag{1.5.10}$$

and

$$G^{p,1}_{p,p} \; [z \exp \; i\pi(1+p-m-n-2\lambda)\} \; \|a_t], \; (t = 1,2,\ldots,n) \tag{1.5.11}$$

For these functions to be linearly independent, it is not necessary that the condition (1.4.9) is satisfied. In this case this condition may be replaced by the less stringent condition (1.4.8).

On the other hand if we have to deal with less than q-p functions $G^{q,0}_{p,q}(z)$, namely with the $\sigma(\sigma < q-p)$ functions.

$$G^{q,0}_{p,q}(z[\exp (q-m-n-2\psi)i\pi]) \tag{1.5.12}$$

$$(\psi = \omega, \omega + 1, \ldots, \omega + \sigma - 1)$$

then evidently the condition (1.5.4) need not be satisfied for $\psi = \omega, \omega+1, \ldots$ $\ldots, \omega + q-p-1$ but only for $\psi = \omega, \omega + 1, \ldots, \omega + \sigma - 1$.

Remark 2: Nair [215] introduced the method of differential equations in solving some problems in Statistical Distributions. But from the discussion in Chapter VI it will be seen that the distributions under consideration can be written in terms of G-functions. Hence the topics discussed in this section are directly applicable in such statistical problems. Recently some authors have laboured a lot in obtaining differential equations for some statistical density functions. But these are directly available from (1.5.1). A detailed discussion of the different techniques used in Statistics including the method of differential equations can be seen from Mathai [185].

1.6. ANALYTIC CONTINUATION OF $G^{m,n}_{p,q}(z)$.

We have already seen that

$$G^{m,n}_{p,q}(z|{}^{a_p}_{b_q}) = G^{n,m}_{q,p}(\tfrac{1}{z}\,|{}^{1-b_q}_{1-a_p}), \quad \arg(\tfrac{1}{z}) = -\arg z . \tag{1.6.1}$$

As remarked earlier this result is of importance in the study of the G-function and without loss of generality, we may assume $p \leq q$. If $p = q$ and $0 < |z| < 1$ in (1.6.1) then the contour C is given by (ii) in Section 1.1 whereas if $|z| > 1$ then contour C is defined by (iii) in Section 1.1. In case C is a contour described in (i) of Section 1.1, we have a representation of the function $G^{m,n}_{p,q}(z)$ valid for all $z, z \neq 0$ provided that $m + n - p \geq 1$ and $|\arg z| < (m+n-p)\pi$. When $|z| < 1$, we may without altering the value of the integral, bend round the contour in (i) so that it coincides with that of (ii).

From (1.5.1) it readily follows that the linear differential equation of order p, namely

$$[(-1)^{p-m-n}z \prod_{j=1}^{p} (\theta - a_j + 1) - \prod_{j=1}^{p} (\theta - b_j)] u = 0, \tag{1.6.2}$$

where θ denotes the operator $z\frac{d}{dz}$, is satisfied by the function $u = G^{m,n}_{p,p}(z)$. It can be easily seen that every point in the finite part of the z-plane is an ordinary point of (1.6.2) with the exception of $z = 0$ and $z = (-1)^{m+n-p}$, being regular singularities of the differential equation. A linear representation of the function $G^{m,n}_{p,p}(z)$ in terms of linear combinations of the type $z^{\chi}\ {}_pF_{p-1}[(-1)^{p-m-n}z]$ can be seen from (1.1.9) for $|z| < 1$.

In order to define the function $G^{m,n}_{p,p}(z)$ outside the unit circle we introduce a cross-cut in the z-plane along the straight line which joins

$$(-1)^{m+n-p} \text{ to } (-1)^{m+n-p}\ (1+ \infty\ e^{i\mu}),\ -\frac{\pi}{2} \leq \mu \leq \frac{\pi}{2}\ ,$$

where μ is usually taken as zero. Then in the cut plane $G^{m,n}_{p,p}(z)$ has no singularity except the branch point at the origin. Now, if $m+n-p \geq 2$, the point $(-1)^{m+n-p}$ lies in the interior of the sector $|\arg z| < (m+n-p)\pi$. So if we confine ourselves to values of z for which $|\arg z| < (m+n-p)\pi$, the function $G^{m,n}_{p,p}(z)$ with $m+n-p \geq 2$ has no singularity at $z = (-1)^{m+n-p}$. Hence if $m+n-p \geq 2$ and $|\arg z| < (m+n-p)\pi$, the cross-cut is not necessary. This shows that the function $G^{m,n}_{p,p}(z)$ with $m+n-p \geq 1$ and $|\arg z| < (m+n-p)\pi$ can be continued analytically from the inside of the unit circle with centre at the origin to outside the unit circle by means of (1.1.9) with $q = p$.

In all other cases, in continuing the function $G^{m,n}_{p,p}(z)$ outside the unit circle, we will make the above mentioned cross-cut.

1.7 SOME EXPANSION FORMULAS OF G-FUNCTIONS AND THEIR PARTICULAR CASES.

In this section we discuss four important expansion formulas of the G-function which express $G^{m,n}_{p,q}(z)$ as a finite series of related G-functions of the type $G^{k,\rho,n}_{p,q}(z\|a_t)$. These expansion formulas are the most powerful tool in the investigation of the asymptotic expansion of the G-function and in studying its various properties. For $\rho = 1$ and $k = q$ each expansion formula reduces to one important theorem which in turn gives rise to one or more associated theorems which express the conditions under which an expansion formula represents $G^{m,n}_{p,q}(z)$ in terms of the fundamental solutions of (1.5.1) valid near $z = \infty$. These results in conjunction with the asymptotic expansion of the G-functions $G^{q,1}_{p,q}(z)$ and $G^{q,0}_{p,q}(z)$ discussed in Section 1.4 yield the various asymptotic expansions of the G-function discussed in Section 1.8. Summation formulas involving finite or infinite number of G-functions are given in Chapter IV.

1.7.1 Definitions and Notations Used.

In all the definitions that follow it is assumed that the parameters are such that the definitions make sense. An empty product will be regarded as unity and an empty sum as zero.

In Sections 1.7 and 1.8 the following notations will be used.

$\sigma^* = m+n-k-\rho$

$\nu^* = m+n-p,\ \mu^* = q-m-n$

$$c^* = m+n-\frac{p}{2}-\frac{q}{2} = \frac{\nu^*-\mu^*}{2}$$

$$d^* = \frac{3q}{2}-\frac{p}{2}-m-n+2 \qquad (1.7.1)$$

$$\epsilon^* = \frac{1}{2} \text{ if } q-p = 1 \text{ and } \epsilon^* = 1 \text{ if } q-p > 1.$$

Let m,n,p and q be integers satisfying the conditions $0 \leq m \leq q, 0 \leq n \leq p$. We give the following definitions.

$$A = A_q^{m,n} = (-2\pi i)^{-\mu^*} e^{i\pi(\sum_{j=1}^{n} a_j - \sum_{j=m+1}^{q} b_j)} ,$$

$$B = B_p^{m,n} = (-2\pi i)^{\nu^*} e^{i\pi(\sum_{j=1}^{m} b_j - \sum_{j=n+1}^{p} a_j)} , \qquad (1.7.2)$$

$$\overline{A} = \overline{A}_q^{m,n} = (2\pi i)^{-\mu^*} e^{-i\pi(\sum_{j=1}^{n} a_j - \sum_{j=n+1}^{q} b_j)} ,$$

$$\overline{B} = \overline{B}_q^{m,n} = (2\pi i)^{\nu^*} e^{-i\pi(\sum_{j=1}^{m} b_j - \sum_{j=n+1}^{p} a_j)} . \qquad (1.7.3)$$

The function $G_{p,q}^{k,\rho,n}(z\|a_t)$ is defined as follows:

Suppose $1 \leq \rho \leq n \leq p \leq q,\ 1 \leq t \leq n-\rho+1,\ 0 \leq k \leq q,$

then $G_{p,q}^{k,\rho}[z|{a_t,a_{n-\rho+2},\ldots,a_p,a_1,\ldots,a_{t-1},a_{t+1},\ldots,a_{n-\rho+1} \atop b_1,\ldots,b_q}] = G_{p,q}^{k,\rho,n}(z\|a_t).$

We also define

$$G_{p,q}^{k,\rho-1}[z|{a_{n-\rho+2},\ldots,a_p,a_1,\ldots,a_{n-\rho+1} \atop b_1,\ldots,b_q}] = G_{p,q}^{k,\rho-1,n}(z) ,$$

where $q \geq 1,\ 0 \leq \rho-1 \leq n \leq p \leq q$ and $0 \leq k \leq q$.

As $G_{p,q}^{k,\lambda}(z)$ is a symmetric function of $a_{\lambda+1},\ldots,a_p$, the functions $G_{p,q}^{k,\rho}(z)$ and $G_{p,q}^{k,\rho-1}(z)$ defined above are independent of n if $\rho = 1$.

Hence it is seen that

$$G_{p,q}^{k,0,n}(z) = G_{p,q}^{k,0}(z|{a_1,\ldots,a_p \atop b_1,\ldots,b_q}) = G_{p,q}^{k,0}(z) \qquad (1.7.4)$$

and

$$G_{p,q}^{k,1,n}(z\|a_t) = G_{p,q}^{k,1}(z|{a_t,a_1,\ldots,a_{t-1},a_{t+1},\ldots,a_p \atop b_1,\ldots,b_q}).$$

Hence the symbol $G_{p,q}^{k,1}(z\|a_t)$ will be used to denote

$$G^{k,1}_{p,q}\left(z\,\middle|\,\begin{matrix}a_t, a_1, \ldots, a_{t-1}, a_{t+1}, \ldots, a_p\\ b_1, \ldots, b_q\end{matrix}\right) \quad (1 \le t \le p)$$

Consequently,

$$G^{k,1,n}_{p,q}(z\|a_t) = G^{k,1}_{p,q}(z\|a_t)\ .$$

If in any expression i is replaced by -i, then the symbol representing that expression will be written with bar over it.

The symbol $\omega^{m,n}_{q}(t)$, $L^{m,n}_{p}(\lambda)$, $\varphi^{m,n}_{p,q}(h,\lambda)$, $\psi^{m,n}_{p,q}(h,\lambda)$, $\theta^{m,n}_{p}(\rho,r)$, are defined repectively as the coefficient of x^t in the expansion of

$$\frac{\prod_{j=m+1}^{q} [1-x\exp(2\pi i b_j)]}{\prod_{j=1}^{n} [1-x\exp(2\pi i a_j)]}\ , \tag{1.7.5}$$

coefficient of x^{λ} in the expansion of

$$\frac{\prod_{j=n+1}^{p} [1-x\exp(2\pi i a_j)]}{\prod_{j=1}^{m} [1-x\exp(2\pi i b_j)]}\ , \tag{1.7.6}$$

coefficient of $x^{h+\lambda-1}$ in the expansion of

$$\left[\sum_{r=o}^{h-1} \omega^{0,p}_{q}(r)x^r\right]\left[\frac{\prod_{j=n+1}^{p} [1-x\exp(2\pi i a_j)]}{\prod_{j=1}^{m} [1-x\exp(2\pi i b_j)]}\right]\ . \tag{1.7.7}$$

coefficient of x^{λ} in the expansion of

$$\left\{\sum_{r=1}^{h} \bar{\omega}^{0,p}_{q}(h-r)x^r\right\}\left\{\prod_{j=n+1}^{p} \frac{[1-x\exp(2\pi i a_j)]}{[1-x\exp(2\pi i b_j)]}\right\}, \tag{1.7.8}$$

and the coefficient of x^r in the expansion of

$$\prod_{j=n+1}^{p} \frac{[1-x\exp(2\pi i a_j)]}{[1-x\exp(2\pi i a_\rho)\ \prod_{j=1}^{m} [1-x\exp(2\pi i b_j)]}\ . \tag{1.7.9}$$

We now state some lemmas given by Meijer. For the proofs, the reader is referred to the original work of Meijer [208]. There are several properties of these symbols but for the sake of brevity, we are giving here only a few interesting ones.

Lemma 1.7.1. Let m,n and q be integers with $0 \le m \le q$, $n \ge 0$ and r is an arbitrary integer, the condition

$$a_j - a_t \neq 0, \pm 1, \pm 2, \ldots, \quad (j = 1,\ldots,n;\ t = 1,\ldots,n;\ j \neq t)$$

is satisfied, then

$$\sum_{t=1}^{n} \exp\,[i\pi a_t(2r - \mu^*)]\Delta_q^{m,n}(t)$$
$$= -\frac{1}{2\pi i}\{A_q^{m,n}\,\omega_q^{m,n}(r) - \bar{A}_q^{m,n}\,\bar{\omega}_q^{m,n}(\mu^* - r)\}. \qquad (1.7.10)$$

Lemma 1.7.2. Let m,n,q and λ be integers with $0 \le m \le q$, $0 \le \lambda \le \mu^* - 1$, and $n \ge 1$ holds and

$$a_t - \beta \neq 0, \pm 1, \pm 2, \ldots \quad (t = 1,\ldots,n)$$

then

$$\sum_{t=1}^{n} \exp\,[-i\pi a_t(2\lambda + \mu + 1)]\,\frac{\Delta(t)}{\sin(\beta - a_t)\pi}$$
$$= -\pi^{-\mu^*-1}\exp[-i\pi\beta(2\lambda + \mu + 1)]\,\frac{\prod_{j=m+1}^{q}\sin(b_j - \beta)\pi}{\prod_{j=1}^{n}\sin(a_j - \beta)\pi} \qquad (1.7.11)$$

Taking $a_{n+1} = \beta$ in (1.7.10) and noting that $\Delta_q^{m,n+1}(t)$ with $1 \le t \le n$ and $a_{n+1} = \beta$ is equal to $\frac{\pi\,\Delta(t)}{\sin(\beta - a_t)\pi}$ the result (1.7.11) follows.

An interesting particular case of (1.7.11) is given below.

Lemma 1.7.3. Let m,n,q,h and λ be integers with $1 \le m+1 \le h \le q$ and $0 \le \lambda \le -\mu^* - 1$ and the conditions (1.4.8) and $a_t - b_h \neq 0, \pm 1, \pm 2, \ldots$ $(t = 1,\ldots,n;\ m+1 \le h \le q)$ are satisfied, then

$$\sum_{t=1}^{n} \frac{\exp[-i\pi a_t(2\lambda + \mu^* + 1)]\Delta(t)}{\sin(b_h - a_t)\pi} = 0\,. \qquad (1.7.12)$$

The result (1.7.12) follows from (1.7.11).

Now the coefficient $D_{p,q}^{m,n}(\lambda)$ is defined as

$$D_{p,q}^{m,n}(\lambda) = (-2\pi i)^{p-q}\exp\{i\pi(\sum_{j=1}^{p} a_j - \sum_{j=1}^{q} b_j\}$$
$$\times [B\,L_p^{m,n}(\lambda) - B\,\bar{L}_q^{m,n}(\lambda - \nu^*)]\,, \qquad (1.7.13)$$

where λ is an arbitrary integer, $0 \le m \le q$ and $0 \le n \le p$.

When

$$b_j - b_s \neq 0, \pm 1, \pm 2, \ldots \quad (j = 1, \ldots, m; \; s = 1, \ldots, m; \; j \neq s)$$

$D^{m,n}_{p,q}(\lambda)$ may be put in the form

$$D^{m,n}_{p,q}(\lambda) = (-2i)^{p-q+1} \pi^{-\mu^*} \exp[i\pi(\sum_{j=1}^{p} a_j - \sum_{j=1}^{q} b_j)]$$

$$\times \sum_{r=1}^{m} \exp[i\pi b_r(2\lambda + \nu^*)] \frac{\prod_{j=n+1}^{p} \sin(a_j - a_r)\pi}{\prod_{\substack{j=1 \\ j\neq r}}^{m} \sin(b_j - b_r)\pi}. \tag{1.7.14}$$

Finally the coefficients $R^{m,n}_{p,q}(h,\lambda)$ and $T^{m,n}_{p,q}(\rho,\lambda)$ are defined in the following manner.

$$R^{m,n}_{p,q}(h,\lambda) = A^{m,n}_{q} \phi^{m,n}_{p,q}(h,\lambda) - A^{0,p}_{q} \overline{B}^{m,n}_{p} \overline{\psi}^{m,n}_{p,q}(h,-\lambda-\nu^* + 1). \tag{1.7.15}$$

$$T^{m,n}_{p,q}(\rho,\lambda) = -\{\exp(i\pi a_\rho) B^{m,n}_{p} \Theta^{m,n}_{p}(\rho,\lambda-1)$$

$$+ \exp[-i\pi a_\rho] \overline{B}^{m,n}_{p} \overline{\Theta}^{m,n}_{p}(\rho, -\nu^*-\lambda)\} \Delta^{0,p}_{q}(\rho)$$

$$+ \exp[i\pi a_\rho(\nu^* + 2\lambda - 1)] \Delta^{m,n}_{q}(\rho), \quad 1 \leq \rho \leq n. \tag{1.7.16}$$

$$T^{m,n}_{p,q}(\rho,\lambda) = -\{\exp(i\pi a_\rho) B^{m,n}_{p} \Theta^{m,n}_{p}(\rho,\lambda-1)$$

$$+ \exp[-i\pi a_\rho] \overline{B}^{m,n}_{p} \overline{\Theta}^{m,n}_{p}(\rho,-\nu^*-\lambda)\} \Delta^{0,p}_{q}(\rho), \quad n+1 \leq \rho \leq p. \tag{1.7.17}$$

When $b_j - b_r \neq 0, \pm 1, \pm 2, \ldots$ $(j = 1, \ldots m; \; r = 1, \ldots, m; \; j \neq r)$, we have the following series representation for the coefficient $T^{m,n}_{p,q}(\rho,\lambda)$.

$$T^{m,n}_{p,q}(\rho,\lambda) = \frac{\pi^{-\mu^*-1}}{\prod_{\substack{j=1 \\ j\neq\rho}}^{p} \sin(a_j - a_\rho)\pi} \sum_{r=1}^{m} \exp[i\pi b_r(\nu^* + 2\lambda - 1)]$$

$$\times \frac{\{\prod_{j=n+1}^{p} \sin(a_j - a_r)\pi\}\{\prod_{\substack{j=1 \\ j\neq r}}^{q} \sin(b_j - a_\rho)\pi\}}{\prod_{\substack{j=1 \\ j\neq r}}^{m} \sin(b_j - b_r)\pi} \tag{1.7.18}$$

The following are the properties of the various symbols defined in this section which follow directly from their definitions.

$$\omega_q^{m,n}(t) = \bar{\omega}_q^{m,n}(t) = 0, \quad \text{for } t = -1,-2,-3,\ldots;$$

$$\omega_q^{m,n}(0) = \bar{\omega}_q^{m,n}(0) = 1 \qquad (1.7.19)$$

$$L_p^{m,n}(\lambda) = \bar{L}_p^{m,n}(\lambda) = 0 \quad \text{for } \lambda = -1,-2,-3,\ldots;$$

$$L_p^{m,n}(0) = \bar{L}_p^{m,n}(0) = 1. \qquad (1.7.20)$$

$$\Phi_{p,q}^{m,n}(h,\lambda) = \bar{\Phi}_{p,q}^{m,n}(h,\lambda) = 0, \quad \text{for } \lambda = -h,\ -h-1,\ -h-2,\ldots;$$

$$\psi_{p,q}^{m,n}(h,\lambda) = \bar{\psi}_{p,q}^{m,n}(\lambda) = 0, \quad \text{for } \lambda = 0,\ -1,-2,\ldots\ . \qquad (1.7.21)$$

$$\theta_p^{m,n}(\rho,r) = \bar{\theta}_p^{m,n}(\rho,r) = 0 \quad \text{for } r = -1,-2,-3,\ldots; \qquad (1.7.22)$$

and

$$\theta_p^{m,n}(\rho,0) = \bar{\theta}_q^{m,n}(\rho,0) = 1$$

$$D_{p,q}^{m,n}(0) = A_q^{m,n} \quad \text{for } \nu^* \geq 1. \qquad (1.7.23)$$

$$D_{p,q}^{m,n}(-\nu^*) = (-1)^{p-q+1} \exp\{2\pi i(\sum_{j=1}^{p} a_j - \sum_{j=1}^{q} b_j)\}\, \bar{A}_q^{m,n}, \qquad (1.7.24)$$

for $\nu^* \geq 1$.

$$D_{p,q}^{m,n}(-1) = 0 \quad \text{if } \nu^* > 1. \qquad (1.7.25)$$

$$D_{p,q}^{m,n}(-1) = (-1)^{p-q+1} \exp\{2\pi i(\sum_{j=1}^{p} a_j - \sum_{j=1}^{q} b_j)\}\, \bar{A}_q^{m,n} \qquad (1.7.26)$$

for $\nu^* = 1$.

We also have

$$T_{p,q}^{m,n}(\sigma,0) = \exp[i\pi a_\sigma(\nu^*-1)]\,\Delta_q^{m,n}(\sigma) \text{ if } \nu^* \geq 1 \text{ and } 1 \leq \sigma \leq n.$$

(1.7.27)

$$T_{p,q}^{m,n}(\sigma,0) = 0, \quad \text{if } \nu^* \geq 1 \text{ and } n+1 \leq \sigma \leq p\ . \qquad (1.7.28)$$

$$T_{p,q+1}^{m,n}(\sigma,\lambda) = \exp[i\pi a_\sigma(\nu^* + 2\lambda - 1)]\,\Delta_{p+1}^{m,n}(\sigma),$$

where $1 \leq \sigma \leq n$ and $1-\nu^* \leq \lambda \leq 0$.

$$T^{m,n}_{p,p+1}(\sigma,\lambda) = 0, \qquad (1.7.29)$$

where $n+1 \le \sigma \le p$ and $1-\nu^{*} \le \lambda \le 0$.

$$T^{m,n}_{p,p}(\sigma,\lambda) = \exp[i\pi a_{\sigma}(2\lambda + \nu^{*} -1]\, \Delta^{m,n}_{p}(\sigma) \text{ for } 1 \le \sigma \le n \text{ and}$$

$$T^{m,n}_{p,p}(\sigma,\lambda) = 0 \text{ for } n+1 \le \sigma \le p. \qquad (1.7.30)$$

1.7.2. A Few Expansion Formulas.

The following are the conditions among the parameters a_j and b_h which are needed in the presentation of theorems that follow.

$$a_j - b_h \ne 1,2,3,\ldots \; (j = 1,\ldots,n;\; h = 1,\ldots,m) \qquad (1.7.31)$$

$$a_j - a_h \ne 0, \pm 1, \pm 2,\ldots \; (j = 1,\ldots,n;\; h = 1,\ldots,n; j \ne h) \qquad (1.7.32)$$

$$a_j - a_h \ne 0, \pm 1, \pm 2,\ldots,(j = 1,\ldots,p; h = 1,\ldots,p; j \ne h) \qquad (1.7.33)$$

$$a_j - b_h \ne 1,2,3,\ldots(j = n-\rho + 2,\ldots,n;\; h = m+1,\ldots,k) \qquad (1.7.34)$$

$$a_j - a_h \ne 0, \pm 1, \pm 2,\ldots,(j = 1,\ldots,n-\rho+1; h = 1,\ldots,n-\rho+1; j \ne h). \qquad (1.7.35)$$

Firstly we state a theorem and then its particular cases are given. The particular cases of the expansion formulas are useful in obtaining the asymptotic expansion of the general G-function under certain restrictions on its parameters. Besides this use it is also possible to represent the function $G^{m,n}_{p,q}(z)$ by means of these expansion formulas in terms of fundamental solutions of the differential Equation (1.5.1) valid near $z = \infty$.

Proof of only Theorem 1.7.1 is given, for the sake of brevity. For further details and complete proofs see the papers of Meijer [208].

<u>Theorem 1.7.1</u>. Let m,n,p,q,k and ρ be integers satisfying the conditions

$$1 \le \rho \le n \le p \le q, \quad 0 \le m \le k \le q, \quad \sigma^{*} \ge 0 \qquad (1.7.36)$$

and (1.7.31), (1.7.34) and (1.7.35) hold, then

$$G^{m,n}_{p,q}(z) = \sum_{r=1}^{n-\rho+1} \exp[(\sigma^{*}-2\lambda)\pi i a_r]\Delta^{m,n-\rho+1}_{k}(r)$$

$$\cdot\, G^{k,\rho,n}_{p,q}(z \exp[(2\lambda-\sigma^{*})\pi i] \,\|\, a_r), \qquad (1.7.37)$$

where λ is an arbitrary integer such that $0 \leq \lambda \leq \sigma^*$, where σ^* is defined in (1.7.1).

<u>Proof</u>: Besides the conditions mentioned above, let

$$a_j - b_h \neq 1,2,3,.. \quad (j = 1,\dots,n-\rho+1;\ h = m+1,\dots,k)$$

$$b_j - b_h \neq 0, \pm 1, \pm 2,\dots, \quad (j = 1,\dots,k;\ h = 1,\dots,k;\ j \neq h)$$

$$a_j - b_h \neq 0, -1, -2,\dots, (j = 1,\dots,n-\rho+1; h = 1,\dots,k).$$

These conditions are assumed as they are involved during the proof but can be waived by an appeal to the considerations of continuity and following the techniques given in Chapter V.

From (1.1.9), we easily get

$$G^{m,\rho,n}_{p,q}[z \exp\{(2\lambda-\sigma^*)\pi i\} \| a_r] = \sum_{j=1}^{k} \frac{C_j}{\sin(b_j-a_r)\pi}$$

where

$$C_j = -\pi \frac{\prod_{h=n-\rho+2}^{n} \Gamma(1+b_j-a_h) \prod_{h=1, h\neq j}^{k} \Gamma(b_h-b_j)}{\prod_{h=n+1}^{p} \Gamma(a_h-b_j) \prod_{h=1}^{n-\rho+1} \Gamma(a_h-b_j) \prod_{h=k+1}^{q} \Gamma(1+b_j-b_h)}$$

$$\times z^{b_j} \exp[\pi i b_j(2\lambda-\sigma^*)]\ {}_pF_{q-1}\left({1+b_j-a_1,\dots,1+b_j-a_p \atop 1+b_j-b_1,\dots*\dots,1+b_j-b_q} ; (-1)^{p-m-n} z \right).$$

Substituting the above value of $G^{k,\rho,n}_{p,q}(\cdot)$ on the R.H.S. of (1.7.37), we see that

$$\sum_{r=1}^{n-\rho+1} \Delta^{m,n-\rho+1}_{k}(r) \exp[(\sigma^*-2\lambda)\pi i a_r] \sum_{j=1}^{k} \frac{C_j}{\sin(b_j-a_r)\pi}$$

$$= \sum_{j=1}^{k} C_j \sum_{r=1}^{n-\rho+1} \frac{\Delta^{m,n-\rho+1}_{k}(r)}{\sin(b_j-a_r)\pi} \exp\ (\sigma^*-2\lambda)\pi i a_\rho$$

$$= \sum_{j=1}^{m} C_j \sum_{r=1}^{n-\rho+1} \frac{\Delta^{m,n-\rho+1}_{k}(r)}{\sin(b_j-a_r)\pi} \exp[(\sigma^*-2\lambda)\pi i a_\rho]$$

by virtue of Lemma 1.7.3.

$$= -\pi^{\sigma^*} \sum_{j=1}^{m} C_j \exp[\pi i b_j(\sigma^*-2\lambda)] \frac{\prod_{h=m+1}^{k} \sin(b_h-b_j)\pi}{\prod_{h=1}^{n-\rho+1} \sin(a_h-b_j)\pi}$$

by virtue of Lemma 1.7.2.

The above equation is equal to (1.1.9) and hence the result follows.

Theorem 1.7.1a. Let m,n,p and q be integers satisfying the conditions $1 \leq n \leq p \leq q$, $1 \leq m \leq q$ and $\mu^* \leq 1$ and (1.7.31) and (1.7.32) hold, then

$$G^{m,n}_{p,q}(z) = \sum_{r=1}^{n} \exp[-(\mu^* + 2\lambda + 1)\pi i a_r]\, \Delta(r)$$

$$\times\, G^{q,1}_{p,q}(z \exp[(2\lambda + 1 + \mu^*)\pi i] \| a_r), \tag{1.7.38}$$

where λ is an arbitrary integer such that $0 \leq \lambda \leq -\mu^* - 1$.

For $\rho = 1$, $k = q$, (1.7.37) reduces to (1.7.38) by virtue of the result (1.7.4).

Theorem 1.7.1b. Let m,n,p and q be integers such that

$$1 \leq n \leq p < q, \quad 2 \leq m \leq q, \quad \mu^* + 1 \leq 0, \tag{1.7.39}$$

the numbers a_j and b_h satisfy (1.7.31) and (1.7.32), $|\arg z| < c^*\pi$; $c^* = m+n- \frac{p}{2} - \frac{q}{2}$; and λ is an arbitrary integer which satisfies the condition $0 \leq \lambda \leq -\mu^* -1$ and $-d^*\pi - \arg z < 2\pi\lambda < c^*\pi - \arg z$, then $G^{m,n}_{p,q}(z)$ can be represented by means of (1.7.38) in terms of fundamental solutions valid near $z = \infty$.

(1.7.40)

Remark: The value of the product

$$\Delta^{m,n-\rho+1}_{k}(r)\; G^{k,\rho,n}_{p,q}(\xi \| a_r), \quad (1 \leq r \leq n-\rho+1;\ m < k) \quad \text{when}$$

$a_r - b_h = 1,2,3,\ldots,(m+1 \leq h \leq k)$, can be obtained by adopting a procedure given in Chapter V, when the poles are of higher orders.

Theorem 1.7.2. Let m,n,p,q,k and ρ be integers satisfying the conditions $q \geq 1$, $0 \leq \rho-1 \leq n \leq p \leq q$ and $0 \leq m \leq k \leq q$, (1.7.31), (1.7.34) and (1.7.35) hold, r is an arbitrary integer such that $r \geq \max(0,-\sigma^*)$, then

$$G^{m,n}_{p,q}(z) = A^{m,n-\rho+1}_{k} \sum_{s=o}^{r-1} \omega^{m,n-\rho+1}_{k}(s)$$

$$\times\, G^{k,\rho-1,n}_{p,q}(z \exp[-(2s + \sigma^* + 1)i\pi])$$

$$+ \sum_{t=1}^{n-\rho+1} \exp[i\pi a_t(2r + \sigma^*)]\, \Delta^{m,n-\rho+1}_{k}(t)$$

$$\times\, G^{k,\rho,n}_{p,q}[z \exp\{-i\pi(2r + \sigma^*)\} \| a_t]\,. \tag{1.7.41}$$

The formula (1.7.41) also holds if i is replaced by -i.

Theorem 1.7.2a. Let m,n,p and q be integers with $0 \leq n \leq p \leq q$, $q \geq 1$, $0 \leq m \leq q$, (1.7.31) and (1.7.32) hold, r is an arbitrary integer satisfying the condition

$$r \geq \max(0, 1 + \mu^*), \qquad (1.7.42)$$

then,

$$G^{m,n}_{p,q}(z) = A^{m,n}_q \sum_{s=0}^{r-1} \omega^{m,n}_q(s) G^{q,0}_{p,q}[z \exp\{(\mu^* - 2s)\pi i\}]$$

$$+ \sum_{t=1}^{n} \exp[i\pi a_t(2r-\mu^*-1)] \Delta^{m,n}_q(t)\, G^{q,1}_{p,q}[z \exp i\pi(\mu^*+1-2r)\}\| a_t] \qquad (1.7.43)$$

The formula (1.7.43) follows from (1.7.41) by taking $\rho = 1$ and $k = q$ and using (1.7.4).

It can be easily seen that theorem 1.7.2a still holds if i is replaced by -i.

Theorem 1.7.2b. Let m,n,p and q be integers with $1 \leq n \leq p < q$, $2 \leq m \leq q$, $\mu^* + 1 \leq 0$, $-d^*\pi < \arg z < (\nu^* + \epsilon^*)\pi$, the conditions (1.7.31) and (1.7.32) hold, r is an arbitrary integer such that $r \geq 0$, and

$$\arg z - c^*\pi < 2r\pi < \arg z + d^*\pi ,$$

then (1.7.43) represents the function $G^{m,n}_{p,q}(z)$ in terms of fundamental solutions which hold near $z = \infty$. (1.7.44)

Theorem 1.7.2c. Let m,n,p and q be integers satisfying the condition (1.7.39), a_j and b_h satisfy (1.7.31) and (1.7.32) and r is an arbitrary integer, such that $r \geq 0$,

$$-c^*\pi - \arg z < 2r\pi < d^*\pi - \arg z$$

and $-(\nu^* + \epsilon^*)\pi < \arg z < d^*\pi$, then (1.7.43), with i replaced by -i expresses the function $G^{m,n}_{p,q}(z)$ in terms of fundamental solutions valid near $z = \infty$. (1.7.45)

Theorem 1.7.2d. Let the parameters a_j and b_h satisfy (1.7.31) and (1.7.32). Further let m,n,p and q be integers with $0 \leq n \leq p < q$, $1 \leq m \leq q$ and

$$\frac{p}{4} + \frac{q}{4} - \frac{\epsilon^*}{2} < m+n \leq q+1, \quad -c^*\pi < \arg z < (\nu^* + \epsilon^*)\pi,$$

r is an arbitrary integer such that

$$r \geq 1 + \mu^* \qquad (1.7.46)$$

and

$$(-c^*\pi) + \arg z < 2r\pi < d^*\pi + \arg z,$$

then (1.7.43) represents the function $G^{m,n}_{p,q}(z)$ in terms of fundamental solutions valid in the neighbourhood of $z = \infty$. (1.7.47)

Theorem 1.7.2e. Let m,n,p and q be integers satisfying the conditions

$0 \leq n \leq p < q,\ 1 \leq m \leq q,\ \frac{p}{4} + \frac{q}{2} - \frac{\epsilon^*}{2} < m+n \leq q+1$

$$-(\nu^* + \epsilon^*)\pi < \arg z < c^*\pi,$$

(1.7.31) and (1.7.32) hold, r is an arbitrary integer such that the condition (1.7.46) is satisfied and

$$-c^*\pi - \arg z < 2r\pi < d^*\pi + \arg z ,$$

then the function $G^{m,n}_{p,q}(z)$ can be represented by means of (1.7.43) with i replaced by -i in terms of fundamental solutions valid in the vicinity of $z = \infty$. (1.7.48)

Theorem 1.7.3. Let m,n,p,q,k and ρ be integers with $q \geq 1$, $0 \leq \rho-1 \leq n \leq p \leq q$ and $0 \leq m \leq k \leq q$ and the conditions (1.7.31), (1.7.34) and (1.7.35) are fulfilled, then

$$G^{m,n}_{p,q}(z) = A^{m,n-\rho+1}_{k} \sum_{s=0}^{r-1} \omega^{m,n-\rho+1}_{k}(s)\, G^{k,\rho-1,n}_{p,q}[z \exp\{i\pi(-2s-1-\sigma^*)\}]$$

$$+ \bar{A}^{m,n-\rho+1}_{k} \sum_{\tau=0}^{-\sigma^*-r-1} \bar{\omega}^{m,n-\rho+1}_{k}(\tau)\, G^{k,\rho-1,n}_{p,q}[z \exp\{(2\tau+1+\sigma^*)i\pi\}]$$

$$+ \sum_{t=1}^{n-\rho+1} \exp\{(\sigma^* + 2r)\pi i a_t\}\, \Delta^{m,n-\rho+1}_{k}(t)$$

$$\times G^{k,\rho,n}_{p,q}[z \exp\{(-2r-\sigma^*)\pi i\} \| a_t], \qquad (1.7.49)$$

where r is an arbitrary integer, positive, negative or zero.

The following particular cases of this theorem are worth mentioning.

(i) If $\sigma^* \geq 0$, $r \geq 0$, (1.7.49) reduces to (1.7.41).

(ii) On the other hand if $\sigma^* \geq 0$, $-\sigma^* \leq r \leq 0$, (1.7.49) gives rise to (1.7.37) with $\lambda = -r$.

(iii) Next, if we take $\sigma^* \geq 0$, $r \leq -\sigma^*$ or $\sigma^* \leq 0$, $r \leq 0$, then (1.7.49) reduces to (1.7.41) with i replaced by -i, r replaced by $-r-\sigma^*$ and s by τ.

(iv) When $\sigma^* \leq 0$, $r \geq -\sigma^*$, (1.7.49) reduces to (1.7.41).

The formula (1.7.49) also holds if $\sigma^* \leq 0$ and $0 \leq r \leq -\sigma^*$.

Theorem 1.7.3a. If m,n,p and q are integers with $q \geq 1$, $0 \leq m \leq q$, $0 \leq n \leq p \leq q$ and $m+n \leq q+1$, the numbers a_j and b_h satisfy (1.7.31) and (1.7. 32), then

$$G_{p,q}^{m,n}(z) = A_q^{m,n} \sum_{s=0}^{r-1} \omega_q^{m,n}(s)\, G_{p,q}^{q,0}[z \exp\{(\mu^* - 2s)\pi i\}]$$

$$+ \bar{A}_q^{m,n} \sum_{\tau=0}^{\mu^*-r} \bar{\omega}_q^{m,n}(\tau)\, G_{p,q}^{q,0}[z \exp\{(2\tau - \mu^*)\pi i\}]$$

$$+ \sum_{t=1}^{n} \exp\{(2r-\mu^*-1)\pi i a_t\}\, \Delta_q^{m,n}(t)\, G_{p,q}^{q,1}[z \exp\{(\mu^*+1-2r)\pi i\} \| a_t], \tag{1.7.50}$$

where r is an arbitrary integer satisfying the condition $0 \leq r \leq \mu^*+1$.

When $\rho = 1$, $k = q$ and $\nu = p$, (1.7.49) reduces to (1.7.50).

Theorem 1.7.3b. If the number z satisfies the inequality, $|\arg z| < (\nu^* + \epsilon^*)\pi$ and the numbers a_j and b_h the conditions (1.7.31) and (1.7.32), m,n,p and q are integers with $0 \leq n \leq p < q$, $1 \leq m \leq q$, $p+1 \leq m+n \leq \frac{3q}{4} + \frac{p}{4} - \frac{\epsilon^*}{2} + 1$ and r an arbitrary integer which fulfil the conditions $0 \leq r \leq \mu^* + 1$, $\arg z - c^*\pi < 2r\pi < d^*\pi + \arg z$, then (1.7.50) represents the function $G_{p,q}^{m,n}(z)$ in the vicinity of $z = \infty$. (1.7.51)

Theorem 1.7.4. If m,n,p,q,k,ρ and ν are integers with

$$q \geq 1,\ \rho \geq 1,\ 0 \leq n-\rho+1 \leq \nu \leq k,\ \rho+\nu-1 \leq p \leq q$$

and $0 \leq m \leq k \leq q$, and the following conditions are satisfied

$$a_j - b_h \neq 1,2,3,\ldots (j = \nu+1,\ldots,\rho+\nu-1;\ h = 1,\ldots,k).$$

$$a_j - b_h \neq 1,2,3,\ldots (j = 1,\ldots,n-\rho+1;\ h = 1,\ldots,m),$$

$$a_j - a_t \neq 0, \pm 1, \pm 2,\ldots (j = 1,\ldots,\nu;\ t = 1,\ldots,\nu;\ j \neq t);$$

then

$$G_{p,q}^{m,n}\left[z \middle| \begin{matrix} a_1,\ldots,a_{n-\rho+1}, a_{\nu+1},\ldots,a_p, a_{n-\rho+2},\ldots,a_\nu \\ b_1,\ldots,b_q \end{matrix}\right]$$

$$= \sum_{h=1}^{k-\nu-\mu} R_{\nu,k}^{m,n-\rho+1}(h,\lambda)\, G_{p,q}^{k,\rho-1,\rho+\nu-1}[z \exp\{-i\pi(\sigma^*+2h+2\lambda-1)\}]$$

$$+ \sum_{e=1}^{\mu} \bar{R}_{\nu,k}^{m,n-\rho+1}(e,\rho-m-n-\lambda+\nu)$$

$$\times G_{p,q}^{k,\rho-1,\rho+\nu-1}[z \exp\{i\pi(\rho-k-m-n+2e-2\lambda+2\nu-1)\}]$$

$$+ \sum_{t=1}^{\nu} \exp[i\pi a_t(k-\nu-2\mu)]\, T_{\nu,k}^{m,n-\rho+1}(t,\lambda)$$

$$\times G_{p,q}^{k,\rho,\rho+\nu-1}[z \exp\{i\pi(\rho-k-m-n-2\lambda+2\mu+2\nu)\} \| a_t] \tag{1.7.52}$$

where λ and μ are arbitrary integers and $0 \leq \mu \leq k-\nu$.

Theorem 1.7.4a. If m,n,p,q and k are integers such that $0 \leq m \leq k \leq q$, $0 \leq n \leq p \leq q$, $q \geq 1$, the condition

$$a_j - b_h \neq 1,2,3,\ldots \quad (j = 1,\ldots,n;\ h = 1,\ldots,k)$$

is satisfied, then

$$G^{m,n}_{p,q}(z) = \sum_{h=1}^{k} R^{m,0}_{0,k}(h,\lambda)$$

$$G^{k,n}_{p,q}[z \exp\{(k-m-2h-2\lambda+2\ )\pi i\}] \tag{1.7.53}$$

where λ is an arbitrary integer.

For $\rho = n+1$ and $\nu = 0$, (1.7.52) reduces to (1.7.53).

Theorem 1.7.4b. If m,n,p and q are integers such that $q \geq 1$, $0 \leq n \leq p \leq q$, $0 \leq m \leq q$, (1.7.31) and (1.7.33) hold, then

$$G^{m,n}_{p,q}(z) = \sum_{h=1}^{q-p-\mu} R^{m,n}_{p,q}(h,\lambda)\ G^{q,0}_{p,q}[z \exp\{(\mu^{*}-2h-2\lambda+2\ )\pi i\}\ ]$$

$$+ \sum_{e=1}^{\mu} \bar{R}^{m,n}_{p,q}(e,-\nu^{*}-\lambda+1)\ \ G^{q,0}_{p,q}[z \exp\{(2p+\mu^{*}+2e-2\lambda-2q)\pi i\ \}]$$

$$+ \sum_{t=1}^{p} \exp\{(q-p-2\mu)\pi i a_t\}\ T^{m,n}_{p,q}(t,\lambda)$$

$$\times\ G^{q,1}_{p,q}\ [z \exp\{(2p-2q+\mu^{*}-2\lambda+2\mu+1)\pi i\}\|a_t], \tag{1.7.54}$$

where λ and μ are arbitrary integers and $0 \leq \mu \leq q-p$.

For $\rho = 1$, $k = q$, $\nu = p$, (1.7.52) yields (1.7.54).

Theorem 1.7.4c. Let m,n,p and q be integers with $0 \leq m \leq q$ and $0 \leq n \leq p < q$, (1.7.31) and (1.7.33) hold, λ and μ be arbitrary integers satisfying

$$(\nu^{*} + \epsilon^{*} + 2\lambda-2\ )\pi \leq \arg z < (\nu^{*} + \epsilon^{*} + 2\lambda)\pi$$

and

$$(m+n-\tfrac{3p}{2} + \tfrac{q}{2} + 2\lambda-2\ )\pi - \arg z < 2\mu\pi < (m+n-\tfrac{5p}{2} + \tfrac{3q}{2} + 2\lambda)\pi - \arg z,$$

then the function $G^{m,n}_{p,q}(z)$ can be expressed in terms of fundamental solutions by means of (1.7.54) valid near $z = \infty$. (1.7.55)

1.8 SOME ASYMPTOTIC EXPANSIONS OF $G^{m,n}_{p,q}(z), (p < q)$.

We now proceed to investigate the behaviour of $G^{m,n}_{p,q}(z)$, $(p < q)$ for all values of m,n,p,q and arg z as $|z| \to \infty$. The theorems of the preceding section enable us to represent $G^{m,n}_{p,q}(z)$ linearly in terms of the functions $G^{q,0}_{p,q}$ and $G^{q,1}_{p,q}$ of which the asymptotic expansions can be readily deduced from the theorems 1.4.1 and 1.4.3 of Section 1.4. Meijer [208] has investigated all the functions of the type $G^{q,0}_{p,q}$ and $G^{q,1}_{p,q}$ on the right hand side of the expansion formulae discussed in Section 1.7 and determined the dominant or dominants among them.

<u>Definition</u>. <u>Dominance.</u> It is useful to define the concept of dominance. A function f(z) is said to be dominant compared with another function g(z) if the leading term of the asymptotic expansion of g(z) is of an order less than the error term of the asymptotic expansion of f(z).

For example, consider the asymptotic expansions

$$f_1(z) \sim e^z \sum_{r=0}^{\infty} a_{1r}\, z^{-r}, \qquad f_2(z) \sim z^5 \sum_{r=0}^{\infty} a_{2r}\, z^{-r},$$

$$f_3(z) \sim z^{-\frac{1}{2}} \sum_{r=0}^{\infty} a_{3r}\, z^{-r}, \qquad f_4(z) \sim e^{iz} z^{-2} \sum_{r=0}^{\infty} a_{4r}\, z^{-r},$$

$$f_5(z) \sim e^{-z} \sum_{r=0}^{\infty} a_{5r}\, z^{-r}, \qquad f_6(z) \sim e^{-2z} \sum_{r=0}^{\infty} a_{6r}\, z^{-r},$$

where z is positive and it is assumed that none of a_{io} for i = 1,2,...,6, is zero. Then evidently $f_1(z)$ is dominant compared with $f_2(z),\ldots,f_6(z)$. $f_2(z)$, $f_3(z)$ and $f_4(z)$ are dominant compared with $f_5(z)$ and $f_6(z)$. Compared with $f_6(z)$, $f_5(z)$ is dominant but $f_2(z)$ is not dominant compared with $f_3(z)$ and $f_4(z)$. Thus among the functions $f_2(z),\ldots,f_6(z)$ there are three dominants, namly $f_2(z)$, $f_3(z)$ and $f_4(z)$.

In order to obtain the asymptotic expansions of $G^{m,n}_{p,q}(z)$ we need only to retain the dominant term or terms in the right hand sides of the formulas in Section 1.7, unless the coefficients of all the dominant functions vanish. Now these coefficients are functions of the parameters a_j and b_j and in general do not vanish. Such a function is only zero if the parameters a_j and b_j satisfy certain equations. As these parameters are mutually independent there exists no relation between them.

We, therefore, assume that if there is only one dominant function, then the coefficient of this function is not zero. When there are two or more dominant functions, it is assumed that at least one of them possesses a non-zero coefficient. If the coefficients of all the dominant functions vanish, it is necessary to make a further investigation, which is being omitted.

The following discussion is based on the work of Meijer [208]. Accordingly we give only the dominant terms in the various asymptotic expansions. The expansion theorems from which these results follow are also indicated in order that the complete expansions may be deduced if so desired.

Theorem 1.8.1. Let m,n,p and q be integers satisfying the conditions $0 \le n \le p \le q-2$, $p+1 \le m+n \le \frac{p}{2}+\frac{q}{2}$ and $-(1+\nu^*)\pi < \arg z < (1+\nu^*)\pi$. Further let the parameters a_j and b_h satisfy the conditions (1.7.31) and (1.7.32) for (1.8.3) and the condition (1.7.31) for (1.8.1), (1.8.2) and (1.8.4) given below. We then have the following results.

(i) The function $G^{m,n}_{p,q}(z)$ admits for large values of $|z|$ with $0 < \arg z < (1+\nu^*)\pi$, the asymptotic expansion

$$G^{m,n}_{q,p}(z) \sim A^{m,n}_{q} H_{p,q} [z \exp(i\pi\mu^*)] . \tag{1.8.1}$$

(ii) The function $G^{m,n}_{p,q}(z)$ admits the asymptotic expansion

$$G^{m,n}_{p,q}(z) \sim \bar{A}^{m,n}_{q} H_{p,q}[z \exp(-i\pi\mu^*)] \tag{1.8.2}$$

for large values of $|z|$ with $-(\nu^* + 1)\pi < \arg z < 0$.

(iii) The asymptotic expansion

$$G^{m,n}_{p,q}(z) \sim \sum_{i,-i} A^{m,n}_{q} H_{p,q} [z \exp(i\pi\mu^*)]$$

$$+ \sum_{t=1}^{n} \Delta^{m,n}_{q}(t) \exp[-(1+\mu^*)i\pi a_t] E_{p,q}\{\exp[i\pi(1+\mu^*)\|a_t\}], \tag{1.8.3}$$

holds for large values of z with $\arg z = 0$ and $m+1 = \frac{p}{2}+\frac{q}{2}$. Here the symbol $\sum_{i,-i}$ indicates that to the expression following it, a similar expression is to be added with i replaced by $-i$.

(iv) The asymptotic expansion

$$G^{m,n}_{p,q}(z) \sim \sum_{i,-i} A^{m,n}_{q} H_{p,q}[z \exp(i\pi\mu^*)] , \tag{1.8.4}$$

holds for large values of z with $\arg z = 0$ and $p+1 \le m+n < \frac{p}{2}+\frac{q}{2}$.

Theorem 1.8.2. Let m,n and q be integers satisfying the condition $0 \le p \le q-2$ and $p+1 \le m \le q-1$ and $(p-m-1)\pi < \arg z < (m-p+1)\pi$, then the following results hold

(i)
$$G^{m,0}_{p,q}(z) \sim \sum_{i,-i} A^{m,0}_{q} H_{p,q} [z \exp\{i\pi(q-m)\}], \tag{1.8.5}$$

for large values of z with $\arg z = 0$.

(ii) $$G^{m,0}_{p,q}(z) \sim A^{m,0}_{q} H_{p,q}[z \exp\{i\pi(q-m)\}] \tag{1.8.6}$$

for large values of $|z|$ with $0 < \arg z < (m-p+1)\pi$.

(iii) $$G^{m,0}_{p,q}(z) \sim \bar{A}^{m,0}_{q} H_{p,q}[z \exp\{-i\pi(q-m)\}], \tag{1.8.7}$$

for large values of $|z|$ with $(p-m-1)\pi < \arg z < 0$.

Theorem 1.8.3. Let m,n,p and q be integers satisfying the conditions $c^* = m+n- \frac{p}{2} - \frac{q}{2} > 0$, $1 \le n \le p < q$, $1 \le m \le q$, and the parameters a_j and b_h fulfil the conditions (1.7.31) and (1.7.32). Then

$$G^{m,n}_{p,q}(z) \sim \sum_{t=1}^{n} \exp\{-\pi i a_t(1+\mu^*\} \Delta^{m,n}_{q}(t)$$

$$\times E_{p,q}[z \exp\{\pi i(1+ \mu)\}\|a_t], \tag{1.8.8}$$

for large values of $|z|$ with $|\arg z| < c^*\pi$.

Theorem 1.8.4. Let m,n,p and q be integers satisfying the conditions $0 \le n \le p < q$, $1 \le m \le q$ and $c^* > 0$ and $c^*\pi \le \arg z < (\nu^* + \epsilon^*)\pi$. Further the numbers a_j and b_h fulfil the conditions (1.7.31) in the statements of the asymptotic expansions under (i) and (ii) and the conditions (1.7.31) and (1.7.32) in the asymptotic expansions under (iii) and (iv) given below. Then the following results hold.

(i) The asymptotic expansion given by (1.8.1) also holds for large values of $|z|$ with $c^*\pi \le |\arg z| <(\nu^* + \epsilon^*)\pi$. (1.8.9)

(ii) The asymptotic expansion given by (1.8.2) also holds for large values of $|z|$ with $-(\nu^* + \epsilon^*)\pi < \arg z <-c^*\pi$. (1.8.10)

(iii) The asymptotic expansion

$$G^{m,n}_{p,q}(z) \sim A^{m,n}_{q} H_{p,q}[z \exp(i\pi\mu^*)] + \sum_{t=1}^{n} \exp\{-i\pi(\mu^*+1)a_t\}$$

$$\times E_{p,q}[z \exp\{(1+\mu^*)\pi i\}\|a_t] \tag{1.8.11}$$

holds for large values of $|z|$ with $\arg z = c^*\pi$.

(iv) The asymptotic expansion

$$G^{m,n}_{p,q}(z) \sim \bar{A}^{m,n}_{q} H_{p,q}[z \exp(-i\pi\mu^*)] + \sum_{t=1}^{n} \exp\{i\pi(\mu^* + 1)a_t\}$$

$$\times E_{p,q}[z \exp\{-(1 + \mu^*)i\pi\}\|a_t], \tag{1.8.12}$$

holds for large values of $|z|$ with $\arg z = -c^*\pi$.

Theorem 1.8.5. Let m,n,p and q be integers such that $0 \leq n \leq p \leq q-2$, $1 \leq m \leq q$. Further suppose that λ is an arbitrary integer or (λ is either an arbitrary integer ≥ 0 or an arbitrary integer $\leq p-m-n$ according as $m+n \leq p+1$ or $m+n \geq p+2$) and the condition (1.7.31) is satisfied in (1.8.13) and (1.8.14). The conditions (1.7.31) and (1.7.33) are satisfied in (1.8.15). Then the following asymptotic expansions hold.

(i)
$$G_{p,q}^{m,n}(z) \sim D_{p,q}^{m,n}(\lambda)\, H_{p,q}\,[z \exp\{i\pi(\mu^* - 2\lambda)\}] \tag{1.8.13}$$

for large of $|z|$ with $(\nu^* + 2\lambda-1)\pi < \arg z < (\nu^* + 2\lambda + 1)\pi$.

(ii)
$$G_{p,q}^{m,n}(z) \sim \sum_{\lambda,\lambda-1} D_{p,q}^{m,n}(\lambda)\, H_{p,q}\,[z \exp\{i\pi(\mu^* - 2\lambda)\}] \tag{1.8.14}$$

for large values of $|z|$ with $\arg z = (\nu^* + 2\lambda-1)\pi$ and $q > p+2$ Here the symbol $\sum_{\lambda,\lambda-1}$ indicates that to the expression following it, a similar expression is to be added in which λ is replaced by $\lambda-1$.

(iii)
$$G_{p,p+2}^{m,n}(z) \sim \sum_{\lambda,\lambda-1} D_{p,p+2}^{m,n}(\lambda)\, H_{p,p+2}[z \exp\{-i\pi(\nu^* + 2\lambda - 2)\}]$$
$$+ \sum_{t=1}^{p} \exp\{-2(1+\lambda)\pi i a_t\}\, T_{p,p+2}^{m,n}(t,\lambda) E_{p,p+2}[z \exp\{i\pi(3-\nu^*)\} \| a_t] \tag{1.8.15}$$

for large values of $|z|$ with $\arg z = (\nu^* + 2\lambda-1)\pi$.

Remark 1: This theorem is also valid if $q \geq p+2$ in the following cases:

(i) $\nu^* \leq 1$ and all values of arg z.

(ii) $\nu^* \geq 2$ and $\arg z \geq (\nu^* - 1)\pi$.

(iii) $\nu^* \geq 2$ and $\arg z < -(\nu^* - 1)\pi$.

The results of the theorem are not applicable when $q \geq p+2$ with $\nu^* \geq 2$ and $-(\nu^*-1)\pi \leq \arg z < (\nu^*-1)\pi$. This case is covered by means of the theorems 1.8.1 to 1.8.4.

Remark 2: By virtue of the properties of $D_{p,q}^{m,n}(\cdot)$ and $T_{p,q}^{m,n}(.,.)$ given in Section 1.7.1, namely the results (1.7.23) to (1.7.28), and the remark preceding (1.4.12) the following interesting particular cases of the theorem are obtained.

(i) For $\lambda = -\nu^*$ and $\nu^* \geq 1$, (1.8.13) reduces to (1.8.2).

(ii) For $\lambda = 0$ and $\nu^* \geq 1$, (1.8.13) reduces to (1.8.1).

(iii) For $\lambda = 0$ and $\nu^* > 1$, (1.8.14) reduces to (1.8.1).

(iv) For $\lambda = 0$ and $\nu^* = 1$, (1.8.14) reduces to (1.8.4).

(v) For $\lambda = 0$ and $\nu^* > 1$, (1.8.15) reduces to (1.8.11) with $q = p+2$.

(vi) For $\lambda = 0$ and $\nu^* = 1$, (1.8.15) reduces to (1.8.3) with $q = p+2$.

Theorem 1.8.6. Let m,n and p be integers such that $p \geq 1$, $0 \leq n \leq p$ and $1 \leq m \leq p+1$. Further assume that both the conditions (1.7.31) and (1.7.33) are fulfilled in case of the results (1.8.16), (1.8.18) and (1.8.19) and the condition (1.7.31) holds in case of (1.8.17).

$$\text{Also } (\nu^* + 2\lambda - \tfrac{3}{2})\pi \leq \arg z < (\nu^* + 2\lambda + \tfrac{1}{2})\pi.$$

Then the following are the asymptotic expansions of the function $G^{m,n}_{p,p+1}(z)$ for large values of z .

(i)
$$G^{m,n}_{p,p+1}(z) \sim \sum_{t=1}^{p} \exp\{-i\pi(2\lambda + 1)a_t\}$$

$$\times\, T^{m,n}_{p,p+1}(t,\lambda)\, E_{p,p+1}[z \exp\{i\pi(2-\nu^*)\} \| a_t], \tag{1.8.16}$$

with $(\nu^* + 2\lambda - \frac{3}{2})\pi < \arg z < (\nu^* + 2\lambda - \frac{1}{2})\pi$.

Remark: It is interesting to observe that the result (1.8.16) also holds under the following conditions:

(i) $\nu^* \geq 2$, $-\nu^* < \lambda < 0$;

$$(\nu^* + 2\lambda - \tfrac{1}{2})\pi \leq \arg z < (\nu^* + 2\lambda + \tfrac{1}{2})\pi$$

(ii) $(\nu^* \geq 2$, $\nu^* + 1 < \lambda < 1$, $\arg z = (\nu^* + 2\lambda - \frac{3}{2})\pi$.

The result (1.8.16) is not, however, true if $n = \lambda = 0$ and $m = p + 1$.

(ii) $$G^{m,n}_{p,p+1}(z) \sim D^{m,n}_{p,p+1}(\lambda)\, H_{p,p+1}[z\, \exp(1-2\lambda-\nu^*)\pi i], \qquad (1.8.17)$$

provided $\nu^* \geq 2$, λ is either an arbitrary integer ≥ 0 or an arbitrary integer $\leq -\nu^*$ and $(\nu^* + 2\lambda - \frac{1}{2})\pi < \arg z < (\nu^* + 2\lambda + \frac{1}{2})\pi$.

Remark: The result (1.8.17) is also valid under the same conditions on arg z as in (1.8.17) provided $\nu^* \leq 1$ and λ is an arbitrary integer.

(iii) $$G^{m,n}_{p,p+1}(z) \sim D^{m,n}_{p,p+1}(\lambda) H_{p,p+1}[z\, \exp\{-i\pi(\nu + 2\lambda-1)\}]$$
$$+ \sum_{t=1}^{p} \exp[-i\pi(2\lambda+1)a_t]\, T^{m,n}_{p,p+1}(t,\lambda)$$
$$\times E_{p,p+1}[z\, \exp\{(2-\nu^*)\pi i\}\| a_t], \qquad (1.8.18)$$

provided $\nu^* \geq 2$ and λ is either an arbitrary integer ≥ 0 or an arbitrary integer $\leq -\nu^*$ and $\arg z = (\nu^* + 2\lambda - \frac{1}{2})\pi$.

Remark: The expansion (1.8.18) is also true if $\nu^* \leq 1$, $\arg z = (\nu^* + 2\lambda - \frac{1}{2})\pi$ and λ is an arbitrary integer.

(iv) $$G^{m,n}_{p,p+1}(z) \sim D^{m,n}_{p,p+1}(\lambda-1)\, H_{p,p+1}[z\, \exp\{(-\nu^*-2\lambda+3)i\pi\}]$$
$$+ \sum_{t=1}^{n} \exp[-(2\lambda+1)i\pi\, a_t]\, T^{m,n}_{p,p+1}(t,\lambda)$$
$$\times E_{p,p+1}[z\, \exp\{(-\nu^*+2)\pi i\}\| a_t], \qquad (1.8.19)$$

provided $\nu^* \geq 2$, $\arg z = (\nu^* + 2\lambda - \frac{3}{2})\pi$ and λ is either an arbitrary integer ≥ 1 or an arbitrary integer $\leq 1-\nu^*$.

Remark 1: The expansion (1.8.19) is also valid if $\nu^* \leq 1$, $\arg z = (\nu^* + 2\lambda - \frac{3}{2})\pi$ and λ is an arbitrary integer.

Remark 2: By virtue of the results given in (1.7.29), the formula (1.8.16) yields (1.8.8), when $q = p+1$, $n \geq 1$, $\nu^* \geq 1$, $1-\nu^* \leq \lambda \leq 0$.

Briefly we indicate the expansion formulas from which the various asymptotic expansions discussed in this section are obtained. For the details of the proofs of these theorems the reader is referred to the original papers by Meijer [208].

Theorem 1.8.1. follows from the formulas (1.7.50) and (1.7.51).

Theorem 1.8.2. can be obtained by the application of the expansion formula (1.7.43) with $n = 0$.

Theorem 1.8.3. can be established with the help of the expansion formula (1.7.38) and theorem 1.4.1.

Use is made of the expansion formula (1.7.43) to prove theorem 1.8.4. Apply the theorems 1.7.4b and 1.7.4c to get theorem 1.8.5.

Finally theorem 1.8.6 can be obtained by the application of the formulas (1.7.54) and (1.7.55).

We recall here that the p-functions defined by

$$G^{q,1}_{p,q}(z \exp [i\pi(\mu^* - 2\lambda+1)] \| a_t)$$

for t = 1,...,p, form a system of fundamental solutions of the differential equation (1.6.2) satisfied by $G^{m,n}_{p,p}(z)$, valid near $z = \infty$. Here we give an expression for the function $G^{m,n}_{p,p}(z)$ in terms of these fundamental solutions which enables us to obtain its analytic continuation in the general case outside the circle $|z| = 1$ by the application of the theorem 1.4.4a.

Theorem 1.8.7. Let m,n and p be integers satisfying the conditions $p \geq 1$, $0 \leq n \leq p$, $0 \leq m \leq q$, (1.7.31) and (1.7.33) hold, then the function $G^{m,n}_{p,p}(z)$ can be expressed in terms of fundamental solutions valid near $z = \infty$, by means of the expression

$$G^{m,n}_{p,p}(z) = \sum_{r=1}^{p} T^{m,n}_{p,p}(r,\lambda)\, G^{p,1}_{p,p}[z \exp i\pi(1-\nu^*-2\lambda) \| a_t], \qquad (1.8.20)$$

where λ is an arbitrary integer and

$$(\nu^* + 2\lambda-2)\pi < \arg z < (\nu^* + 2\lambda)\pi . \qquad (1.8.21)$$

Further the function $G^{m,n}_{p,p}(z)$ admits in the sector (1.8.21) an analytic continuation outside the unit circle $|z| = 1$ which can be represented as

$$G^{m,n}_{p,p}(z) = \sum_{r=1}^{p} \exp[-2i\pi\lambda a_r]\, T^{m,n}_{p,p}(r,\lambda)$$

$$\times E_{p,p}[z \exp\{-i\pi(\nu^*-1)\} \| a_r]. \qquad (1.8.22)$$

The result (1.8.20) follows from (1.7.54) for $\mu = 0$ and $q = p$, whereas the formula (1.8.22) is obtained from (1.8.21), (1.4.4) and (1.4.16).

In particular if we set $\nu^* \geq 1$, $1-\nu^* \leq \lambda \leq 0$ and make use of the properties of $T^{m,n}_{p,p}(t,\lambda)$ given in (1.7.30) then it readily follows that the analytic continuation of $G^{m,n}_{p,p}(z)$ outside the unit circle can be put in the form

$$G^{m,n}_{p,p}(z) = \sum_{r=1}^{n} \exp[i\pi a_r(\nu^*-1)]\Delta^{m,n}_{p}(r)$$

$$\times E_{p,p} [z \exp\{i\pi(-\nu^*+1)\} \| a_r]. \qquad (1.8.23)$$

Remark: (1.8.23) is equivalent to theorem 1.4.4 though slightly less general, since we have now excluded the values

$$\arg z = (2-\nu^*)\pi, \quad (4-\nu^*)\pi, \ldots, (\nu^*-2)\pi.$$

General Remark: As a concluding remark it is useful to note that in these asymptotic expansion formulas we have assumed that the parameters satisfy one or two of the conditions (1.7.31), (1.7.32) and (1.7.33). If these conditions are not fulfilled the right hand side of these expansion formulas assumes an indeterminate form and must be replaced by its limit. A general series expansion of the G-function when its parameters a_j and b_h differ in any manner, due to Mathai [75], is applicable. A detailed account of this will be found in Chapter V.

EXERCISES

1.1 Show that

(i) $$G_{0,1}^{1,0}\left(px^{\alpha}\middle|\, \frac{\beta}{\alpha}\right) = p^{\frac{\beta}{\alpha}} x^{\beta} \exp(-px^{\alpha}).$$

(ii) $$G_{1,1}^{1,0}\left(x\middle|\, {\alpha+\beta+1 \atop \alpha}\right) = \frac{x^{\alpha}(1-x)^{\beta}}{\Gamma(\beta+1)}, \quad 0 < x < 1 .$$

(iii) $$G_{1,1}^{1,1}\left(-x\middle|\, {1-\alpha \atop 0}\right) = \Gamma(\alpha)\,(1-x)^{-\alpha}, \quad 0 < x < 1 .$$

(iv) $$G_{1,2}^{1,1}\left(x\middle|\, {1-\alpha \atop 0,1-\gamma}\right) = \frac{\Gamma(\alpha)}{\Gamma(\gamma)}\, {}_1F_1(\alpha;\gamma;x)$$

Note: These functions are associated with statistical distributions.

(Kaufman, Mathai and Saxena, 1969, [134]) .

1.2 Prove that

$$G_{2,2}^{2,2}\left(z\middle|\, {1-a,1-b \atop 0,c-a-b}\right) = \frac{\Gamma(a)\,\Gamma(b)\,\Gamma(c-a)\,\Gamma(c-b)}{\Gamma(c)}\, {}_2F_1(a,b;c;1-z),$$

provided $c-a, c-b \neq 0, -1, -2$.

Hence or otherwise deduce

$$G_{2,2}^{2,2}\left(1\middle|\, {1-\alpha,1-\beta \atop \gamma\,,\,\delta}\right) = \frac{\Gamma(\gamma+\alpha)\,\Gamma(\delta+\alpha)\,\Gamma(\gamma+\beta)\,\Gamma(\delta+\beta)}{\Gamma(\alpha+\beta+\gamma+\delta)}$$

provided $\gamma+\alpha, \delta+\alpha, \gamma+\beta, \delta+\beta \neq 0, -1, -2, \ldots$

(Meijer, 1956, p.78, [209]).

1.3 Prove that

$$\frac{1}{\Gamma(\alpha)\,\Gamma(\beta)} \; G^{2,1}_{2,2}\left(z\Big|{}^{1,\gamma}_{\alpha,\beta}\right) = \left(\frac{z}{1+z}\right)^{\beta} \; {}_2\Phi_1\left(\gamma-\alpha,\ \beta,\gamma;\ \frac{1}{1+z}\right).$$

(Meijer 1954, p.279, [209]).

1.4 Prove that

$$\frac{d^k}{dx^k}\, x^{r(a_1-1)} \quad G^{m,n}_{p,q}\left[x^{-r}\Big|{}^{a_p}_{b_q}\right]$$

$$= (-n)^k\, x^{r(a_1-1)-k} \quad G^{m,n}_{p,q}\left(x^{-r}\Big|\begin{matrix} a_1-\frac{k}{r},\ldots,a_r-\frac{k}{r},\ a_{r+1},\ldots,a_p \\ b_1,\ldots,b_q\end{matrix}\right)$$

provided $r < n$ and the parameters $a_1,\ldots,a_r$ are in arithmetic progression (A.P) with common difference $-\frac{1}{r}$.

Hence or otherwise deduce

$$\frac{d^k}{dx^k}\left\{x^{-r\,b_1}\quad G^{m,n}_{p,q}\left(x\Big|{}^{a_p}_{b_q}\right)\right\} = (-r)^k\, x^{-rb_1-k}$$

$$\times\, G^{m,n}_{p,q}\left(x^{r}\Big|\begin{matrix} a_1,\ldots,a_p \\ b_1+\frac{k}{r},\ldots,b_r+\frac{k}{r},\ b_{r+1},\ldots,b_q\end{matrix}\right)$$

provided $r < m$ and the parameters $b_1, b_2, \ldots, b_r$ are in A.P. with common difference $\frac{1}{r}$.

Also prove that

$$\frac{d^k}{dx^k}\left\{x^{r}(a_{p-r+1}-\tfrac{1}{r})\ G^{m,n}_{p,q}\left(x^{-r}\Big|{}^{a_p}_{b_q}\right)\right\} = r^k\, x^{r(a_{p-r+1}-\frac{1}{r})-k}$$

$$\times\, G^{m,n}_{p,q}\left(x^{-r}\Big|\begin{matrix} a_1,\ldots,a_{p-r},a_{p-r+1}-\frac{k}{r},\ldots,a_p-\frac{k}{r} \\ b_1,\ldots,b_q\end{matrix}\right)$$

provided that $p-r+1 > n$ and the parameters $a_{p-r+1},\ldots,a_p$ are in A.P. with common difference $\frac{1}{r}$.(Sundararajan, 1967, [325]).

1.5 Prove that

$$G^{p,1}_{q+1,p}\left(z\Big|{}^{1,b_q}_{a_p}\right) = E(a_1,\ldots,a_p;\ b_1,\ldots,b_q;\ z)$$

where E denotes MacRobert's E-function.

1.6 Show that

$$2\pi\, G^{p,0}_{q+1,p}\left(z\,\Big|\,{1,b_q \atop a_p}\right) = \sum_{i,-i} \frac{1}{i}\exp(i\pi b_1)\, E(a_p;\, b_q;\, ze^{-i\pi})$$

where the symbol $\sum_{i,-i}$ indicates that to the expression following it, a similar expression with i interchanged by -i is to be added.

1.7 Prove the expansion

$$G^{m+1,n}_{p+1,q+1}\left(x\,\Big|\,{a_p, a_1+\nu-k \atop a_1+\nu,\, b_q}\right)$$

$$= \sum_{r=o}^{k} (-1)^r \binom{k}{r} \prod_{\rho=o}^{k-r-1} (\nu-\rho)\; G^{m,n}_{p,q}\left(x\,\Big|\,{a_1-r, a_2, \ldots, a_p \atop b_1, \ldots, b_q}\right)$$

where k is a non-negative integer and $1 \leq n \leq p-1$.

(Srivastava and Gupta 1969, [310]).

1.8 Show that

$$(a_p-a_1)\; G^{m+1,n}_{p+1,q+1}\left(x\,\Big|\,{1+a_1, a_2, \ldots, a_{p-1}, 1+a_p, a_1+\nu \atop 1+a_1+\nu,\, b_1, \ldots, b_q}\right)$$

$$= G^{m,n}_{p,q}\left(x\,\Big|\,{1+a_1, a_2, \ldots, a_p}\right) + (\nu-q_p+a_1)\; G^{m,n}_{p,q}\left[x\,\Big|\,{a_1, \ldots, a_{p-1}, 1+a_p \atop b_1, \ldots, b_q}\right]$$

where $1 \leq n \leq p-1$, and

$$\frac{d^r}{dx^r}\left\{x^{a_p-1}\, G^{m,n}_{p,q}\left(\frac{z}{x}\Big|{a_p \atop b_q}\right)\right\} = x^{a_p-r-1}\, G^{m,n}_{p,q}\left(\frac{z}{x}\,\Big|\,{a_1, \ldots, a_{p-1}, a_p-r \atop b_1, \ldots, b_q}\right)$$

where $1 \leq n \leq p-1$ and r is any positive integer.

(Srivastava and Gupta, 1969, [310]).

1.9 Show that

$${}_{p+1}F_p\left({a_{p+1} \atop b_p};\, z\right) = \frac{\prod_{j=1}^{p}\Gamma(b_j)}{\prod_{j=1}^{p+1}\Gamma(a_j)}$$

$$\times \sum_{k=1}^{p+1} \frac{\Gamma(a_k) \prod_{j=1}^{p+1} \Gamma(a_j-a_k)\, (e^{i\pi}/z)^{a_k}}{\prod_{j=1}^{p}\Gamma(b_j-a_k)}$$

$$\times\ {}_{p+1}F_p\left(\begin{matrix}1+a_k-b_1,\ldots,1+a_k-b_p\\ 1+a_k-a_1,\ldots,*,\ldots,1+a_k-a_{p+1}\end{matrix};\frac{1}{z}\right)$$

where $0 < \arg z < 2\pi$. Also prove that this formula provides the analytic continuation of the function ${}_{p+1}F_p$ on the left from inside the unit disc to the outside of this disc.

1.10 Establish the following expansions for the G-function.

$$G^{m,n}_{p,q}\left(z\left|\begin{matrix}a_p\\ b_q\end{matrix}\right.\right) = \sum_{h=1}^{m} \frac{[\prod'_{j=1} \{\Gamma(b_j-b_h)\Gamma(1+b_h-b_j)\}\exp[-i\pi b_h(p+1-m-n)]}{\prod_{j=n+1}^{n+1} \Gamma(a_j-b_h)\ \Gamma(1+b_h-a_j)}$$

$$\times\ G^{1,p}_{p,q}[(-1)^{p+1-m-n}\ z\left|\begin{matrix}a_1,\ldots,a_p\\ b_h,\ldots*\ldots,b_q\end{matrix}\right.]$$

$$= \sum_{h=1}^{n} \frac{\prod'^{n}_{j=1} \{\Gamma(a_h-a_j)\Gamma(1+a_j-a_h)\}\exp[-i\pi(q+1-m-n)(a_h-1)]}{\prod_{j=m+1}^{q} \{\Gamma(a_h-b_j)\Gamma(1+b_j-a_h)\}}$$

$$\times\ G^{q,1}_{p,q}[(-1)^{q+1-m-n}\ z\left|\begin{matrix}a_h,\ldots*\ldots,a_p\\ b_1,\ldots,b_q\end{matrix}\right.] .$$

CHAPTER II

PARTICULAR CASES OF MEIJER'S G-FUNCTION

2.1 INTRODUCTION

In the previous chapter we have seen that the G-function can be expressed in terms of a generalized hypergeometric series in special cases. Since generalized hypergeometric series include, as particular cases, nearly all the elementary Special Functions occurring in Mathematical Physics, it is possible to express them by means of the G-function. Keeping this in mind we give mainly two types of relations, one in which the elementary Special Functions are expressed in terms of the G-function and the second in which G-function is represented in terms of elementary Special Functions. The material presented in this chapter is based mostly on the papers by Meijer ([204] to [209]) and Kaufman, Mathai and Saxena [134]. The results presented here will find applications not only in the following chapters but also in various problems of Physical, Biological and Sociological Sciences. The definitions of the various Special Functions are also given in order that the reader may be able to get representations for products of elementary Special Functions in terms of G-functions. A detailed discussion of the various properties of the elementary Special Functions can be found in the Monograph by Magnus, Oberhettinger and Soni [167]. These definitions can also be used in obtaining the relations that exist between the elementary Special Functions and generalized hypergeometric functions and vice-versa.

It is important to note that the results listed here are the key formulas from which a large number of results for the various Special Functions can be derived. Because of the importance of the results involving G-functions, we repeat here a few results given in the earlier chapter.

Algebraic and Logarithmic functions given in Sections 2.6 and 2.7 occur in Statistical problems. The reader is advised to look into the Mellin-Barnes type integral representations given in Sections 2.2 to 2.5 for the elementary Special Functions whenever there is any confusion to identify simple and logarithmic cases.

2.2 GAMMA FUNCTION AND RELATED FUNCTIONS

2.2.1. Gamma Function.

$$\Gamma(z) = \int_0^\infty e^{-x} x^{z-1} dx, \quad R(z) > 0.$$

$$\frac{1}{\Gamma(z)} = \frac{1}{2\pi i} \int_{c-i\infty}^{c+i\infty} e^t t^{-z} dt, \quad c > 0, \quad R(z) > 0 .$$

2.2.2 Psi-Function.

$$\psi(z) = \frac{d}{dz}\log\Gamma(z) = \frac{\Gamma'(z)}{\Gamma(z)} = \int_0^\infty [t^{-1}e^{-t} - (1-e^{-t})^{-1}e^{-tz}]dt, R(z) > 0,$$

$$= -\gamma + (z-1)\sum_{k=0}^{\infty}[(k+1)(z+k)]^{-1}, \quad \gamma \approx 0.5772156649\ldots$$

2.2.3 Beta Function.

$$B(x,y) = \frac{\Gamma(x)\,\Gamma(y)}{\Gamma(x+y)} = \int_0^1 t^{x-1}(1-t)^{y-1}dt. \qquad R(x) > 0,\ R(y) > 0.$$

2.2.4 Euler's Dilogarithm

$$L_2(z) = \sum_{n=1}^{\infty}\frac{z^n}{n^2} = -\int_0^z \frac{\log(1-z)}{z}\,dz .$$

2.2.5 Riemann's Zeta Function and Related Functions.

$$\zeta(s) = \sum_{n=1}^{\infty} n^{-s} .$$

$$\xi(t) = -\frac{1}{2}(t^2+\frac{1}{4})\pi^{-\frac{it}{2}-\frac{1}{4}}\Gamma(\frac{it}{2}+\frac{1}{4})\,\zeta(it+\frac{1}{2}) .$$

$$\zeta(z,a) = \sum_{n=0}^{\infty}(n+a)^{-z} .$$

$$\underset{\sim}{\phi}(z,s,\nu) = \sum_{n=0}^{\infty}\frac{z^n}{(\nu+n)^s} .$$

2.3 HYPERGEOMETRIC FUNCTIONS

2.3.1 Gauss's Hypergeometric Function:

$${}_2F_1(a,b;c;z) = \sum_{r=0}^{\infty}\frac{(a)_r(b)_r}{(c)_r\, r!}z^r ,$$

where c is not a negative integer or zero, $|z| < 1$; $z = 1$ and $R(c-a-b) > 0$; $z = -1$ and $R(c-a-b+1) > 0$.

$$= \frac{\Gamma(c)}{\Gamma(a)\,\Gamma(b)}\frac{1}{2\pi i}\int_{-i\infty}^{+i\infty}\frac{\Gamma(a+s)\,\Gamma(b+s)\,\Gamma(-s)}{\Gamma(c+s)}(-z)^s ds$$

where $|\arg(-z)| < \pi$ and the path of integration, is indented, if necessary, in such a manner so as to separate the poles at $s = 0,1,2,\ldots$ from the poles of $s = -a-n$, $s = -b-n (n = 0,1,2,\ldots,)$ of the integrand.

2.3.2 Generalized Hypergeometric Series.

$$ {}_pF_q(a_1,\dots,a_p;b_1,\dots,b_q;z) = \sum_{r=0}^{\infty} \frac{\prod_{j=1}^{p}(a_j)_r}{\prod_{j=1}^{q}(b_j)_r} \frac{z^r}{r!} $$

$p \leq q$; $p = q+1$ and $|z| < 1$.

$$ {}_pF_q(a_1,\dots,a_p;\, b_1,\dots,b_q;z) = \frac{\prod_{j=1}^{q}\Gamma(b_j)}{\prod_{j=1}^{p}\Gamma(a_j)} $$

$$ \times \frac{1}{2\pi i} \int_{-i\infty}^{+i\infty} \frac{\prod_{j=1}^{p}\Gamma(a_j+s)\,\Gamma(-s)}{\prod_{j=1}^{q}\Gamma(b_j+s)} (-z)^s\, ds, $$

where $|\arg(-z)| < \pi$ and the path of integration is indented if necessary in such a manner that the poles of $\Gamma(-s)$ are separated from those of $\Gamma(a_j+s)$ for $j = 1,\dots,p$.

$$ S_n(b_1,b_2,b_3,b_4;z) = \sum_{r=1}^{n} \frac{\prod_{j=1}^{n}{}'\,\Gamma(b_j-b_r)}{\prod_{j=n+1}^{4}\Gamma(1+b_r-b_j)} z^{1+2b_r} $$

$$ \times {}_0F_3(1+b_r-b_1,\dots *\dots,1+b_r-b_4;\,(-1)^n z^2) , $$

where the prime in Π' and the asterisk in ${}_0F_3$ indicate that the term containing b_h-b_h is to be omitted.

An empty product is always interpreted as unity.

2.3.3 Incomplete Beta Function.

(i) $$ B_x(p,q) = \int_0^x t^{p-1}(1-t)^{q-1}dt $$

$$ = p^{-1} x^p \, {}_2F_1(p,1-q;\, p+1;x) . $$

(ii) $$ I_x(p,q) = \frac{B_x(p,q)}{B_1(p,q)} . $$

2.3.4 Confluent Hypergeometric Functions.

An account of these functions can be found in the books by Whittaker and Watson [356], Tricomi [336] and Slater [301] .

2.3.4a. Whittaker Functions.

$$M_{k,m}(z) = z^{m+\frac{1}{2}} e^{-\frac{z}{2}} \; {}_1F_1(\tfrac{1}{2} - k+m;\ 2m+1; z)$$

$$= z^{m+\frac{1}{2}} e^{\frac{z}{2}} \; {}_1F_1(\tfrac{1}{2} + k+m;\ 2m+1;\ -z)$$

$$= \frac{\Gamma(1+2m)}{\Gamma(\frac{1}{2}+m+k)\ \Gamma(\frac{1}{2}+m-k)} e^{-\frac{z}{2}} z^{m+\frac{1}{2}}$$

$$\times \int_0^1 e^{-zt}\, t^{m-k-\frac{1}{2}} (1-t)^{m+k-\frac{1}{2}}\, dt\ ,$$

$$R(\tfrac{1}{2} + m \pm k) > 0, \quad |\arg z| < \pi,$$

$$= \frac{\Gamma(1+2m)}{\Gamma(\frac{1}{2}+m-k)} e^{-\frac{z}{2}} z^{m+\frac{1}{2}} \ .$$

$$\times \int_{c-i\infty}^{c+i\infty} \frac{\Gamma(-t)\ \Gamma(\frac{1}{2}+m-k+t)}{\Gamma(1+2m+t)} (-z)^t\, dt$$

$$|\arg z| < \frac{\pi}{2}\ , \quad 2m \neq 0, -1, -2, \ldots$$

$$W_{k,m}(z) = \sum_{m,-m} \frac{\Gamma(-2m)}{\Gamma(\frac{1}{2}-k-m)} M_{k,m}(z)\ ,$$

where the symbol $\sum_{m,-m}$ indicates that to the expression following it an expression in which m has been replaced by -m is to be added.

$$W_{k,m}(z) = \frac{2z^{\frac{1}{2}} e^{-\frac{z}{2}}}{\Gamma(\frac{1}{2}+m-k)\ \Gamma(\frac{1}{2}-m-k)} \int_0^\infty e^{-t}\, t^{-k-\frac{1}{2}} K_{2m}(2(zt)^{\frac{1}{2}})dt,$$

$R(\frac{1}{2} \pm m-k) > 0$ and $K_\nu(\cdot)$ is a Bessel function defined in Section 2.4.

$$= \frac{z^k \exp(-\frac{z}{2})}{\Gamma(\frac{1}{2}+m-k)\Gamma(\frac{1}{2}-m-k)} \frac{1}{2\pi i} \int_{c-i\infty}^{c+i\infty} \Gamma(-t)\Gamma(\tfrac{1}{2}+m-k+t)\Gamma(\tfrac{1}{2}-m-k+t) z^{-t} dt,$$

$$|\arg z| < \frac{3\pi}{2}\ , \quad \frac{1}{2} + k \pm m \neq 0, 1, 2, \ldots$$

2.3.4b. Parabolic Cylinder Function

$$D_\nu(z) = 2^{\frac{\nu}{2}+\frac{1}{4}} z^{-\frac{1}{2}} W_{\frac{\nu}{2}+\frac{1}{4},\frac{1}{4}}(\frac{z^2}{2}) = (-1)^n \exp(\frac{z^2}{4}) \frac{d^n}{dz^n} (e^{-\frac{z^2}{2}}).$$

$$D_\nu(z) = 2^{\frac{\nu}{2}-\frac{1}{4}} z^{\frac{1}{2}} e^{-\frac{3z^2}{4}} \frac{1}{2\pi i} \int_{-i\infty}^{i\infty} \frac{\Gamma(s)\Gamma(\frac{\nu}{2}+\frac{1}{4}-s)\Gamma(\frac{\nu}{2}-\frac{1}{4}-s)}{\Gamma(\frac{\nu}{2}+\frac{1}{4})\,\Gamma(\frac{\nu}{2}-\frac{1}{4})} (\frac{z^2}{2})^s \, ds,$$

$|\arg z| < \frac{3\pi}{4}$, $\nu \neq 1, 2, -\frac{1}{2}, -\frac{3}{2}, \ldots$

2.3.4c Bateman's Function

$$k_{2\nu}(z) = \frac{1}{\Gamma(\nu+1)} W_{\nu,\frac{1}{2}}(2z) .$$

2.3.4d The Exponential Integral and Related Functions.

$$-Ei(-x) = E_1(x) = \int_x^\infty \frac{e^{-t}\,dt}{t} = \Gamma(0,x), \; -\pi < \arg x < \pi .$$

$$\overset{+}{Ei}(x) = E_i(x+i0), \; \bar{Ei}(x) = Ei(x-i0), \quad x > 0 .$$

$$\overline{Ei(x)} = \frac{1}{2} [\overset{+}{Ei}(x) + \bar{Ei}(x)] , \quad x > 0$$

$$\ell i(z) = \int_o^z \frac{dt}{\log t} = Ei(\log z) .$$

$$si(x) = -\int_x^\infty \frac{\sin t}{t} dt = \frac{1}{2i} [Ei(ix) - Ei(-ix)].$$

$$Si(x) = \int_o^x \frac{\sin t}{t} dt = \frac{\pi}{2} + si(x) .$$

$$Ci(x) = -\int_x^\infty \frac{\cos t}{t} dt = -Ci(x) = \frac{1}{2} [Ei(ix) + Ei(-ix)] .$$

2.3.4e. Error Functions and Related Functions.

$$\mathrm{Erf}(x) = 2\pi^{-\frac{1}{2}} \int_0^x e^{-t^2} dt = \frac{2x}{\sqrt{\pi}} \; {}_1F_1(\tfrac{1}{2}; \tfrac{3}{2}; -x^2) .$$

$$\mathrm{Erfc}(x) = 2\pi^{-\frac{1}{2}} \int_x^\infty e^{-t^2} dt = 1-\mathrm{Erf}(x)$$

$$= (\pi x)^{-\frac{1}{2}} e^{-\frac{x^2}{2}} W_{-\frac{1}{4},\frac{1}{4}}(x^2) .$$

$$C(x) = (2\pi)^{-\frac{1}{2}} \int_0^x t^{-\frac{1}{2}} \cos t \, dt .$$

$$S(x) = (2\pi)^{-\frac{1}{2}} \int_0^x t^{-\frac{1}{2}} \sin t \, dt.$$

2.3.4f. Incomplete Gamma Functions

$$\gamma(a,x) = \int_0^x e^{-t} t^{\alpha-1} dt = a^{-1} x^a \; {}_1F_1(a;a+1;-x) .$$

$$\Gamma(a,x) = \int_x^\infty e^{-t} t^{\alpha-1} dt = \Gamma(a) - \gamma(a,x)$$

$$= x^{\frac{a-1}{2}} e^{-\frac{x}{2}} W_{\frac{a-1}{2},\frac{a}{2}}(x) .$$

2.3.4g. Coulomb Wave Functions.

$$F_L(\eta,\sigma) = C_L(\eta)\sigma^{L+1} e^{-i\sigma} \; {}_1F_1(\tfrac{L+1-i\eta}{2L+2} ; 2i\sigma) ,$$

$$C_L(\eta) = \frac{2^L e^{-\frac{\pi\eta}{2}} |\Gamma(L+1+i\eta)|}{(2L+1)!}$$

By virtue of Kummer's transformation namely,

$${}_1F_1(\alpha; \beta, z) = e^z \; {}_1F_1(\beta-\alpha; \beta; -z)$$

it follows that, $F_L(\eta,\sigma)$ is real if σ,η and L are real. In the applications we usually take L to be a positive integer or zero. An account of these functions can be found in the work of Abramowitz and Stegun [2] .

2.4 BESSEL FUNCTIONS AND ASSOCIATED FUNCTIONS

$$J_\nu(z) = \sum_{r=0}^{\infty} \frac{(-1)^r \left(\frac{z}{2}\right)^{\nu+2r}}{r!\,\Gamma(\nu+r+1)}$$

$$= \frac{\left(\frac{z}{2}\right)^\nu}{\Gamma(\nu+1)} \quad {}_0F_1\left(- \,;\, 1+\nu;\, -\frac{z^2}{4}\right) .$$

$$J_\nu(x) = \frac{1}{4\pi} \int_{c-i\infty}^{c+i\infty} \frac{\Gamma\left(\frac{\nu+s}{2}\right)}{\Gamma\left(1+\frac{\nu-s}{2}\right)} \left(\frac{x}{2}\right)^{-s} ds ,$$

$$x > 0,\ -R(\nu) < c < 1 .$$

$$I_\nu(z) = \sum_{r=0}^{\infty} \frac{\left(\frac{z}{2}\right)^{\nu+2r}}{r!\,\Gamma(\nu+r+1)}$$

$$= \frac{\left(\frac{z}{2}\right)^\nu}{\Gamma(\nu+1)} \quad {}_0F_1\left(- \,;\, 1+\nu;\, \frac{z^2}{4}\right)$$

$$= e^{-\frac{i\nu\pi}{2}} J_\nu\left(z e^{\frac{i\pi}{2}}\right), \quad -\pi < \arg z \le \frac{\pi}{2} .$$

$$M_{o,m}(z) = 2^{2m}\,\Gamma(m+1)\, z^{\frac{1}{2}}\, I_m\left(\frac{z}{2}\right) .$$

$$Y_\nu(z) = \operatorname{cosec}(\nu\pi)\,[J_\nu(z)\cos(\nu\pi) - J_{-\nu}(z)] ,$$

$$Y_\nu(x) = \frac{1}{2\pi i} \int_{c-i\infty}^{c+i\infty} \frac{\Gamma\left(s-\frac{\nu}{2}\right)\Gamma\left(s+\frac{\nu}{2}\right)}{\Gamma\left(s-\frac{\nu+1}{2}\right)\Gamma\left(\frac{3+\nu}{2}-s\right)} \left(\frac{x^2}{4}\right)^{-s} ds,$$

$$R(s\pm\nu) > 0 .$$

$$K_\nu(z) = \frac{\pi}{2} \frac{(I_{-\nu}(z) - I_\nu(z))}{\sin \nu\pi}$$

$$= \frac{1}{2} \sum_{\nu,-\nu} \Gamma(-\nu)\,\Gamma(1+\nu)\, I_\nu(z)$$

$$= \left(\frac{2z}{\pi}\right)^{-\frac{1}{2}} W_{0,\nu}(2z) .$$

$$K_\nu(x) = \frac{1}{4\pi i} \int_{c-i\infty}^{c+i\infty} \Gamma(s \pm \frac{\nu}{2}) (\frac{x^2}{4})^{-s} \, ds \, .$$

$$H_\nu^{(1)}(z) = J_\nu(z) + iY_\nu(z) \, .$$

$$H_\nu^{(2)}(z) = J_\nu(z) - iY_\nu(z) \, .$$

$$H_{i_\nu}(z) = \int_\infty^x J_\nu(t) \frac{dt}{t} \, .$$

$$2\pi i \, J_\nu(x) = \int_{-i\infty}^{+i\infty} \Gamma(-s)[\Gamma(\nu+s+1)]^{-1} (\frac{x}{2})^{\nu+2s} \, ds \, , \quad x > 0, \ R(\nu) > 0 \, .$$

$$2\pi^2 \, H_\nu^{(1)}(z) = -e^{-\frac{i\pi}{2}} \int_{c-i\infty}^{-c+i\infty} \Gamma(-s) \, \Gamma(-\nu-s) \, (-\frac{iz}{2})^{\nu+2s} \, ds \, ,$$

$$-\frac{\pi}{2} < \arg(-iz) < \frac{\pi}{2} \, .$$

$$2\pi^2 \, H_\nu^{(2)}(z) = e^{\frac{i\nu\pi}{2}} \int_{-c-i\infty}^{-c+i\infty} \Gamma(-s) \, \Gamma(-\nu-s) \, (\frac{iz}{2})^{\nu+2s} \, ds \, ,$$

$$-\frac{\pi}{2} < \arg(iz) < \frac{\pi}{2} \, .$$

2.4.1 Kelvin's and Associated Functions.

$$\mathrm{ber}_\nu(z) + i \, \mathrm{bei}_\nu(z) = J_\nu(z e^{\frac{3\pi i}{4}}) \, .$$

$$\mathrm{ber}_\nu(z) - i \, \mathrm{bei}_\nu(z) = J_\nu(z e^{\frac{-3\pi i}{4}}) \, .$$

$$\mathrm{ker}_\nu(z) + i \, \mathrm{kei}_\nu(z) = K_\nu(z e^{\frac{\pi i}{4}}) \, .$$

$$\mathrm{ker}_\nu(z) - i \, \mathrm{kei}_\nu(z) = K_\nu(z e^{-\frac{\pi i}{4}}) \, .$$

$$\mathrm{ber}(z) = \mathrm{ber}_0(z), \quad \mathrm{bei}(z) = \mathrm{bei}_0(z) \, .$$

$$\mathrm{ker}(z) = \mathrm{ker}_0(z) \, , \quad \mathrm{kei}(z) = \mathrm{kei}_0(z) \, .$$

$$\chi_\nu^{(b)}(z) = \mathrm{ber}_\nu^2(z) + \mathrm{bei}_\nu^2(z) .$$

$$V_\nu^{(b)}(z) = [\mathrm{ber}'_\nu(z)]^2 + [\mathrm{bei}'_\nu(z)]^2 .$$

$$W_\nu^{(b)}(z) = \mathrm{ber}_\nu(z)\,\mathrm{bei}'_\nu(z) - \mathrm{bei}_\nu(z)\,\mathrm{ber}'_\nu(z) .$$

$$\frac{1}{2} z_\nu^{(b)}(z) = \mathrm{ber}_\nu(z)\mathrm{bei}'_\nu(z) + \mathrm{ber}_\nu(z)\mathrm{ber}'_\nu(z).$$

2.4.2 Neumann Polynomials.

$$O_o(x) = \frac{1}{x} ; \quad O_n(x) = \frac{1}{4} \sum_{r=o}^{\leq \frac{n}{2}} \frac{n(n-r-1)!}{r!\, (\frac{x}{2})^{n-2r+1}} , \quad n = 1,2,\ldots .$$

$$\underline{O}_n(x) = (-1)^n\, O_n(x), \quad n = 1,2,\ldots .$$

2.4.3. Anger-Weber Functions.

$$\underset{\sim}{J}_\nu(z) = \pi^{-1} \int_o^\pi \cos(\nu\theta - z \sin\theta)d\theta.$$

$$\underset{\sim}{E}_\nu(z) = \pi^{-1} \int_o^\pi \sin(\nu\theta - z \sin\theta)d\theta .$$

2.4.4 Struve's Functions.

$$H_\nu(z) = \sum_{r=o}^{\infty} \frac{(-1)^r \, (\frac{z}{2})^{\nu+2r+1}}{\Gamma(r+\frac{3}{2})\,\Gamma(\nu+r+\frac{3}{2})}$$

$$= \frac{(\frac{z}{2})^{\nu+1}}{\Gamma(\frac{3}{2})\,\Gamma(\nu+\frac{3}{2})} \; {}_1F_2(1; \frac{3}{2}, \nu+\frac{3}{2} ; -\frac{z^2}{4})$$

$$= \frac{2^{1-\nu}}{\Gamma(\frac{1}{2})\,\Gamma(\nu+\frac{1}{2})} \; s_{\mu,\nu}(z) ,$$

where $s_{\mu,\nu}$ is a Lommel's function.

$$\underset{\sim}{L}_\nu(z) = \exp[-\frac{i\pi}{2}(\nu+1)]\, \underset{\sim}{H}_\nu(z \exp(\frac{i\pi}{2})) .$$

$$\underset{\sim}{H}_\nu(x) = \frac{2^{\mu-2}}{\pi i\, \Gamma(\frac{1}{2} - \frac{\mu}{2} \pm \frac{\nu}{2})} \int_{c-i\infty}^{c+i\infty} \Gamma(\frac{1+\mu}{2} + s)\, \Gamma(s\pm\frac{\nu}{2})\, \Gamma(\frac{1-\mu}{2} - s)(\frac{x^2}{4})^{-s} ds,$$

$R(s \pm \nu) > 0$, $R(s + \mu + 1) > 0$, $R(s) < R(1-\mu)$.

2.4.5. Lommel's Functions

$$s_{\mu,\nu}(z) = \frac{z^{\mu+1}}{(\mu-\nu+1)(\mu+\nu+1)} \; {}_1F_2\left(1; \frac{\mu-\nu+3}{2}, \frac{\mu+\nu+3}{2}; \frac{z^2}{4}\right).$$

$$S_{\mu,\nu}(z) = s_{\mu,\nu}(z) + 2^{\mu-1}\,\Gamma\left(\frac{\mu+\nu+1}{2}\right)\Gamma\left(\frac{\mu-\nu+1}{2}\right)$$

$$\times \left[\operatorname{Sin}\left\{\left(\frac{\mu-\nu}{2}\right)\pi\right\} J_\nu(z) - \cos\left\{\left(\frac{\mu-\nu}{2}\right)\pi\right\} Y_\nu(z) \right].$$

2.4.6. Elliptic Functions and Integrals

Complete Elliptic Integrals.

$$K(k) = \int_0^{\pi/2} (1-k^2\sin^2\emptyset)^{-\frac{1}{2}}\, d\emptyset = \frac{\pi}{2}\; {}_2F_1\left(\frac{1}{2}, \frac{1}{2}; 1; k^2\right).$$

$$E(k) = \int_0^{\pi/2} (1-k^2\sin^2\emptyset)^{\frac{1}{2}} = \frac{\pi}{2}\; {}_2F_1\left(-\frac{1}{2}, \frac{1}{2}; 1; k^2\right).$$

2.4.7 Theta Functions.

$$\Theta_0(\nu|\tau) = (-i\tau)^{-\frac{1}{2}} \sum_{r=-\infty}^{\infty} \exp\left[-i\pi \frac{(\nu-\frac{1}{2}+r)^2}{\tau}\right].$$

$$\Theta_1(\nu|\tau) = (-i\tau)^{-\frac{1}{2}} \sum_{r=-\infty}^{\infty} (-1)^r \exp \frac{[-i\pi(\nu+r-\frac{1}{2})^2}{\tau}].$$

$$\Theta_2(\nu|\tau) = (-i\tau)^{-1/2} \sum_{r=-\infty}^{\infty} (-1)^r \exp\left[\frac{-i\pi(\nu+r)^2}{\tau}\right].$$

$$\Theta_3(\nu|\tau) = (-i\tau)^{-1/2} \sum_{r=-\infty}^{\infty} \exp\left[\frac{-i\pi(\nu+r)^2}{\tau}\right].$$

$$\Theta_4(\nu|\tau) = \Theta_0(\nu|\tau).$$

2.4.8. Modified Theta Functions.

$$\hat{\Theta}_0(\nu|\tau) \doteq (-i\tau)^{-1/2} \left[\sum_{r=0}^{\infty} \exp\left(\frac{-i\pi(\nu+\frac{1}{2}+r)^2}{\tau}\right) - \sum_{r=-\infty}^{\infty} \exp\left[\frac{-i\pi(\nu+r+\frac{1}{2})^2}{\tau}\right]\right].$$

$$\hat{\theta}_2(\nu|\tau) = (-i\tau)^{-1/2}\Big[\sum_{r=o}^{\infty}(-1)^r \exp\Big(\frac{-i\pi(\nu+r)^2}{\tau}\Big) - \sum_{r=-1}^{-\infty}(-1)^r \exp\Big(\frac{-i\pi(\nu+r)^2}{\tau}\Big)\Big].$$

$$\hat{\theta}_3(\nu|\tau) = (-i\tau)^{-1/2}\Big[\sum_{r=o}^{\infty} \exp\Big\{\frac{-i\pi(\nu+r)^2}{\tau}\Big\} - \sum_{n=-1}^{\infty} \exp\Big\{\frac{-i\pi(\nu+r)^2}{\tau}\Big\}\Big].$$

For the definitions of MacRobert's E-function and Meijer's G-function, see Chapter I.

2.5. ORTHOGONAL POLYNOMIALS.

2.5.1. Legendre Polynomial.

$$P_n(x) = \frac{1}{2^n\, n!} \frac{d^n}{dx^n}(x^2 - 1)^n .$$

2.5.2. Gegenbauer Polynomial.

$$C_n^{\nu}(x) = \frac{(-2)^n(\nu)_n}{n!(2\nu+n)_n}(1-x)^{-\nu+\frac{1}{2}} \frac{d^n}{dx^n}(1-x^2)^{\nu+n-\frac{1}{2}} .$$

2.5.3. Tchebechef Polynomials.

$$T_n(x) = \cos(n \cos^{-1}x).$$

$$U_n(x) = \frac{[\sin (n+1) \cos^{-1}x]}{\sin (\cos^{-1}x)} .$$

2.5.4. Jacobi Polynomial.

$$P_n^{(\alpha,\beta)}(x) = \frac{(-1)^n}{2^n\, n!} (1-x)^{-\alpha} (1+x)^{-\beta} \frac{d^n}{dx^n} [(1-x)^{n+\alpha}(1+x)^{n+\beta}].$$

For the applications of these polynomials to non-linear differential equations, see the work of Saxena and Kushwaha ([274],[275]).

2.5.5. Laguerre Polynomial.

$$L_n^{(\alpha)}(z) = \frac{e^z z^{-\alpha}}{n!} \frac{d^n}{dx^n} (e^{-z} z^{n+\alpha}).$$

$$L_n^{o}(z) = L_n(z).$$

2.5.6. Hermite Polynomials.

$$He_n(x) = (-1)^n e^{\frac{x^2}{2}} \frac{d^n}{dx^n} (e^{-\frac{x^2}{2}}).$$

$$H_n(x) = (-1)^n e^{x^2} \frac{d^n}{dx^n} (e^{-x^2}).$$

2.5.7. Charlier Polynomial.

$$P_n(x;a) = n!\, a^{-n} L_n^{x-n}(a).$$

2.5.8. Legendre Functions.

$$P_\nu^\mu(z) = \frac{1}{\Gamma(1-\mu)} \left(\frac{z+1}{z-1}\right)^{\frac{\mu}{2}} {}_2F_1(-\nu,\nu+1;\, 1-\mu;\, \frac{1-z}{2}).$$

$$Q_\nu^\mu(z) = \frac{e^{\mu\pi i}\,\Gamma(\frac{1}{2})\,\Gamma(\mu+\nu+1)}{2^{\nu+1}\,\Gamma(\nu+\frac{3}{2})} z^{-\mu-\nu-1} (z^2-1)^{\frac{\mu}{2}}$$
$$\times\, {}_2F_1\left(\frac{\mu+\nu+1}{2}, \frac{\mu+\nu+2}{2};\, \nu+\frac{3}{2};\, \frac{1}{z^2}\right)$$

z in the complex plane cut along the real axis from -1 to 1.

$$P_\nu^\mu(x) = \frac{1}{\Gamma(1-\mu)} \left(\frac{1+x}{1-x}\right)^{\frac{\mu}{2}} {}_2F_1(-\nu,\ \nu+1;1-\mu\ ;\ \frac{1-x}{2}),\quad -1 < x < 1\ .$$

$$Q_\nu^\mu(x) = \frac{e^{-i\mu\pi}}{2} \left[e^{-\frac{\mu\pi i}{2}} Q_\nu^\mu(x+i0) + e^{\frac{\pi i}{2}} Q_\nu^\mu(x-i0)\right],\quad -1 < x < 1\ .$$

$$P_\nu(z) = P_\nu^0(z),\quad Q_\nu(x) = Q_\nu^0(z)\ .$$

2.6. ELEMENTARY SPECIAL FUNCTIONS EXPRESSED IN TERMS OF MEIJER'S G-FUNCTION.

$$z^a e^{-z} = G_{0,1}^{1,0}(z|\,a)\ .$$

$$\sin z = \left(\frac{\pi z}{2}\right)^{1/2} G_{0,2}^{1,0}\left(\frac{z^2}{4}\middle|\ \frac{1}{4}\ ,\ -\frac{1}{2}\right) = \pi^{1/2}\ G_{0,2}^{1,0}\left(\frac{z^2}{4}\middle|\frac{1}{2}\ ,0\right)\ .$$

$$\cos z = \left(\frac{\pi z}{2}\right)^{1/2} G_{0,2}^{1,0}\left(\frac{z^2}{4}\middle|\ -\frac{1}{4},\ \frac{1}{4}\right) = \pi^{\frac{1}{2}}\ G_{0,2}^{1,0}\left(\frac{z^2}{4}\ \middle|\ 0,\ \frac{1}{2}\right).$$

$$\sinh z = \frac{\pi^{1/2}}{i}\ G_{0,2}^{1,0}\left(-\ \frac{z^2}{4}\middle|\ \frac{1}{2}\ ,\ 0\right).$$

$$\cosh z = \pi^{1/2}\ G_{0,2}^{1,0}\left(-\ \frac{z^2}{4}\middle|\ 0,\ \frac{1}{2}\right).$$

$$\log(1 \pm z) = \pm\ G_{2,2}^{1,0}\left(\pm\ z\middle|\ \begin{matrix}1,1\\ 1,0\end{matrix}\right)\ .$$

$$\log\left(\frac{1+z}{1-z}\right) = z\ G_{2,2}^{1,2}\left(-\ z^2\middle|\ \begin{matrix}\frac{1}{2}\ ,\ 0\\ 0\ ,\ \frac{1}{2}\end{matrix}\right)\ .$$

$$\text{arc sin } z = \frac{1}{2}\ G_{2,2}^{1,2}\left(-\ z^2\middle|\ \begin{matrix}\frac{1}{2}\ ,\ \frac{1}{2}\\ 0\ \ -\frac{1}{2}\end{matrix}\right)\ .$$

$$\text{arc tan } z = \frac{1}{2}\ G_{2,2}^{1,2}\left(z^2\middle|\ \begin{matrix}1,\ \frac{1}{2}\\ \frac{1}{2}\ ,\ 0\end{matrix}\right)\ .$$

$$(1+z^2)^{-1/2}\ (1+(1+z^2)^{\frac{1}{2}})^{-2a} = \pi^{-\frac{1}{2}}\ G_{2,2}^{1,2}\left(z^2\middle|\ \begin{matrix}-a,\ \frac{1}{2}\ -a\\ 0,\ -2a\end{matrix}\right)\ .$$

$$\log(z + \sqrt{1+z^2}) = \frac{1}{2\pi^{1/2}}\ G_{2,2}^{1,2}\left(z^2\middle|\ \begin{matrix}1,1\\ \frac{1}{2}\ ,0\end{matrix}\right)\ .$$

$$(1+z)^{-2a} + (1-z)^{-2a} = \frac{1}{2^{2a} \, \Gamma(2a)} \quad G_{2,2}^{1,2} \left(-z^2 \Big| \begin{array}{c} 1-a, \ \frac{1}{2}-a \\ 0, \ \frac{1}{2} \end{array} \right) .$$

$$(1+z)^{1-2a} - (1-z)^{1-2a} = \frac{z(1-2a)}{2^{2a-1} \, \Gamma(2a)} \quad G_{2,2}^{1,2} \left(-z^2 \Big| \begin{array}{c} 1-a, \ \frac{1}{2}-a \\ 0, \ -\frac{1}{2} \end{array} \right)$$

$$z^{\alpha}(1-z)^{\beta} = \Gamma(\beta+1) \ G_{1,1}^{1,0} \left(z \Big| \begin{array}{c} \beta+\alpha+1 \\ \alpha \end{array} \right) .$$

$$\frac{z^{\beta}}{1+az^{\alpha}} = a^{-\frac{\beta}{\alpha}} \ G_{1,1}^{1,1} \left(az^{\alpha} \Big| \begin{array}{c} \frac{\beta}{\alpha} \\ \frac{\beta}{\alpha} \end{array} \right) .$$

$$(1-z)^{a-1} = \Gamma(a) \ G_{1,1}^{1,1} \left(z \Big| \begin{array}{c} a \\ 0 \end{array} \right) .$$

$$(1-z)^{-a} = \frac{1}{\Gamma(a)} \ G_{1,1}^{1,1} \left(-z \Big| \begin{array}{c} 1-a \\ 0 \end{array} \right) .$$

$$z^a \, J_{\nu}(z) = 2^a \ G_{0,2}^{1,0} \left(\frac{z^2}{4} \Big| \frac{a+\nu}{2} \ , \ \frac{a-\nu}{2} \right)$$

$$= 2^{2a} \ G_{0,4}^{2,0} \left(\frac{z^4}{256} \Big| \Delta(2, \frac{a+\nu}{2}) , \ \Delta(2, \frac{a-\nu}{2}) \right) .$$

$$z^a \, Y_{\nu}(z) = z^a \ G_{1,3}^{2,0} \left(\frac{z^2}{4} \Big| \begin{array}{c} \frac{a-\nu-1}{2} \\ \frac{a+\nu}{2}, \ \frac{a-\nu}{2}, \ \frac{a-\nu-1}{2} \end{array} \right) .$$

$$z^a K_{\nu}(z) = 2^{a-1} \ G_{0,2}^{2,0} \left(\frac{z^2}{4} \Big| \frac{a+\nu}{2} \ , \ \frac{a-\nu}{2} \right)$$

$$= 2^{2a-2} \, \pi^{-1}$$

$$\times \ G_{0,4}^{4,0} \left(\frac{z^4}{256} \Big| \Delta(2, \frac{a+\nu}{2}) , \ \Delta(2, \frac{a-\nu}{2}) \right).$$

$$(2z)^a \, e^{-z} I_{\nu}(z) = \pi^{-\frac{1}{2}} \ G_{1,2}^{1,1} \left(2z \Big| \begin{array}{c} \frac{1}{2}+a \\ a+\nu \ , \ a-\nu \end{array} \right) .$$

$$(2z)^a e^{-z} K_\nu(z) = \pi^{\frac{1}{2}} G^{2,0}_{1,2}\left(2z \,\middle|\, \begin{matrix} \frac{1}{2}+a \\ a+\nu,\, a-\nu \end{matrix}\right).$$

$$(2a)^a e^{z} K_\nu(z) = \pi^{-\frac{1}{2}} \cos(\nu\pi)\, G^{2,1}_{1,2}\left(2z \,\middle|\, \begin{matrix} \frac{1}{2}+a \\ a+\nu,\, a-\nu \end{matrix}\right).$$

$$(\tfrac{z}{2})^a s_{\mu,\nu}(z) = 2^{\mu-1}\,\Gamma\left(\frac{\mu+\nu+1}{2}\right)\Gamma\left(\frac{\mu-\nu+1}{2}\right)$$

$$\times\, G^{1,1}_{1,3}\left(\frac{z^2}{4} \,\middle|\, \begin{matrix} \frac{a+\mu+\nu}{2} \\ \frac{a+\mu+1}{2},\, \frac{a+\nu}{2},\, \frac{a-\nu}{2} \end{matrix}\right).$$

$$(\tfrac{z}{2})^a \mathbf{H}_\nu(z) = G^{1,1}_{1,3}\left(\frac{z^2}{4} \,\middle|\, \begin{matrix} \frac{a+\nu+1}{2} \\ \frac{a+\nu+1}{2},\, \frac{a+\nu}{2},\, \frac{a-\nu}{2} \end{matrix}\right).$$

$$(\tfrac{z}{2})^a [\mathbf{H}_\nu(z) - Y_\nu(z)] = \frac{\cos(\nu\pi)}{\pi^2} G^{3,1}_{1,3}\left(\frac{z^2}{4} \,\middle|\, \begin{matrix} \frac{a+\nu+1}{2} \\ \frac{a+\nu+1}{2},\, \frac{a+\nu}{2},\, \frac{a-\nu}{2} \end{matrix}\right).$$

$$(\tfrac{z}{2})^a [I_\nu(z) - \mathbf{L}_\nu(z)] = \frac{1}{\pi} G^{2,1}_{1,3}\left(\frac{z^2}{4} \,\middle|\, \begin{matrix} \frac{a+\nu+1}{2} \\ \frac{a+\nu+1}{2}\;\; \frac{a+\nu}{2}\;\; \frac{a-\nu}{2} \end{matrix}\right).$$

$$(\tfrac{z}{2})^a [I_\nu(z) - \mathbf{L}_\nu(z)] = \pi^{-1} \cos(\nu\pi) G^{2,1}_{1,3}\left(\frac{z^2}{4} \,\middle|\, \begin{matrix} \frac{a+\nu+1}{2} \\ \frac{a+\nu+1}{2}\;\; \frac{a-\nu}{2}\;\; \frac{a+\nu}{2} \end{matrix}\right).$$

$$(\tfrac{z}{2})^a S_{\mu,\nu}(z) = \frac{2^{\mu-1}}{\Gamma\left(\frac{1-\mu+\nu}{2}\right)\Gamma\left(\frac{1-\mu-\nu}{2}\right)}$$

$$\times\, G^{3,1}_{1,3}\left(\frac{z^2}{4} \,\middle|\, \begin{matrix} \frac{a+\mu+1}{2} \\ \frac{a+\mu+1}{2},\, \frac{a+\nu}{2},\, \frac{a-\nu}{2} \end{matrix}\right).$$

$$\mathrm{Ci}(z) + i\,\mathrm{Si}(z) = -\frac{z^{-1}e^{iz}}{\pi^{1/2}}\left\{2z^{-1}\,G^{3,1}_{1,3}\left(\frac{z^2}{4}\,\middle|\,\begin{matrix}1\\ \frac{3}{2},1,1\end{matrix}\right)\right.$$

$$\left.+\,i\,G^{3,1}_{1,3}\left(\frac{z^2}{4}\,\middle|\,\begin{matrix}1\\ \frac{1}{2},1,1\end{matrix}\right)\right\}.$$

$$C(Z) + iS(z) = \frac{1+i}{2} - \frac{(2\pi z)^{-1/2}\,e^{iz}}{2^{1/2}}$$

$$\times\left\{\left(\frac{2}{z}\right)G^{3,1}_{1,3}\left(\frac{z^2}{4}\,\middle|\,\begin{matrix}1\\ \frac{3}{4},\frac{5}{4},1\end{matrix}\right)\right.$$

$$\left.+\;i\,G^{3,1}_{1,3}\left(\frac{z^2}{4}\,\middle|\,\begin{matrix}1\\ \frac{1}{4},\frac{3}{4},1\end{matrix}\right)\right\}.$$

$$z^{\sigma}e^{\frac{z}{2}}M_{k,m}(z) = z^{\frac{1}{2}}\,\Gamma(2m+1)\,\Gamma(\tfrac{1}{2}+k-m)\,G^{1,0}_{0,2}\left(z\,\middle|\,\begin{matrix}\frac{1}{2}+\sigma+k\\ \sigma+m,\sigma-m\end{matrix}\right)$$

$$= \frac{(2\pi z)^{\frac{1}{2}}\Gamma\,(2m+1)\,\Gamma(k-m+\tfrac{1}{2})}{2^{k-\sigma}}$$

$$\times\,G^{2,0}_{2,4}\left(\frac{z^2}{4}\,\middle|\,\begin{matrix}\Delta(2,\frac{1}{2}+\sigma+k)\\ \Delta(2,\sigma+m),\ \Delta(2,\sigma-m)\end{matrix}\right).$$

$$z^{\sigma}e^{-\frac{z}{2}}M_{k,m}(z) = \frac{z^{\frac{1}{2}}\,\Gamma(2m+1)}{\Gamma(\frac{1}{2}+k+m}\,G^{1,1}_{1,2}\left(z\,\middle|\,\begin{matrix}\frac{1}{2}+\sigma-k\\ \sigma+m,\sigma-m\end{matrix}\right)$$

$$= \frac{(\frac{z}{2\pi})^{\frac{1}{2}}\,\Gamma(2m+1)2^{k+\sigma}}{\Gamma(\frac{1}{2}+k+m)}$$

$$\times\,G^{2,2}_{2,4}\left(\frac{z^2}{4}\,\middle|\,\begin{matrix}\Delta(2,\frac{1}{2}+\sigma-k)\\ \Delta(2,\sigma+m),\ \Delta(2,\sigma-m)\end{matrix}\right).$$

$$z^{\sigma} e^{\frac{z}{2}} W_{k,m}(z) = \frac{1}{\Gamma(\frac{1}{2} - k+m)\,\Gamma(\frac{1}{2} - k-m)}$$

$$\times G^{2,1}_{1,2}\left(z \,\Big|\, \begin{matrix} 1+\sigma+k \\ \frac{1}{2}+\sigma+m,\ \frac{1}{2}+\sigma-m \end{matrix}\right)$$

$$= \frac{z^{\frac{1}{2}}\, 2^{\sigma-k}}{(2\pi)^{\frac{3}{2}}\,\Gamma(\frac{1}{2} - k+m)\,\Gamma(\frac{1}{2} - k-m)}$$

$$\times G^{4,2}_{2,4}\left(\frac{z^2}{4} \,\Big|\, \begin{matrix} \Delta(2, \frac{1}{2}+\sigma+k) \\ \Delta(2,\sigma+m),\ \Delta(2,\sigma-m) \end{matrix}\right).$$

$$z^{\sigma} e^{-\frac{z}{2}} W_{k,m}(z) = z^{\frac{1}{2}} G^{2,0}_{1,2}\left(z \,\Big|\, \begin{matrix} \frac{1}{2}+\sigma-k \\ \sigma+m,\sigma-m \end{matrix}\right) = \left(\frac{z}{2\pi}\right)^{\frac{1}{2}} 2^{\sigma+k}$$

$$\times G^{4,0}_{2,4}\left(\frac{z^2}{4} \,\Big|\, \begin{matrix} \Delta(2, \frac{1}{2}+\sigma-k) \\ \Delta(2, \sigma+m),\ \Delta(2,\sigma-m) \end{matrix}\right).$$

$$W_{k,m}(z)\, M_{-k,m}(z) = \frac{\Gamma(1+2m)}{\pi^{\frac{1}{2}}\,\Gamma(\frac{1}{2} - k+m)} G^{3,1}_{2,4}\left(\frac{z^2}{4} \,\Big|\, \begin{matrix} 1+k,\ 1-k \\ \frac{1}{2},\ 1,\ \frac{1}{2}+m,\ \frac{1}{2}-m \end{matrix}\right)$$

$$z^2 W_{k,m}(2iz) W_{k,m}(-2iz) = \frac{z\pi^{-\frac{1}{2}}}{\Gamma(\frac{1}{2}+m-k)\,\Gamma(\frac{1}{2}-m-k)}$$

$$\times G^{4,1}_{2,4}\left(z^a \,\Big|\, \begin{matrix} \frac{1}{2}+\frac{\lambda}{2}+k,\ \frac{1}{2}+\frac{\lambda}{2}-k \\ \frac{\lambda}{2},\ \frac{1+\lambda}{2},\ \frac{\lambda}{2}+m,\ \frac{\lambda}{2}-m \end{matrix}\right).$$

$$W_{k,m}(z)\, W_{-k,m}(z) = \pi^{-\frac{1}{2}} G^{4,0}_{2,4}\left(\frac{z^2}{4} \,\Big|\, \begin{matrix} k+1,\ 1-k \\ \frac{1}{2},\ 1,\ m+\frac{1}{2},\ -m+\frac{1}{2} \end{matrix}\right).$$

$$z^a M_{k,m}(iz) M_{k,m}(-iz) = \frac{2^{a+1}\pi^{\frac{1}{2}} \Gamma^2(2m+1)}{\Gamma(\frac{1}{2}+k+m)\,\Gamma(\frac{1}{2}-k-m)}$$

$$\times\, G^{1,2}_{2,4}\left(\frac{z^2}{4}\,\middle|\,\begin{matrix}\frac{a}{2}+1+k,\ \frac{a}{2}+1-k\\ \frac{a+1}{2}+m,\ \frac{a+1}{2}-m,\ \frac{a}{2}+1,\ \frac{a+1}{2}\end{matrix}\right).$$

$$z^{\lambda} J_{\mu}(z)\, J_{\nu}(z) = \pi^{-\frac{1}{2}} G^{1,2}_{2,4}\left(z^2\,\middle|\,\begin{matrix}\frac{\lambda}{2},\ \frac{\lambda+1}{2}\\ \frac{\lambda+\mu+\nu}{2},\ \frac{\lambda-\mu+\nu}{2}\ \frac{\lambda+\mu-\nu}{2}\ \frac{\lambda-\mu-\nu}{2}\end{matrix}\right).$$

$$z^{\lambda} J^2_{\nu}(z) = \pi^{-\frac{1}{2}} G^{1,1}_{1,3}\left(z^2\,\middle|\,\begin{matrix}\frac{\lambda+1}{2}\\ \frac{\lambda}{2}+\nu,\ \frac{\lambda}{2}-\nu,\ \frac{\lambda}{2}\end{matrix}\right).$$

$$z^{\lambda} J_{-\nu}(z)\, J_{\nu}(z) = \pi^{-\frac{1}{2}} G^{1,1}_{1,3}\left(z^2\,\middle|\,\begin{matrix}\frac{\lambda+1}{2}\\ \frac{\lambda}{2},\ \frac{\lambda}{2}+\nu,\ \frac{\lambda}{2}-\nu\end{matrix}\right).$$

$$z^{\lambda} J_{\nu}(z)\, I_{\nu}(z) = \pi^{\frac{1}{2}}\, 2^{\frac{3\lambda}{2}}\, G^{1,0}_{0,4}\left(\frac{z^4}{64}\,\middle|\,\frac{\lambda+2\nu}{4},\frac{\lambda-2\nu}{4},\frac{\lambda}{4},\frac{\lambda+2}{4}\right).$$

$$z^{\lambda} J_{\nu}(z)\, Y_{\nu}(z) = -\pi^{-\frac{1}{2}} G^{2,0}_{1,3}\left(z^2\,\middle|\,\begin{matrix}\frac{\lambda+1}{2}\\ \frac{\lambda}{2},\ \frac{\lambda}{2}+\nu,\ \frac{\lambda}{2}-\nu\end{matrix}\right).$$

$$\left(\frac{z^2}{8}\right)^{\lambda} I_{\nu}(z)\, J_{-\nu}(z) = \pi^{\frac{1}{2}} \cos\left(\frac{\nu\pi}{2}\right) G^{1,0}_{0,4}\left(\frac{z^4}{64}\,\middle|\,\frac{\lambda}{2},\ \frac{\lambda+1}{2},\ \frac{\lambda+\nu}{2},\frac{\lambda-\nu}{2}\right) - \pi^{\frac{1}{2}} \sin\left(\frac{\nu\pi}{2}\right)$$

$$\times\, G^{1,0}_{0,4}\left(\frac{z^4}{64}\,\middle|\,\frac{\lambda+1}{2},\ \frac{\lambda}{2},\ \frac{\lambda+\nu}{2},\ \frac{\lambda-\nu}{2}\right).$$

$$I_{\nu}(z)\, K_{\nu}(z) = (4\pi)^{-\frac{1}{2}} G^{2,1}_{1,3}\left(z^2\,\middle|\,\begin{matrix}\frac{1}{2}\\ \nu,\ 0,\ -\nu\end{matrix}\right).$$

$$z^{\lambda} I_{\nu}(z)K_{\mu}(z) = (4\pi)^{-\frac{1}{2}} G^{2,2}_{2,4}\left(z^2 \middle| \begin{matrix} \frac{\lambda}{2}, \frac{\lambda+1}{2} \\ \frac{\lambda+\mu+\nu}{2} \; \frac{\lambda-\mu+\nu}{2} \; \frac{\lambda+\mu-\nu}{2} \; \frac{\lambda-\mu-\nu}{2} \end{matrix}\right).$$

$$z^{\lambda} K_{\nu}(z)\, J_{\nu}(z) = \pi^{-\frac{1}{2}}\, 2^{\frac{3\lambda-1}{2}}\, G^{3,0}_{0,4}\left(\frac{z^4}{64} \middle| \begin{matrix} \\ \frac{\lambda}{4}+\frac{\nu}{2}, \frac{\lambda}{4}+\frac{1}{2}, \frac{\lambda}{4}, \frac{\lambda}{4}-\frac{\nu}{2} \end{matrix}\right).$$

$$z^{\lambda} K_{\nu}(z)\, K_{\mu}(z) = 2^{-1}\pi^{\frac{1}{2}}\, G^{4,0}_{2,4}\left(z^2 \middle| \begin{matrix} \frac{\lambda}{2}, \frac{\lambda}{2}+\frac{1}{2} \\ \frac{\lambda+\mu+\nu}{2} \; \frac{\lambda-\mu+\nu}{2} \; \frac{\lambda+\mu-\nu}{2} \; \frac{\lambda-\mu-\nu}{2} \end{matrix}\right).$$

$$z^{\lambda} K^{2}_{\nu}(z) = 2^{-1}\pi^{\frac{1}{2}}\, G^{3,0}_{1,3}\left(z^2 \middle| \begin{matrix} \frac{\lambda+1}{2} \\ \nu+\frac{\lambda}{2}, \frac{\lambda}{2}-\nu, \frac{\lambda}{2} \end{matrix}\right).$$

$$z^{\lambda} H^{(1)}_{\nu}(z)\, H^{(2)}_{\nu}(z) = 2\pi^{-\frac{5}{2}} \cos(\nu\pi)\, G^{3,1}_{1,3}\left(z^2 \middle| \begin{matrix} \frac{\lambda+1}{2} \\ \frac{\lambda}{2}+\nu, \frac{\lambda}{2}-\nu, \frac{\lambda}{2} \end{matrix}\right).$$

$$z^{2\lambda} K_{2\nu}(ze^{\frac{\pi i}{4}})\, K_{2\nu}(ze^{-\frac{\pi i}{4}}) = 2^{3\lambda-3}\pi^{-\frac{1}{2}}\, G^{4,0}_{0,4}\left(\frac{z^4}{64} \middle| \frac{\lambda}{2}, \frac{\lambda+1}{2}, \frac{\lambda}{2}+\nu, \frac{\lambda}{2}-\nu\right).$$

$$z^{\lambda}[J_{\mu}(z)\, J_{\nu}(z) + J_{-\mu}(z)\, J_{-\nu}(z)] = \{\pi^{-\frac{1}{2}}[2\cos\frac{\pi(\mu+\nu)}{2}]\}$$

$$\times G^{2,1}_{2,4}\left(z^2 \middle| \begin{matrix} \frac{\lambda+1}{2}, \frac{\lambda}{2} \\ \frac{\lambda+\mu+\nu}{2}, \frac{\lambda-\mu-\nu}{2}, \frac{\lambda-\mu+\nu}{2}, \frac{\lambda+\mu-\nu}{2} \end{matrix}\right).$$

$$z^{\lambda}[J_{\mu}(z)J_{\nu}(z) - J_{-\mu}(z)J_{-\nu}(z)] = -\{\pi^{-\frac{1}{2}}[2\sin\frac{\pi(\mu+\nu)}{2}]\}$$

$$\times G^{2,1}_{2,4}\left(z^2 \middle| \begin{matrix} \frac{\lambda}{2}, \frac{\lambda+1}{2} \\ \frac{\lambda+\mu+\nu}{2}, \frac{\lambda-\mu-\nu}{2}, \frac{\lambda-\mu+\nu}{2}, \frac{\lambda+\mu-\nu}{2} \end{matrix}\right).$$

$$z^{\lambda}[H^{(1)}_{\mu}(z)H^{(1)}_{\nu}(z) - H^{(2)}_{\nu}(z)H^{(2)}_{\nu}(z)]$$

$$= -4i\pi^{-\frac{1}{2}}\, G^{3,0}_{2,4}\left(z^2 \middle| \begin{matrix} \frac{\lambda}{2}, \frac{\lambda+1}{2} \\ \frac{\lambda+\mu+\nu}{2}, \frac{\lambda-\mu+\nu}{2}, \frac{\lambda+\mu-\nu}{2}, \frac{\lambda-\mu-\nu}{2} \end{matrix}\right).$$

$$\left(\frac{z^{\lambda}}{2i}\right)\left[\exp\left\{\frac{i\pi}{2}(\nu-\mu)\right\} H_{\nu}^{(1)}(z)\, H_{\mu}^{(2)}(z) - \exp\left\{\frac{i\pi}{2}(\mu-\nu)\right\} H_{\mu}^{(1)}(z) H_{\nu}^{(2)}(z)\right] = \pi^{-\frac{5}{2}}[\cos(\mu\pi)-\cos(\nu\pi)]$$

$$\times\, G_{2,4}^{4,1}\left(z^{2}\,\middle|\,\begin{matrix}\frac{\lambda}{2}, \frac{\lambda+1}{2}\\ \frac{\lambda+\mu+\nu}{2}, \frac{\lambda-\mu+\nu}{2}, \frac{\lambda+\mu-\nu}{2}, \frac{\lambda-\mu-\nu}{2}\end{matrix}\right).$$

$$\frac{z^{\lambda}}{2}\left[\exp\left\{\frac{i\pi}{2}(\nu-\mu)\right\} H_{\nu}^{(1)}(z)\, H_{\mu}^{(2)}(z) + \exp\left\{\frac{i\pi}{2}(\mu-\nu)\right\} H_{\mu}^{(1)}(z) H_{\nu}^{(2)}(z)\right] = \pi^{-\frac{5}{2}}\,[\cos(\mu\pi)+\cos(\nu\pi)]$$

$$\times\, G_{2,4}^{4,1}\left(z^{2}\,\middle|\,\begin{matrix}\frac{\lambda}{2}, \frac{\lambda+1}{2}\\ \frac{\lambda+\mu+\nu}{2}\ \ \frac{\lambda-\mu+\nu}{2}\ \ \frac{\lambda+\mu-\nu}{2}\ \ \frac{\lambda-\mu-\nu}{2}\end{matrix}\right).$$

$$z^{\lambda}[I_{\mu}(z)I_{\nu}(z) - I_{-\mu}(z)(I_{-\nu}(z)] = -\frac{\sin(\mu+\nu)\pi}{\pi^{\frac{3}{2}}}$$

$$\times\, G_{2,4}^{2,2}\left(z^{2}\,\middle|\,\begin{matrix}\frac{\lambda}{2}, \frac{\lambda+1}{2}\\ \frac{\lambda+\mu+\nu}{2}, \frac{\lambda-\mu-\nu}{2}, \frac{\lambda-\mu+\nu}{2}, \frac{\lambda+\mu-\nu}{2}\end{matrix}\right).$$

$$z^{\lambda}\, {}_{2}F_{1}(a,b;c;-z) = \frac{\Gamma(c)}{\Gamma(a)\,\Gamma(b)}\; G_{2,2}^{1,2}\left(z\,\middle|\,\begin{matrix}1+\lambda-a, & 1+\lambda-b\\ \lambda & , 1+\lambda-c\end{matrix}\right).$$

$$z^{\lambda}\ {}_pF_q(a_p;b_q;-z) = \frac{\prod_{j=1}^{q}\Gamma(b_j)}{\prod_{j=1}^{p}\Gamma(a_j)}\ G^{1,p}_{p,p+1}\left(z\,\middle|\,\begin{matrix}1+\lambda-(a_p)\\ \lambda,1+\lambda-(b_q)\end{matrix}\right),$$

$p \le q$ or $p = q+1$ and $|z| < 1.$

$$z^{\lambda}\ {}_pF_q(a_p;b_q;-z) = \frac{\prod_{j=1}^{q}\Gamma(b_j)}{\prod_{j=1}^{p}\Gamma(a_j)}\ G^{p,1}_{q+1,p}\left(z^{-1}\,\middle|\,\begin{matrix}\lambda+1,\lambda+(b_q)\\ \lambda+(a_p)\end{matrix}\right),$$

$p \le q$ or $p = q+1$ and $|z| < 1.$

$${}_2F_1(a,b;c;1-z) = \frac{\Gamma(c)}{\Gamma(a)\Gamma(b)\Gamma(c-a)\Gamma(c-b)}\ G^{2,2}_{2,2}\left(z\,\middle|\,\begin{matrix}1-a,1-b\\ 0,\ c-a-b\end{matrix}\right),$$

provided $c-a,\ c-b \ne 0,-1,-2,\dots$.

Various other elementary Special Functions, for example, combinations of Legendre functions and also combinations of generalized hypergeometric functions can be expressed in terms of the G-function. In this connection, see the original papers of Meijer given in the references.

2.7. THE G-FUNCTION EXPRESSED IN TERMS OF ELEMENTARY SPECIAL FUNCTIONS

$$G^{1,0}_{0,1}(z|a) = z^a e^{-z}.$$

$$G^{1,0}_{1,1}\left(z\,\middle|\,\begin{matrix}a+\frac{1}{2}\\ a\end{matrix}\right) = \pi^{-\frac{1}{2}} z^a (1-z)^{-\frac{1}{2}},\quad |z| < 1.$$

$$G^{1,0}_{0,2}(z|\alpha,\beta) = z^{\frac{\alpha+\beta}{2}} J_{\alpha-\beta}(2z^{\frac{1}{2}}).$$

$$G^{1,0}_{1,2}\left(z\,\middle|\,\begin{matrix}\frac{1}{2}\\ a,-a\end{matrix}\right) = \pi^{-\frac{1}{2}} \cos(a\pi) e^{\frac{z}{2}} I_a(\tfrac{z}{2}).$$

$$G^{2,0}_{0,2}(z|\alpha,\beta) = 2\pi^{\frac{\alpha+\beta}{2}} K_{\alpha-\beta}(2z^{\frac{1}{2}}).$$

$$G^{1,1}_{1,2}\left(z\,\middle|\,\begin{matrix}\frac{1}{2}\\ \beta,-\beta\end{matrix}\right) = \pi^{\frac{1}{2}}\, e^{-\frac{z}{2}}\, I_\beta\left(\frac{z}{2}\right).$$

$$G^{1,1}_{1,2}\left(z\,\middle|\,\begin{matrix}\alpha\\ \beta,\gamma\end{matrix}\right) = \frac{\Gamma(1-\alpha+\beta)}{\Gamma(1-\gamma+\beta)}\, z^{\beta}\, {}_1F_1(1-\alpha+\beta;\ 1+\beta-\gamma;\ -z).$$

$$G^{2,0}_{1,2}\left(z\,\middle|\,\begin{matrix}\frac{1}{2}\\ \beta,-\beta\end{matrix}\right) = \pi^{-\frac{1}{2}}\, e^{-\frac{z}{2}}\, K_\beta\left(\frac{z}{2}\right).$$

$$G^{2,0}_{1,2}\left(z\,\middle|\,\begin{matrix}\alpha\\ \beta,\gamma\end{matrix}\right) = z^{\frac{\beta+\gamma-1}{2}}\, e^{-\frac{z}{2}}\, W_{\frac{1+\beta+\gamma}{2}-\alpha,\ \frac{\beta-\gamma}{2}}(z).$$

$$G^{2,0}_{2,2}\left(z\,\middle|\,\begin{matrix}a+\frac{1}{3},\ a+\frac{2}{3}\\ a,\quad a\end{matrix}\right) = z^{a}\, {}_2F_1\left(\frac{2}{3},\frac{1}{3};\ 1;\ 1-z\right),\ |z| < 1,$$

(logarithmic case)

$$G^{2,0}_{2,2}\left(z\,\middle|\,\begin{matrix}a,a\\ a-\frac{1}{2},\ a-1\end{matrix}\right) = z^{\frac{1}{2}}\, \pi^{\frac{1}{2}}\, z^{a-1} - \left(2z^{a-\frac{1}{2}}\, \pi^{-\frac{1}{2}}\right)\, {}_2F_1\left(\frac{1}{2},\frac{1}{2};\frac{3}{2};z\right)$$

$$= \frac{1}{\Gamma\left(\frac{3}{2}\right)}\, z^{a-1}(1-z)^{\frac{1}{2}}\, {}_2F_1\left(\frac{1}{2},\frac{1}{2};\frac{3}{2};1-z\right),\ |z| < 1,$$

(logarithmic case).

$$G^{2,1}_{1,2}\left(z\,\middle|\,\begin{matrix}\frac{1}{2}\\ \beta,\ -\beta\end{matrix}\right) = \frac{\pi^{\frac{1}{2}}}{\cos(\beta\pi)}\, e^{\frac{z}{2}}\, K_\beta\left(\frac{z}{2}\right).$$

$$G^{2,1}_{1,2}\left(z\,\middle|\,\begin{matrix}\alpha\\ \beta,\gamma\end{matrix}\right) = \Gamma(\beta-\alpha+1)\Gamma(\gamma-\alpha+1)\, z^{\frac{\alpha+\gamma-1}{2}}\, e^{\frac{z}{2}}\, W_{\alpha-\frac{(\beta+\gamma+1)}{2},\ \frac{\beta-\gamma}{2}}(z).$$

$$G^{1,0}_{0,4}\left(z\,\middle|\,\alpha,\beta,2\beta-\alpha,\beta+\frac{1}{2}\right) = \pi^{-\frac{1}{2}}\, I_{2(\alpha-\beta)}\left(2^{\frac{3}{2}} z^{\frac{1}{4}}\right) J_{2(\alpha-\beta)}\left(2^{\frac{3}{2}} z^{\frac{1}{4}}\right) z^{\beta}.$$

$$G^{1,0}_{0,4}\left(z\,\middle|\,\alpha,\alpha+\frac{1}{2},\beta,2\alpha-\beta\right) = \frac{\pi^{-\frac{1}{2}}}{2}\, \sec[(\beta-\alpha)\pi]\, z^{\alpha}\, \Big[J_{2(\alpha-\beta)}\left(2^{\frac{3}{2}} z^{\frac{1}{4}}\right)$$

$$\times J_{2(\beta-\alpha)}\left(2^{\frac{3}{2}} z^{\frac{1}{4}}\right) + I_{2(\beta-\alpha)}\left(2^{\frac{3}{2}} z^{\frac{1}{4}}\right)\ J_{2(\beta-\alpha)}\left(2^{\frac{3}{2}} z^{\frac{1}{4}}\right)\Big].$$

$$G^{1,0}_{0,4}(z\,|\,\alpha+\tfrac{1}{2},\,\alpha,\beta,2\alpha-\beta) = \frac{\pi^{-\frac{1}{2}}\,z^{\alpha}}{2}\ \mathrm{cosec}(\alpha-\beta)\pi$$

$$\times\,[J_{2(\alpha-\beta)}(2^{\frac{3}{2}}\,z^{\frac{1}{4}})\,I_{2(\beta-\alpha)}(2^{\frac{3}{2}}\,z^{\frac{1}{4}})$$

$$-\,I_{2(\alpha-\beta)}(2^{\frac{3}{2}}\,z^{\frac{1}{4}})\,J_{2(\beta-\alpha)}(2^{\frac{3}{2}}\,z^{\frac{1}{4}})].$$

$$G^{1,1}_{1,3}\left(z\,\middle|\,\begin{matrix}\frac{1}{2}\\ \alpha,0,\,-\alpha\end{matrix}\right) = \pi^{\frac{1}{2}}\,J^{2}_{\alpha}\,(z^{\frac{1}{2}})\ .$$

$$G^{1,1}_{1,3}\left(z\,\middle|\,\begin{matrix}\frac{1}{2}\\ 0,\alpha,\,-\alpha\end{matrix}\right) = \pi^{\frac{1}{2}}\,J_{\alpha}(z^{\frac{1}{2}})\,J_{-\alpha}(z^{\frac{1}{2}})\ .$$

$$G^{1,1}_{1,3}\left(z\,\middle|\,\begin{matrix}\alpha\\ \alpha,\beta,\alpha-\frac{1}{2}\end{matrix}\right) = z^{\frac{\alpha+\beta}{2}-\frac{1}{4}}\,\underset{\sim}{H}_{a-b-\frac{1}{2}}\,(2z^{\frac{1}{2}})\ .$$

$$G^{2,0}_{1,3}\left(z\,\middle|\,\begin{matrix}\alpha-\frac{1}{2}\\ \alpha,\beta,\alpha-\frac{1}{2}\end{matrix}\right) = z^{\frac{\alpha+\beta}{2}}\,Y_{\beta-\alpha}(2z^{\frac{1}{2}})\ .$$

$$G^{2,0}_{1,3}\left(z\,\middle|\,\begin{matrix}\alpha+\frac{1}{2}\\ \beta,\alpha,2\alpha-\beta\end{matrix}\right) = -\,\pi^{\frac{1}{2}}\,z^{\alpha}\,J_{\beta-\alpha}(z^{\frac{1}{2}})\,Y_{\beta-\alpha}(z^{\frac{1}{2}})\ .$$

$$G^{2,0}_{1,3}\left(z\,\middle|\,\begin{matrix}\frac{1}{2}\\ \alpha,-\alpha,0\end{matrix}\right) = \frac{\pi^{\frac{1}{2}}}{2}\ \mathrm{cosec}(\alpha\pi)\ [J^{2}_{-\alpha}(z^{\frac{1}{2}}) - J^{2}_{\alpha}(z^{\frac{1}{2}})].$$

$$G^{2,1}_{1,3}\left(z\,\middle|\,\begin{matrix}\frac{1}{2}\\ \alpha,0,-\alpha\end{matrix}\right) = 2\pi^{\frac{1}{2}}\,[I_{\alpha}(z^{\frac{1}{2}})\,K_{\alpha}(z^{\frac{1}{2}})\,]\,.$$

$$G^{2,1}_{1,3}\left(z\,\middle|\,\begin{matrix}\frac{1}{2}\\ \alpha,-\alpha,0\end{matrix}\right) = \pi^{\frac{3}{2}}\mathrm{cosec}\ 2a\pi\ [I^{2}_{-\alpha}(z^{\frac{1}{2}}) - I_{\alpha}(z^{\frac{1}{2}})]\ .$$

$$G^{2,1}_{1,3}\left(z\,\middle|\,\begin{matrix}\alpha+\frac{1}{2}\\ \alpha+\frac{1}{2},\beta,\alpha\end{matrix}\right) = \frac{\pi z^{\frac{\alpha+\beta}{2}}}{\cos(\alpha-\beta)\pi}\ [I_{\beta-\alpha}(2z^{\frac{1}{2}}) - \underset{\sim}{L}_{\alpha-\beta}(2z^{\frac{1}{2}})].$$

$$G^{2,1}_{1,3}\left(z\,\middle|\,\begin{matrix}\alpha+\frac{1}{2}\\ \alpha,\alpha+\frac{1}{2},\beta\end{matrix}\right) = \pi z^{\frac{\alpha+\beta}{2}}\,[I_{\alpha-\beta}(2z^{\frac{1}{2}}) - \mathbf{L}_{\alpha-\beta}(2z^{\frac{1}{2}})\,].$$

$$G^{2,0}_{0,4}(z\,|\,\alpha,\alpha+\tfrac{1}{2},\beta,\beta+\tfrac{1}{2}) = z^{\frac{\alpha+\beta}{2}}\,J_{2(\alpha-\beta)}(4z^{\frac{1}{4}})\,.$$

$$G^{2,0}_{0,4}(z\,|\,\alpha,-\alpha,0,\tfrac{1}{2}) = -\pi^{\frac{1}{2}}\operatorname{cosec}(2\alpha\pi)$$

$$\times\,[\,J_{2\alpha}(ue^{\frac{i\pi}{4}})\,J_{2\alpha}(ue^{-\frac{i\pi}{4}}) - J_{-2\alpha}(ue^{\frac{\pi i}{4}})\,J_{-2\alpha}(ue^{-\frac{\pi i}{4}})\,,\quad u = 2^{\frac{3}{2}}z^{\frac{1}{4}}$$

$$G^{2,0}_{0,4}(z\,|\,0,\tfrac{1}{2},\alpha,-\alpha) = -\pi^{\frac{1}{2}}\operatorname{cosec}(2\alpha\pi)[\exp(2\alpha\pi i)J_{2\alpha}(ue^{-\frac{\pi i}{4}})\,J_{-2\alpha}(ue^{\frac{\pi i}{4}})]$$

$$-\exp(-2\alpha\pi i)J_{2\alpha}(ue^{\frac{\pi i}{4}})\,J_{-2\alpha}(ue^{-\frac{\pi i}{4}}),\quad u = 2^{\frac{3}{2}}\,z^{\frac{1}{4}}\,.$$

$$G^{2,0}_{2,2}\left(z\,\middle|\,\begin{matrix}\alpha_1+\beta_1-1,\alpha_2+\beta_2-1\\ \alpha_1-1,\alpha_2-1\end{matrix}\right) = \frac{z^{\alpha_2-1}(1-z)^{\beta_1+\beta_2-1}}{\Gamma(\beta_1+\beta_2)}\;{}_2F_1\left(\begin{matrix}\alpha_2+\beta_2-\alpha_1,\beta_1;1-z\\ \beta_1+\beta_2\;;\end{matrix}\right)$$

$$|z| < 1\,.$$

$$G^{3,0}_{0,3}\left(z\,\middle|\,\begin{matrix}a+1,a+\frac{3}{2},a+1\\ a,a-\frac{1}{2},a-1\end{matrix}\right) = 4z^{a-1}[\,-2z^{\frac{1}{2}} + \tfrac{2}{3}\,z^{\frac{3}{2}} + \tfrac{1}{3}(1-\log z)z]\,,\quad |z| < 1.$$

$$G^{3,0}_{1,3}\left(z\,\middle|\,\begin{matrix}\alpha+\frac{1}{2}\\ \alpha+\beta,\alpha-\beta,\alpha\end{matrix}\right) = 2\pi^{-\frac{1}{2}}\,z^{\alpha}\,K^2_{\beta}\,(z^{\frac{1}{2}})\,.$$

$$G^{3,1}_{1,3}\left(z\,\middle|\,\begin{matrix}\alpha+\frac{1}{2}\\ \alpha+\frac{1}{2},-\alpha,\alpha\end{matrix}\right) = \frac{\pi^2}{\cos 2\alpha\pi}\,[H_{2\alpha}(2z^{\frac{1}{2}}) - Y_{2\alpha}(2z^{\frac{1}{2}})]\,.$$

$$G^{3,1}_{1,3}\left(z\,\middle|\,\begin{matrix}\alpha\\ \alpha,\beta,-\beta\end{matrix}\right) = 2^{2-2\alpha}\,\Gamma(1-\alpha-\beta)\,\Gamma(1-\alpha+\beta)\;\;S_{2\alpha-1,2\beta}(2z^{\frac{1}{2}})\,.$$

$$G^{3,1}_{1,3}\left(z\,\middle|\,\begin{matrix}\alpha+\frac{1}{2}\\ \beta,2\alpha-\beta,\alpha\end{matrix}\right) = \frac{\pi^{\frac{5}{2}}}{2}\;\sec(\beta-\alpha)\pi\;z^{\alpha}\,H^{(1)}_{\beta-\alpha}(z^{\frac{1}{2}})\,H^{(2)}_{\beta-\alpha}(z^{\frac{1}{2}})\,.$$

$$G^{3,0}_{0,4}(z|3\alpha-\tfrac{1}{2},\ \alpha,-\alpha-\tfrac{1}{2},\ \alpha-\tfrac{1}{2}) = 2\pi^{\frac{1}{2}}\sec(2\alpha\pi)\ z^{\alpha-\frac{1}{2}}\ K_{4\alpha}(2^{\frac{3}{2}} z^{\frac{1}{4}})$$

$$\times [J_{4\alpha}(2^{\frac{3}{2}} z^{\frac{1}{4}}) + J_{-4\alpha}(2^{\frac{3}{2}} z^{\frac{1}{4}})] .$$

$$G^{3,0}_{0,4}(z|\ 0,\alpha-\tfrac{1}{2},\ -\alpha-\tfrac{1}{2},\ -\tfrac{1}{2}) = 4\pi^{\frac{1}{2}}\ z^{-\frac{1}{2}}\ K_{2\alpha}(2^{\frac{3}{2}} z^{\frac{1}{4}})\ [J_{2\alpha}(2^{\frac{3}{2}} z^{\frac{1}{4}})\cos(\alpha\pi)$$

$$-\ Y_{2\alpha}(2^{\frac{3}{2}} z^{\frac{1}{4}})\ \sin(\alpha\pi)] .$$

$$G^{3,0}_{0,4}(z|\ -\tfrac{1}{2},\ \alpha-\tfrac{1}{2},\ -\alpha-\tfrac{1}{2},0) = -4\pi^{\frac{1}{2}}\ z^{-\frac{1}{2}}\ K_{2\alpha}(2^{\frac{3}{2}} z^{\frac{1}{4}})$$

$$\times [\ J_{2\alpha}(2^{\frac{3}{2}} z^{\frac{1}{4}})\ \sin(\alpha\pi) + Y_{2\alpha}(2^{\frac{3}{2}} z^{\frac{1}{4}})\ \cos(\alpha\pi)] .$$

$$G^{3,0}_{0,4}(z|\ \alpha,\beta+\tfrac{1}{2},\ \beta,2\beta-\alpha) = (2\pi)^{\frac{1}{2}}\ z^{\beta}\ K_{2(\alpha-\beta)}(2^{\frac{3}{2}} z^{\frac{1}{4}})\ J_{2(\alpha-\beta)}(2^{\frac{3}{4}} z^{\frac{1}{4}}) .$$

$$G^{4,0}_{0,4}(z|\ \alpha,\alpha+\tfrac{1}{2},\beta,\beta+\tfrac{1}{2}) = 4\pi z^{\frac{\alpha+\beta}{2}}\ K_{2(\alpha-\beta)}(4z^{\frac{1}{4}}) .$$

$$G^{4,0}_{0,4}(z|\ \alpha,\alpha+\tfrac{1}{2},\beta,2\alpha-\beta) = 2^{\frac{3}{2}}\ \pi^{\frac{1}{2}}\ z^{\alpha}\ K_{2(\beta-\alpha)}(2^{\frac{3}{2}} z^{\frac{1}{4}}\ e^{\frac{\pi i}{4}})K_{2(\beta-\alpha)}(2^{\frac{3}{2}}z^{\frac{1}{4}}e^{-\frac{\pi i}{4}}).$$

$$G^{n,0}_{0,4}(z|\ \alpha,\beta,\gamma,\delta) = z^{-\frac{1}{2}}\ S_n(\alpha,\beta,\gamma,\delta;\ z^{\frac{1}{2}}),\quad n = 1,2,3,4 .$$

$$G^{1,2}_{2,2}\left(z\middle|\begin{matrix}-\gamma_1,-\gamma_2\\ \alpha-1,\beta\end{matrix}\right) = \frac{\Gamma(\alpha+\gamma_1)\ \Gamma(\alpha+\gamma_2)}{\Gamma(\alpha+\beta)}\ z^{\alpha-1}\ {}_2F_1(\alpha+\gamma_1,\alpha+\gamma_2;\ \alpha+\beta;\ -z) .$$

$$G^{1,2}_{2,4}\left(z\middle|\begin{matrix}\alpha+\frac{1}{2},\alpha\\ \alpha+\beta,\alpha-\gamma,\alpha+\gamma,\alpha-\beta\end{matrix}\right) = \pi^{\frac{1}{2}}\ z^{\alpha}\ J_{\beta+\gamma}(z^{\frac{1}{2}})\ J_{\beta-\gamma}(z^{\frac{1}{2}}) .$$

$$G^{2,2}_{2,4}\left(z\ \middle|\begin{matrix}\alpha,\alpha+\frac{1}{2}\\ \beta,\gamma,2\alpha-\gamma,2\alpha-\beta\end{matrix}\right) = 2\pi^{\frac{1}{2}}\ z^{\alpha}\ I_{\beta+\gamma+2\alpha}(z^{\frac{1}{2}})\ K_{\beta-\gamma}(z^{\frac{1}{2}}) .$$

$$G^{3,0}_{2,4}\left(z\,\middle|\,\begin{matrix}0,\frac{1}{2}\\ \alpha,\beta,-\beta,-\alpha\end{matrix}\right)=\frac{i}{4}\pi^{\frac{1}{2}}\,[H^{(1)}_{\alpha-\beta}(z^{\frac{1}{2}})H^{(1)}_{\alpha+\beta}(z^{\frac{1}{2}})-H^{(2)}_{\alpha-\beta}(z^{\frac{1}{2}})H^{(2)}_{\alpha+\beta}(z^{\frac{1}{2}})].$$

$$G^{3,1}_{2,4}\left(z\,\middle|\,\begin{matrix}\frac{1}{2}+\alpha,\frac{1}{2}-\alpha\\ 0,\frac{1}{2},\beta,-\beta\end{matrix}\right)=\frac{\pi^{\frac{1}{2}}\,\Gamma(\frac{1}{2}-\alpha+\beta)}{z^{\frac{1}{2}}\Gamma(1+2\alpha)}\,W_{\alpha,\beta}(2z^{\frac{1}{2}})\,M_{-\alpha,\beta}(2z^{\frac{1}{2}})\,.$$

$$G^{4,0}_{2,4}\left(z\,\middle|\,\begin{matrix}\frac{1}{2}+\alpha,\frac{1}{2}-\alpha\\ 0,\frac{1}{2},\beta,-\beta\end{matrix}\right)=\pi^{\frac{1}{2}}z^{-\frac{1}{2}}\,W_{\alpha,\beta}(2z^{\frac{1}{2}})W_{-\alpha,\beta}(2z^{\frac{1}{2}})\,.$$

$$G^{4,0}_{2,4}\left(z\,\middle|\,\begin{matrix}\alpha,\alpha+\frac{1}{2}\\ \beta+\gamma,\beta-\gamma,\beta+\gamma\frac{1}{2},\beta-\gamma+\frac{1}{2}\end{matrix}\right)=\pi^{\frac{1}{2}}\,2^{2\gamma-2\beta-\frac{1}{2}}\,z^{\beta-\frac{1}{4}}e^{-z^{\frac{1}{2}}}\,W_{\frac{1}{2}+2\beta-2\gamma,2\gamma}(2z^{\frac{1}{2}}).$$

$$G^{4,0}_{2,4}\left(z\,\middle|\,\begin{matrix}\alpha,\alpha+\frac{1}{2}\\ \alpha+\beta,\alpha+\gamma,\alpha-\gamma,\alpha-\beta\end{matrix}\right)=2\pi^{-\frac{1}{2}}\,z^{\alpha}\,K_{\beta+\gamma}(z^{\frac{1}{2}})\,K_{\beta-\gamma}(z^{\frac{1}{2}})\,.$$

$$G^{4,1}_{2,4}\left(z\,\middle|\,\begin{matrix}0,\frac{1}{2}\\ \alpha,\beta,-\alpha,-\beta\end{matrix}\right)=\frac{i\,\pi^{\frac{5}{2}}}{4\sin(\alpha\pi)\,\sin(\beta\pi)}\,[\exp(-\beta\pi i)H^{(1)}_{\alpha-\beta}(z^{\frac{1}{2}})\,H^{(2)}_{\alpha+\beta}(z^{\frac{1}{2}})$$

$$-\ \exp(\beta\pi i)\,H^{(1)}_{\alpha+\beta}(z^{\frac{1}{2}})\,H^{(2)}_{\alpha-\beta}(z^{\frac{1}{2}})\,].$$

$$G^{4,1}_{2,4}\left(z\,\middle|\,\begin{matrix}\frac{1}{2},0\\ \alpha,\beta,-\beta,-\alpha\end{matrix}\right)=\frac{\pi^{\frac{5}{2}}}{4\cos(\alpha\pi)\,\cos(\beta\pi)}\,[\exp(-\beta\pi i)H^{(1)}_{\alpha-\beta}(z^{\frac{1}{2}})\,H^{(2)}_{\alpha+\beta}(z^{\frac{1}{2}})$$

$$+\ \exp(\beta\pi i)H^{(1)}_{\alpha+\beta}(z^{\frac{1}{2}})\,H^{(2)}_{\alpha-\beta}(z^{\frac{1}{2}})].$$

$$G^{4,1}_{2,4}\left(z\,\middle|\,\begin{matrix}\frac{1}{2}+\alpha,\frac{1}{2}-\alpha\\ 0,\frac{1}{2},\beta,-\beta\end{matrix}\right)=z^{-\frac{1}{2}}\,\pi^{\frac{1}{2}}\,\Gamma(\tfrac{1}{2}-\alpha+\beta)$$

$$\times\,\Gamma(\tfrac{1}{2}-\alpha-\beta)\,W_{\alpha,\beta}(2iz^{\frac{1}{2}})\,W_{\alpha,\beta}(-2iz^{\frac{1}{2}})\,.$$

$$G^{4,2}_{2,4}\left(z\,\middle|\,\begin{matrix}\alpha,\alpha+\frac{1}{2}\\ \beta+\gamma,\beta-\gamma,\beta+\gamma+\frac{1}{2},\ \beta-\gamma+\frac{1}{2}\end{matrix}\right) = 2^{2\alpha-2\beta+\frac{1}{2}}\,\pi^{\frac{3}{2}}\,\Gamma(1-2\alpha+2\beta+2\gamma)\,\Gamma(1-2\alpha+2\beta-2\gamma)$$

$$\times\, z^{\beta-\frac{1}{4}}\exp(z^{\frac{1}{2}})\,W_{2\alpha-2\beta-\frac{1}{2},\ 2\gamma}(2z^{\frac{1}{2}})\ .$$

$$G^{1,p}_{p,q}\left(z\,\middle|\,\begin{matrix}\alpha_p\\ \beta_q\end{matrix}\right) = \frac{\prod_{j=1}^{p}\Gamma(1+\beta_1-\alpha_j)\,z^{\beta_1}}{\prod_{j=2}^{q}\Gamma(1+\beta_1-\beta_j)}\ {}_pF_{q-1}\left(\begin{matrix}1+\beta_1-\alpha_1,\ldots,1+\beta_1-\alpha_p\ ;\\ 1+\beta_1-\beta_2,\ldots,1+\beta_1-\beta_q\ ;\end{matrix}\ -z\right),\ p\le q\ .$$

$$G^{1,n}_{p,q}\left(z\,\middle|\,\begin{matrix}\alpha_p\\ \beta_q\end{matrix}\right) = \frac{\prod_{j=1}^{n}\Gamma(1+\beta_1-\alpha_j)\,z^{\beta_1}}{\prod_{j=2}^{q}\Gamma(1+\beta_1-\beta_j)\ \prod_{j=n+1}^{p}\Gamma(\alpha_j-\beta_1)}$$

$$\times\ {}_pF_{q-1}\left(\begin{matrix}1+\beta_1-\alpha_1,\ldots,1+\beta_1-\alpha_p\\ 1+\beta_1-\beta_2,\ldots,1+\beta_1-\beta_q\end{matrix}\ ;(-1)^{p-n-1}\,z\right)\ ,\ p\le q\ .$$

$$G^{q,1}_{p,q}\left(z\,\middle|\,\begin{matrix}a_p\\ b_q\end{matrix}\right) = z^{\alpha_1-1}\ E(1-\alpha_1+\beta_1,\ldots,1-\alpha_1+\beta_q;1-\alpha_1+\alpha_2,\ldots,1-\alpha_1+\alpha_p\ ;\ z)\,.$$

For further representations of G-functions in terms of MacRobert's E-function, see Sharma [290] .

$$G^{m,n}_{p,q}\left(z\,\middle|\,\begin{matrix}a_p\\ b_q\end{matrix}\right) = \sum_{h=1}^{m}\frac{\prod'_{j=1}^{m}\Gamma(b_j-b_h)\ \prod_{j=1}^{n}\Gamma(1+b_h-a_j)\,z^{b_h}}{\prod_{j=m+1}^{q}\Gamma(1+b_h-b_j)\ \prod_{j=n+1}^{p}\Gamma(a_j-b_h)}$$

$$\times\ {}_pF_{q-1}(1+b_h-a_1,\ldots,1+b_h-a_p;1+b_h-b_1,\ldots *,\ldots,1+b_h-b_q;(-1)^{p-m-n}\,z)\ ,$$

$p < q$ or $p = q$ and $|z| < 1$, and the poles are simple

$$= \sum_{h=1}^{n} \frac{\prod_{j=1}^{n}{}' \Gamma(a_h-a_j) \prod_{j=1}^{m} \Gamma(b_j-a_h+1) z^{a_h-1}}{\prod_{j=n+1}^{p} \Gamma(1+a_j-a_h) \prod_{j=m+1}^{q} \Gamma(a_h-b_j)}$$

$$* {}_qF_{p-1}(1+b_1-a_h, 1+b_2-a_h, \ldots, 1+b_q-a_h; 1+a_1-a_h, \ldots, * \ldots, 1+a_p-a_h; (-1)^{q-m-n}z).$$

$q < p$ or $q = p$ and $|z| > 1$, and the poles are simple.

Note: See Chapters V and VI for logorithmic cases.

CHAPTER III

INTEGRALS OF G-FUNCTIONS

This chapter deals with the evaluation of integrals involving G-functions. An integral involving products of two G-functions in which the argument of one of the G-functions contains rational exponents is evaluated by Saxena [258]. This result is important as it can readily yield integrals involving products of any two Special Functions with arguments containing positive integers. This is given in Section 3.2.1 and its particular cases are given in Section 3.3.

We have also discussed integrals of G-functions due to Saxena [263] in which the argument of the G-function contains a factor

$$x^k \{x^{\frac{1}{2}} + (1+x)^{\frac{1}{2}}\}^{2\rho},$$

where k and ρ are positive integers and x is the variable of integration.

The results given in this chapter are the key formulas for the elementary Special Functions occurring in Mathematical Physics since the results involving Bessel, Whittaker and Hypergeometric Functions can be derived as their special cases. The results of this chapter will find applications in the following chapters as well as in many branches of Physical and Biological Sciences. A few applications of some of these formulas are given in Chapters VI and VII. These are mainly applications, in Distributions of Multivariate Test Criteria, Null and Non-null Distributions, Characterizations, Bayesian Inference, Prior and Posterior Distributions and Structural setup of Densities in the field of Statistics; Hard-limiting of Sinusoidal and Gaussian Signals in the field of Engineering and Communication Technology; Heat conduction and Heating and Cooling in Physical Sciences; and in the theory of Integral Equations.

3.1. INTEGRALS INVOLVING HYPERGEOMETRIC FUNCTIONS AND RELATED FUNCTIONS

In this section we have given a number of important integrals involving Gauss's hypergeometric functions and related functions which will be required in evaluating the integrals associated with a G-function. For the source of these integrals, see the papers by Sharma, B.L. ([282],[284]), Saxena [264] and the literature cited in the references especially Erdélyi, A. et al [86], Slater ([301],[302]) and Magnus, Oberhttinger and Soni [167].

For the sake of brevity, the pairs of parameters like $\alpha+\beta$, $\alpha-\beta$ will be written as $\alpha\pm\beta$, the set of parameters $\alpha+\beta+\gamma$, $\alpha-\beta+\gamma$, $\alpha+\beta-\gamma$ and $\alpha-\beta-\gamma$ as $\alpha\pm\beta\pm\gamma$, the gamma product $\Gamma(\alpha+\beta)\,\Gamma(\alpha-\beta)$ as $\Gamma(\alpha\pm\beta)$ and the set of gamma product $\Gamma(\alpha+\beta+\gamma)\,\Gamma(\alpha-\beta+\gamma)\,\Gamma(\alpha+\beta-\gamma)$ $(\alpha-\beta-\gamma)$ as $\Gamma(\alpha\pm\beta\pm\gamma)$ and so on.

The symbol $\Delta(n,\alpha)$ will be used to represent the set of n parameters

$$\frac{\alpha}{n}, \frac{\alpha+1}{n}, \ldots, \frac{\alpha+n-1}{n}.$$

Since the results to be presented in this section are not difficult to prove they are given without proofs. Also the derivations are readily available from the papers cited in the references. Due to their importance in practical problems, integrals involving some particular cases of a G-function are given in the following sections. The list is by no means exhaustive because almost all elementary Special Functions qualify to be particular cases of a G-function.

Throughout this chapter the poles of the integrands of the G-functions involved are assumed to be simple.

$$\int_0^1 t^{\alpha-1}(1-t)^{\beta-1}dt = B(\alpha,\beta) = \frac{\Gamma(\alpha)\,\Gamma(\beta)}{\Gamma(\alpha+\beta)}, \quad R(\alpha) > 0, \quad R(\beta) > 0, \tag{3.1.1}$$

$$= 2\int_0^{\frac{\pi}{2}} (\sin\theta)^{2\alpha-1}(\cos\theta)^{2\beta-1}\,d\theta. \tag{3.1.2}$$

$$\int_0^1 t^{\alpha-1}(1-t)^{\beta-1}\,\{1+ct+d(1-t)\}^{-\alpha-\beta}\,dt$$

$$= (1+c)^{-\alpha}(1+d)^{-\beta}\,B(\alpha,\beta), \tag{3.1.3}$$

$R(\alpha) > 0$, $R(\beta) > 0$ and the constants c and d are such that none of the expressions $1+c$, $1+d$, $1+ct+d(1-t)$, where $0 \le t \le 1$, is zero.

$$\int_0^{\frac{\pi}{2}} \exp\,[i(\alpha+\beta)\theta]\,(\sin\theta)^{\alpha-1}\,(\cos\theta)^{\beta-1}\,d\theta$$

$$= \exp(i\,\frac{\pi\alpha}{2})\;B(\alpha,\beta), \tag{3.1.4}$$

$R(\alpha) > 0$, $R(\beta) > 0$.

$$\int_0^{\pi} \sin^{\alpha}t\,\exp(i\beta t)dt = \frac{\Pi\,\exp(i\,\frac{\pi}{2}\,\beta)\,\Gamma(1+\alpha)}{2^{\alpha}\Gamma[1+(\frac{\alpha+\beta}{2})]\Gamma[1+(\frac{\alpha-\beta}{2})]}, \quad R(\alpha) > -1. \tag{3.1.5}$$

$$\int_0^{\frac{\pi}{2}} (1+a\sin^2 t)^{-\alpha-\beta} (\sin t)^{2\alpha-1} (\cos t)^{2\beta-1}\, dt = \frac{1}{2}(1+a)^{-\alpha} B(\alpha,\beta), \quad (3.1.6)$$

$$a > 0,\ R(\alpha) > 0,\ R(\beta) > 0 .$$

$$\int_0^{\frac{\pi}{2}} (\cos^2 t + a\sin^2 t)^{-\alpha-\beta} (\sin t)^{2\alpha-1} (\cos t)^{2\beta-1} dt = \frac{a^{-\alpha}}{2} B(\alpha,\beta),$$

$$a > 0,\ R(\alpha) > 0,\ R(\beta) > 0. \quad (3.1.7)$$

$$\int_0^{\frac{\pi}{2}} e^{i(\alpha+\beta)\theta} (\sin\theta)^{\alpha-1} (\cos\theta)^{\beta-1}\ {}_2F_1(a,b;\beta;e^{i\theta}\cos\theta)\,d\theta$$

$$= e^{i\frac{\pi}{2}\alpha} \frac{\Gamma(\alpha)\,\Gamma(\beta)\,\Gamma(\alpha+\beta-a-b)}{\Gamma(\alpha+\beta-a)\,\Gamma(\alpha+\beta-b)}, \quad (3.1.8)$$

$$R(\alpha) > 0,\quad R(\beta) > 0,\ R(\alpha+\beta-a-b) > 0 \cdot$$

$$\int_0^{\frac{\pi}{2}} e^{i(\alpha+\beta)\theta} (\sin\theta)^{\alpha-1} (\cos\theta)^{\beta-1}\ {}_2F_1[a,b;\alpha;e^{i(\theta-\frac{\pi}{2})}\sin\theta]\ d\theta$$

$$= e^{i\frac{\pi}{2}\alpha} \frac{\Gamma(\alpha)\,\Gamma(\beta)\,\Gamma(\alpha+\beta-a-b)}{\Gamma(\alpha+\beta-a)\,\Gamma(\alpha+\beta-b)}, \quad (3.1.9)$$

$$R(\alpha) > 0,\ R(\beta) > 0,\ R(\alpha+\beta-a-b) > 0 .$$

$$\int_0^1 x^{\rho-1}(1-x)^{\beta-\gamma-n}\ {}_2F_1(-n,\beta;\gamma;x)dx$$

$$= \frac{\Gamma(\gamma)\,\Gamma(\rho)\,\Gamma(\beta-\gamma+1)\,\Gamma(\gamma-\rho+n)}{\Gamma(\gamma+n)\,\Gamma(\gamma-\rho)\,\Gamma(\beta-\gamma+\rho+1)}, \quad (3.1.10)$$

$$R(\rho) > 0,\quad R(\beta-\gamma) > n-1\ ,\quad n = 0,1,2,\ldots$$

$$\int_0^1 t^{\rho-1}\,(1-t)^{\beta-\rho-1}\ {}_2F_1(\alpha,\beta/\gamma;t)dt$$

$$= \frac{\Gamma(\gamma)\,\Gamma(\rho)\,\Gamma(\beta-\rho)\,\Gamma(\gamma-\alpha-\rho)}{\Gamma(\beta)\,\Gamma(\gamma-\alpha)\,\Gamma(\gamma-\rho)}, \qquad (3.1.11)$$

$R(\rho) > 0,\quad R(\beta-\rho) > 0,\quad R(\gamma-\alpha-\rho) > 0$.

$$\int_0^{\frac{\pi}{2}} \cos^{\alpha} t\, \cos(\beta t)dt = \frac{\pi\,\Gamma(1+\alpha)}{2^{\alpha+1}\,\Gamma[1+\frac{1}{2}(\alpha\pm\beta)]}, \quad R(\alpha) > -1 . \qquad (3.1.12)$$

$$\int_{-\frac{\pi}{2}}^{\frac{\pi}{2}} (\cos\theta)^{m+n-2}\, e^{i(m-n)\theta} d\theta = \frac{\pi\,\Gamma(m+n+1)}{2^{m+n-2}\,\Gamma(m)\,\Gamma(n)}, R(m+n) > 1. \qquad (3.1.13)$$

$$\int_0^{\infty} e^{-t}\, t^{\alpha-1}\, dt = \int_0^1 [\log \frac{1}{t}]^{\alpha-1} dt = \Gamma(\alpha)\ ,\ R(\alpha) > 0. \qquad (3.1.14)$$

$$\sigma^t \int_0^{\infty e^{i\gamma}} e^{-t\sigma}\, t^{\alpha-1} dt = \Gamma(\alpha)\ , \qquad (3.1.15)$$

$-\frac{\pi}{2} - \gamma < \arg\sigma < \frac{\pi}{2} - \gamma,\ R(\alpha) > 0$.

$$\int_0^{\infty} t^{\nu-1}\,(t+x)^{-\rho} dx = B(\nu,\rho-\nu)\, x^{\nu-\rho}\ ,\quad R(\nu) > 0,\ R(\rho-\nu) > 0. \qquad (3.1.16)$$

$$\int_0^{\infty} t^{\lambda-1}\,(1+t)^{-\frac{1}{2}}\,[t^{\frac{1}{2}} + (1+t)^{\frac{1}{2}}]^{2\mu} dt$$

$$= 2^{1-2\lambda}\, B(2\lambda, \tfrac{1}{2} - \lambda - \mu), \qquad (3.1.17)$$

$R(\lambda) > 0,\ R(\lambda+\mu) < \frac{1}{2}$.

$$\int_0^1 t^{\lambda-1}\,(1-t)^{\mu-1}\,[1+ct+d(1-t)]^{-\lambda-\mu}$$

$$\times\ {}_2F_1\,[\,\alpha,\beta;\mu;\frac{(1-t)(1+d)}{1+ct+d(1-t)}\,]\, dt$$

$$= \frac{(1+c)^{-\lambda}\,(1+d)^{-\mu}\,\Gamma(\lambda)\,\Gamma(\mu)\,\Gamma(\lambda+\mu-\alpha-\beta)}{\Gamma(\lambda+\mu-\alpha)\,\Gamma(\lambda+\mu-\beta)}\ , \qquad (3.1.18)$$

provided that the constants c and d are such that none of the expressions $1 + c$, $1 + d$, $1 + ct + d(1-t)$ where $0 \leq t \leq 1$, is zero, $R(\lambda) > 0$, $R(\mu) > 0$ and $R(\lambda + \mu - \alpha - \beta) > 0$.

$$\int_0^\infty e^{2\mu\theta} (\sinh \theta)^{2\lambda-1} d\theta = 2^{-2\lambda} B(2\lambda, \tfrac{1}{2} - \lambda - \mu), \tag{3.1.19}$$

$R(\lambda) > 0$, $R(\frac{1}{2} - \lambda - \mu) > 0$.

$$\int_0^\infty e^{2\mu\theta} (\sinh \theta)^{2\lambda-1} {}_2F_1(\lambda-\mu + \tfrac{1}{2}, \beta ; \delta ; 2e^{-\theta} \sinh \theta) d\theta$$

$$= \frac{\Gamma(\lambda)\Gamma(\delta)\Gamma(\lambda + \frac{1}{2})\Gamma(\frac{1}{4} - \frac{\lambda}{2} - \frac{\mu}{2})\, \Gamma(\frac{3}{4} - \frac{\lambda}{2} - \frac{\mu}{2})\, \Gamma(\frac{\delta - \beta}{2} - \lambda)}{2^{\lambda+\mu+\beta+\frac{3}{2}}\, \pi\, \Gamma(\delta - \beta)\, \Gamma(\frac{1+\lambda-\mu}{2})\, \Gamma(\frac{\delta}{2} - \lambda)}$$

$$\cdot \frac{\Gamma(\frac{1+\delta-\beta}{2} - \lambda)}{\Gamma(\frac{1 + \delta}{2} - \lambda)}, \tag{3.1.20}$$

$R(\lambda) > 0$, $R(\frac{1}{2} - \lambda - \mu) > 0$, $R(\delta - \beta - \lambda + \mu) > \frac{1}{2}$.

$$\int_0^1 t^{\rho-1} (1-t)^{\sigma-1} {}_2F_1(\alpha ,\beta ; \gamma; t) dt$$

$$= \frac{\Gamma(\rho)\, \Gamma(\sigma)}{\Gamma(\rho+ \sigma)}\, {}_3F_2 (\alpha,\beta,\rho; \gamma, \rho+\sigma;1), \tag{3.1.21}$$

$R(\rho) > 0$, $R(\sigma) > 0$, $R(\gamma + \sigma - \alpha -\beta) > 0$.

$$\int_0^1 t^{\rho-1} (1-t)^{\rho-1} {}_2F_1(\alpha,\beta; \tfrac{\alpha+\beta+1}{2}; t)\, dt$$

$$= \frac{\pi\, \Gamma(\rho)\, \Gamma(\frac{1}{2} + \frac{\alpha}{2} + \frac{\beta}{2})\, \Gamma(\frac{1}{2} + \rho - \frac{\alpha}{2} - \frac{\beta}{2})}{2^{2\rho-1}\, \Gamma(\frac{1}{2} + \frac{\alpha}{2})\, \Gamma(\frac{1}{2} + \frac{\beta}{2})\, \Gamma(\rho + \frac{1}{2} - \frac{\alpha}{2})\, \Gamma(\rho + \frac{1}{2} - \frac{\beta}{2})},$$

$R(\rho) > 0$, $R(\rho + \frac{1}{2} - \frac{\alpha}{2} - \frac{\beta}{2}) > 0$. (3.1.22)

$$\int_0^1 t^{\rho-1}(1-t)^{\rho-\nu} \; {}_2F_1(\alpha, 1-\alpha ; \nu; t)dt$$

$$= \frac{\pi \; 2^{1-2\rho} \, \Gamma(\nu) \, \Gamma(\rho) \, \Gamma(\rho-\nu+1)}{\Gamma(\frac{\alpha}{2}+\frac{\nu}{2}) \, \Gamma(\frac{1}{2}-\frac{\alpha}{2}+\frac{\nu}{2}) \, \Gamma(\rho+\frac{\alpha}{2}-\frac{\nu}{2}+\frac{1}{2}) \, \Gamma(\rho-\frac{\alpha}{2}-\frac{\nu}{2})},$$

$R(\rho) > 0, \quad R(\rho - \nu + 1) > 0$. (3.1.23)

$$\int_0^t t^{\rho-1} (1-t)^{\sigma-1} \; {}_2F_1 (\alpha,\beta; \gamma; tz)dt$$

$$= B(\rho,\sigma) \; {}_3F_2 (\alpha,\beta,\rho; \gamma, \rho+\sigma;z), \tag{3.1.24}$$

$R(\rho) > 0, \quad R(\sigma) > 0, \; |\arg(1-z)| < \pi$.

$$\int_0^1 t^{\gamma-1}(1-t)^{\rho-1} e^{-tz} \; {}_2F_1(\alpha,\beta;\gamma;t)dt$$

$$= \frac{\Gamma(\gamma) \, \Gamma(\rho) \, \Gamma(\gamma+\rho-\alpha-\beta)}{\Gamma(\gamma+\rho-\alpha) \, \Gamma(\gamma+\rho-\beta)} e^{-z}$$

$$\times \; {}_2F_2 (\rho,\gamma+\rho-\alpha-\beta; \gamma+\rho-\alpha , \gamma+\rho-\beta;z), \tag{3.1.25}$$

$R(\gamma) > 0, \quad R(\rho) > 0, \quad R(\gamma + \rho - \alpha - \beta)$.

$$\int_0^1 t^{\gamma-1}(1+t)^{-\sigma} \; {}_2F_1(\alpha,\beta;\gamma; -t)dt$$

$$= \frac{\Gamma(\gamma) \, \Gamma(\alpha-\gamma+\sigma) \, \Gamma(\beta-\gamma+\sigma)}{\Gamma(\rho) \, \Gamma(\alpha+\beta-\gamma+\sigma)},$$

$R(\gamma) > 0, \quad R(\alpha-\gamma+\sigma) > 0, \quad R(\beta-\gamma+\sigma) > 0$. (3.1.26)

$$\int_0^1 t^{\lambda-1}(1-t^2)^{-\frac{1}{2}\mu} P_\nu^\mu(t)dt$$

$$= \frac{\pi^{\frac{1}{2}}\, 2^{\mu-\lambda}\, \Gamma(\lambda)}{\Gamma(\frac{1}{2}+\frac{\lambda}{2}-\frac{\mu}{2}-\frac{\nu}{2})\,\Gamma(1+\frac{\lambda}{2}-\frac{\mu}{2}+\frac{\nu}{2})}, \quad R(\lambda) > 0, R(\mu) < 1.$$

(3.1.27)

$$\int_0^\pi (\sin\theta)^{\lambda-1}\, T_\nu^{-\mu}(\cos\theta)d\theta$$

$$= \frac{\pi\,\Gamma(\frac{\lambda}{2}+\frac{\mu}{2})\,\Gamma(\frac{\lambda}{2}-\frac{\mu}{2})}{2^\mu\Gamma(\frac{1}{2}+\frac{\mu}{2}-\frac{\nu}{2})\,\Gamma(\frac{1}{2}+\frac{\mu}{2}+\frac{\nu}{2})\,\Gamma(\frac{1}{2}+\frac{\lambda}{2}+\frac{\nu}{2})\,\Gamma(\frac{\lambda}{2}-\frac{\nu}{2})},$$

$R(\lambda \pm \mu) > 0.$ (3.1.28)

$$\int_1^\infty (t^2-1)^{\lambda-1}\, P_\nu^\mu(t)dt$$

$$= \frac{2^{\mu-1}\,\Gamma(\lambda-\frac{\mu}{2})\,\Gamma(1-\lambda+\frac{\nu}{2})\,\Gamma(\frac{1}{2}-\lambda-\frac{\nu}{2})}{\Gamma(1-\frac{\mu}{2}+\frac{\nu}{2})\,\Gamma(\frac{1}{2}-\frac{\mu}{2}-\frac{\nu}{2})\,\Gamma(1-\lambda-\frac{\mu}{2})},$$

$R(\lambda) > R(\mu)$, $R(1-2\lambda-\nu) > 0$, $R(2-2\lambda+\nu) > 0.$ (3.1.29)

$$\int_1^\infty t^{-\rho}(t^2-1)^{-\frac{\mu}{2}}\, P_\nu^\mu(t)dt$$

$$= \frac{2^{\rho+\mu-2}\,\Gamma(\frac{\rho+\mu+\nu}{2})\,\Gamma(\frac{\rho+\mu-\nu-1}{2})}{\pi^{\frac{1}{2}}\,\Gamma(\rho)}, \quad (3.1.30)$$

$R(\mu) < 1$, $R(\rho+\mu+\nu) > 0$, $R(\rho+\mu-\nu) > 1.$

$$\int_1^\infty (t-1)^{\lambda-1}(t^2-1)^{\frac{\mu}{2}}\, P_\nu^\mu(t)dt$$

$$= \frac{2^{\lambda+\mu}\,\Gamma(\lambda)\,\Gamma(-\lambda-\mu-\nu)\,\Gamma(1-\lambda-\mu+\nu)}{\Gamma(1-\mu+\nu)\,\Gamma(-\mu-\nu)\,\Gamma(1-\lambda-\mu)}, \quad (3.1.31)$$

$R(\lambda) > 0$, $R(\lambda+\mu+\nu) < 0$, $R(\lambda+\mu-\nu) < 1$.

$$\int_1^\infty (t-1)^{\lambda-1}\,(t^2-1)^{-\frac{\mu}{2}}\,P_\nu^\mu(t)dt$$

$$= -\frac{2^{\lambda-\mu}\sin(\nu\pi)\,\Gamma(\lambda-\mu)\Gamma(-\lambda+\mu-\nu)\,\Gamma(1-\lambda+\mu-\nu)}{\pi\,\Gamma(1-\lambda)}, \qquad (3.1.32)$$

$R(\lambda-\mu) > 0$, $R(\mu-\lambda-\nu) > 0$, $R(\mu-\lambda+\nu) > -1$.

3.1.1 Integral of Orthogonal Polynomials.

Here we list a number of integrals involving orthogonal polynomials such as Jacobi polynomial, Laguerre polynomial, Gegenbauer Polynomial and Hermite polynomial. These will be used in later sections:

$$\int_{-1}^{1} (1-t)^\alpha\,(1+t)^\sigma\,P_n^{(\alpha,\beta)}(t)dt$$

$$= \frac{2^{\alpha+\sigma+1}\,\Gamma(\sigma+1)\,\Gamma(\alpha+n+1)\,\Gamma(\sigma-\beta+1)}{\Gamma(\sigma-\beta-n+1)\,\Gamma(\alpha+\sigma+n+2)}, \qquad (3.1.33)$$

$R(\alpha) > -1$, $R(\sigma) > -1$.

$$\int_{-1}^{1} P_n^{(\alpha,\beta)}(t)\,P_m^{(\alpha,\beta)}(t)\,(1-t)^\alpha(1+t)^\beta\,dt$$

$$= \frac{\Gamma(\alpha+n+1)\,\Gamma(\beta+n+1)}{n!\,\Gamma(\alpha+\beta+n+1)}\;\frac{2^{\alpha+\beta+1}}{\alpha+\beta+2n+1}\;\delta_{mn}, \qquad (3.1.34)$$

where δ_{mn} is the Kronecker delta, defined by

$$\delta_{mn} = \begin{cases} 0, & \text{if } m \neq n \\ 1, & \text{if } m = n, \end{cases}$$

$R(\alpha) > -1$, $R(\beta) > -1$.

$$\int_0^\infty t^{\beta-1}\,e^{-t}\,L_n^\alpha(t)dt = \frac{\Gamma(\alpha-\beta+n+1)\,\Gamma(\beta)}{n!\;\Gamma(\alpha-\beta+1)},\ R(\beta) > 0. \qquad (3.1.35)$$

$$\int_0^\infty e^{-t}\,t^\alpha\,L_m^\alpha(t)\,L_n^\alpha(t)dt = \frac{\Gamma(\alpha+n+1)}{n!}\,\delta_{mn}\,. \qquad (3.1.36)$$

$$\int_0^1 t^{2\rho+n} (1-t^2)^{\nu-\frac{1}{2}} C_n^{\nu}(t)dt$$

$$= \frac{(2\nu)n\ (2\rho+1)n\ \Gamma(\nu+\frac{1}{2})\ \Gamma(\rho+\frac{1}{2})}{2^{n+1}\ n!\ \ \Gamma(n+\nu+\rho+1)}, \quad R(\rho) > -\frac{1}{2}. \tag{3.1.37}$$

$$\int_{-1}^1 (1-t^2)^{\nu-\frac{1}{2}} (1+t)^{\beta} C_n^{\nu}(t)dt$$

$$= \frac{2^{\beta+\nu+\frac{1}{2}}\Gamma(\beta+1)\ \Gamma(\nu+\frac{1}{2})\ \Gamma(2\nu+n)\ \Gamma(\beta-\nu+\frac{3}{2})}{n!\ \ \Gamma(2\nu)\ \Gamma(\beta-\nu+\frac{3}{2}-n)\ \Gamma(\beta+\nu+\frac{3}{2}+n)}, \tag{3.1.38}$$

$R(\beta) > -1, \quad R(\nu) > -\frac{1}{2}$.

$$\int_{-1}^1 (1-t^2)^{\nu-\frac{1}{2}} C_m^{\nu}(t)\ C_n^{\nu}(t)dt$$

$$= \frac{\pi\ 2^{1-2\nu}\ \Gamma(n+2\nu)}{n!\ (n+\nu)\ \{\Gamma(\nu)\}^2}\ \delta_{mn}, \quad R(\nu) > -\frac{1}{2}. \tag{3.1.39}$$

$$\int_{-\infty}^{\infty} e^{-t^2} H_n(t)\ H_m(t)dt = 2^n\ n!\ \pi^{\frac{1}{2}}\ \delta_{mn}. \tag{3.1.40}$$

$$\int_{-1}^1 P_n^m(t)\ P_k^m(t)dt = \frac{2(m+n)!}{(2n+1)\ (n-m)!}\ \delta_{nk}. \tag{3.1.41}$$

$$\int_{-1}^1 \frac{P_n^m(t)\ P_n^k(t)}{1-t^2}\ dt = \frac{(n+m)!}{\{m(n-m)!\}}\ \delta_{mk}. \tag{3.1.42}$$

3.1.2. Infinite Integrals of Bessel Functions.

$$\int_0^{\infty} t^{s-1} J_{\nu}(\alpha t)dt = \frac{2^{s-1}\alpha^{-s}\Gamma(\frac{\nu+s}{2})}{\Gamma(1+\frac{\nu-s}{2})}, \quad -R(\nu) < R(s) < \frac{3}{2}, \quad \alpha > 0. \tag{3.1.43}$$

$$\int_0^{\infty} t^{s-1} Y_{\mu}(\alpha t)dt = -\ 2^{s-1}\ \pi^{-1}\ \alpha^{-s} \cos\left[\frac{\pi(s-\nu)}{2}\right]\ \Gamma(\frac{\nu \pm s}{2}),$$

$$R(s \pm \nu) > 0, \quad R(s) < \frac{3}{2}, \ \alpha > 0. \tag{3.1.44}$$

$$\int_0^\infty t^{s-1} K_\nu(\alpha t)dt = 2^{s-2}\alpha^{-s}\Gamma(\frac{s\pm\nu}{2}), \quad R(s\pm\nu) > 0,\ R(\alpha) > 0. \tag{3.1.45}$$

$$\int_0^\infty t^{s-1} e^{-\alpha t} K_\nu(\alpha t)dt = \frac{\pi^{\frac{1}{2}}\Gamma(s\pm\nu)}{(2\alpha)^s\Gamma(s+\frac{1}{2})}, \quad R(s) > |R(\nu)|,\ R(\alpha) > 0. \tag{3.1.46}$$

$$\int_0^\infty t^{s-1} J_\mu(\alpha t)\, J_\nu(\alpha t)dt = \frac{2^{s-1}\alpha^{-s}B(1-s, \frac{\mu+\nu+s}{2})}{\Gamma(\frac{\nu-\mu-s}{2}+1)\,\Gamma(\frac{\mu-\nu-s}{2}+1)}, \tag{3.1.47}$$

$-R(\mu+\nu) < R(s) < 1,\ \alpha > 0.$

$$\int_0^\infty t^{s-1} J_\nu(\alpha t)\, Y_\mu(\alpha t)dt = 2^{s-1}\alpha^{-s}\pi^{-1}\sin[\frac{\pi(\nu+s-\mu-1)}{2}]$$

$$\times\frac{\Gamma[(\frac{s+\mu+\nu}{2})]\Gamma[(\frac{s-\mu+\nu}{2})]}{\Gamma[(\frac{\nu+2-s-\mu}{2})]\ \Gamma[(\frac{\nu+\mu-s+2}{2})]}, \quad R(-\nu\pm\mu) < R(s) < 1, \alpha > 0. \tag{3.1.48}$$

$$\int_0^\infty t^{s-1} I_\nu(\alpha t)K_\mu(\alpha t)dt = \frac{\Gamma(\frac{s+\mu+\nu}{2})\ B(1-s, \frac{s-\mu+\nu}{2})}{2^{2-s}\alpha^s\Gamma(\frac{\nu+\mu-s}{2}+1)}, \tag{3.1.49}$$

$R(-\nu\mp\mu) < R(s) < 1,\ R(s) < 1,\ R(\alpha) > 0.$

$$\int_0^\infty t^{s-1} K_\mu(\alpha t)K_\nu(\alpha t)dt = 2^{s-3}\alpha^{-s}\{\Gamma(s)\}^{-1}\ \Gamma[(\frac{s\pm\mu\pm\nu}{2})], \tag{3.1.50}$$

$R(\alpha) > 0,\ R(s) > |R(\mu)| + |R(\nu)|.$

$$\int_0^\infty e^{it}\, t^\mu J_\nu(t)dt = \frac{\exp\{\frac{i\pi(\mu+\nu+1)}{2}\}\ \Gamma(\mu+\nu+1)\ \Gamma(-\mu-\frac{1}{2})}{2^{\mu+1}\ \Gamma(\frac{1}{2})\ \Gamma(\nu-\mu)},$$

$R(\mu+\nu) > -1,\ R(\mu) < -\frac{1}{2}.$ (3.1.51)

$$\int_0^\infty t^{-1} J_{\nu+2n+1}(t)\, J_{\nu+2m+1}(t)dt = 0 \quad \text{if } m \neq n$$

$$= (4n+2\nu+2)^{-1} \quad \text{if } m = n,\ R(\nu)+m+n > -1. \tag{3.1.52}$$

3.1.3. Integrals Involving Whittaker Functions.

$$\int_0^\infty t^{\nu-1}\exp(-\tfrac{t}{2})\,W_{\lambda,\mu}(t)dt = \frac{\Gamma(\frac{1}{2}\pm\mu+\nu)}{\Gamma(1-\lambda+\nu)}, \qquad (3.1.53)$$

$R(\frac{1}{2}\pm\mu+\nu) > 0$.

$$\int_0^\infty t^{\nu-1}\exp(\tfrac{t}{2})\,W_{\lambda,\mu}(t)dt = \frac{\Gamma(\frac{1}{2}\pm\mu+\nu)\,\Gamma(-\lambda-\nu)}{\Gamma(\frac{1}{2}-\lambda\pm\mu)}, \qquad (3.1.54)$$

$R(\frac{1}{2}\pm\mu+\nu) > 0,\quad R(\lambda+\nu) < 0$.

$$\int_0^\infty t^{\nu-1}\exp(-\tfrac{t}{2})\,M_{k,m}(t)dt = \frac{\Gamma(2m+1)\,B(m+\nu+\frac{1}{2},\,k-\nu)}{\Gamma(m-\nu+\frac{1}{2})}, \qquad (3.1.55)$$

$-\frac{1}{2} - R(m) < R(\nu) < R(k)$.

$$\int_0^\infty t^{\nu-1}\,W_{k,m}(t)\,W_{-k,m}(t)dt = \frac{\Gamma(\nu+1)\,\Gamma(\frac{\nu+1}{2}\pm\mu)}{2\,\Gamma(1+\frac{\nu}{2}\pm k)}, \qquad (3.1.56)$$

$R(\nu) > 2|R(m)| - 1$.

3.2. MELLIN TRANSFORM OF THE G-FUNCTION.

In view of the definition (1.1.1) and Mellin inversion theorem (Titchmarsh, [329]) it follows that

$$\int_0^\infty x^{s-1}\,G_{p,q}^{m,n}\left(\omega x\,\middle|\,{a_p \atop b_q}\right)dx = \frac{\omega^{-s}\prod_{j=1}^{m}\Gamma(b_j+s)\prod_{j=1}^{n}\Gamma(1-a_j-s)}{\prod_{j=m+1}^{q}\Gamma(1-b_j-s)\prod_{j=n+1}^{p}\Gamma(a_j+s)}, \qquad (3.2.1)$$

$0\le n\le p<q,\quad 0\le m\le q,\ \omega\neq 0,\quad c^* = m+n-\frac{p}{2}-\frac{q}{2} > 0,\ |\arg\omega| < c^*\pi,$
$-\min R(b_j) < R(s) < 1 - \max R(a_k)$, for $j = 1,\ldots,m;\ k = 1,\ldots,n.$

Evidently the Mellin transform of the G-function is the coefficient of x^{-s} in the integrand of (1.1.1)

Luke [152] has given seven cases of the validity of (3.2.1). In this connection, see Luke [152] and the conditions for the validity of (3.2.2) below.

3.21. Mellin Transform of the Product of Two G-Functions.

Saxena [258] has shown that

$$\int_0^\infty x^{\sigma-1}\, G_{p,q}^{m,n}\left(\omega x \,\Big|\, \begin{matrix} a_p \\ b_q \end{matrix}\right) G_{\gamma,\delta}^{\alpha,\beta}\left(\eta x^{\frac{k}{\rho}} \,\Big|\, \begin{matrix} c_\gamma \\ d_\delta \end{matrix}\right) dx$$

$$= \omega^{-\sigma} (2\pi)^{c^*(1-k) + b^*(1-\rho)}\, k^{U+\sigma(q-p)-1}\, \rho^{V}$$

$$\times\, G_{\rho\gamma+kq,\,\rho\delta+kp}^{\rho\alpha+kn,\,\rho\beta+km}\left[W \,\Big|\, \begin{matrix} \Delta(\rho,c_1),\ldots,\ (\rho,c_\beta),\ \Delta(k,1-b_q-\sigma),\ \Delta(\rho,c_{\beta+1}),\ \ldots,\ \Delta(\rho,c_\gamma) \\ \Delta(\rho,d_1),\ldots,\ \Delta(\rho,d_\alpha),\ \Delta(k,1-a_p-\sigma),\ \Delta(\rho,d_{\alpha+1}),\ \ldots,\ \Delta(\rho,d_\delta) \end{matrix}\right], \tag{3.2.2}$$

where k and ρ are positive integers, $c^* = m+n - \frac{p}{2} - \frac{q}{2}$, $b^* = \alpha+\beta - \frac{\gamma}{2} - \frac{\delta}{2}$,

$U = \sum_{j=1}^{q} b_j - \sum_{j=1}^{p} a_j + \frac{p}{2} - \frac{q}{2} + 1$, $V = \sum_{j=1}^{\delta} d_j - \sum_{j=1}^{\gamma} c_j + \frac{\gamma}{2} - \frac{\delta}{2} + 1$ and

$$W = \{\eta^{\rho}\, \rho^{\rho(\gamma-\delta)}\} / \{\omega^{k}\, k^{k(p-q)}\}. \tag{3.2.3}$$

The following conditions are useful in the discussion of the conditions of validity of (3.2.2).

$$\omega \neq 0, \quad \eta \neq 0. \tag{3.2.4}$$

$$R(\sigma + \min b_j + \frac{k}{\rho} \min dh) > 0, \text{ for } j = 1,\ldots,m;\ h = 1,\ldots,\alpha. \tag{3.2.5}$$

$$R(\sigma + a_j + \frac{k}{\rho} c_h) > \frac{k}{\rho} + 1, \quad \text{for } j = 1,\ldots,\beta;\ h = 1,\ldots,n. \tag{3.2.6}$$

$$a_j - b_h \neq 1,2,3,\ldots\ ;\ j = 1,\ldots,n \text{ and } h = 1,\ldots,m. \tag{3.2.7}$$

$$c_j - d_h \neq 1,2,3,\ldots;\ j = 1,\ldots,\beta \text{ and } h = 1,\ldots,\alpha. \tag{3.2.8}$$

$$R\,[\{\sum_{j=1}^{\gamma} c_j - \sum_{j=1}^{\delta} d_j\} + (\frac{\delta-\gamma+1}{2}) + \frac{\rho}{k}\,\{R(a_h) + \sigma - 1)(\gamma-\delta)\}] > 0,$$

for $h = 1,2,\ldots,n.$ (3.2.9)

$$R\,[\{\sum_{j=1}^{p} a_j - \sum_{j=1}^{q} b_j\} + (\frac{q-p+1}{2}) + \frac{k}{\rho}\,[(R(c_h) + \sigma - 1)(p-q)] > 0,$$

for $h = 1,2,\ldots,\beta$. (3.2.10)

To prove (3.2.2) replace the function $G^{\alpha,\beta}_{\gamma,\delta}(\eta\, x^{\frac{k}{\rho}})$ by its equivalent contour integral from the definition (1.1.1) and interchange the order of integration which is permissible since the integrals involved are absolutely convergent. The inner integral is readily evaluated by means of (3.2.1). The result now follows on making use of the multiplication formula for the gamma functions (1.2.6) and interpreting the result with the help of (1.1.1) .

The conditions of the validity of (3.2.2) are as follows.

(i) $1 \leq m \leq q$, $1 \leq n \leq p < q$; (or $1 \leq m \leq q = p+1$, $0 \leq n \leq p$, $p \geq 1$, (but excluding $m = p+1$ and $n = 0$); or $0 \leq m \leq q = p$, $p \geq 1$, $0 \leq n \leq p$, provided that $|\arg \omega| = (c^* - 2j)\pi$ for $j = 0,1,\ldots, [\frac{c^*}{2}]$ is excluded); $1 \leq \alpha \leq \delta$, $1 \leq \beta \leq \gamma < \delta$; (or $\gamma \geq 1$, $0 \leq \gamma \leq \delta$, $1 \leq \alpha \leq \delta = \gamma+1$ (but excluding $\alpha = \gamma+1$ and $\beta = 0$); or $\gamma \geq 1$, $0 \leq \beta \leq \gamma$, $0 \leq \alpha \leq \delta = \gamma$ provided that $|\arg \eta| = (b^*-2j)\pi$ for $j = 0,1,\ldots, [\frac{b^*}{2}]$ is excluded); and the conditions (3.2.4) to (3.2.8), $c^* > 0$, $|\arg \omega| < c^*\pi$, $b^* > 0$ and $|\arg \eta| < b^*\pi$ are fulfilled.

(ii) Let $\alpha,\beta,\gamma,\delta$ be as in (i) above, $b^* > 0$, $|\arg \eta| < b^*\pi$; $1 \leq m \leq q$, $0 \leq n \leq p < q$, $c^* > 0$, $|\arg \omega| = c^*\pi$ (or $\arg \omega = 0$, $c^* = 0$, $0 \leq n \leq p \leq q - 2$) and the conditions (3.2.4) to (3.2.8) and (3.2.10) are satisfied.

(iii) Let m,n,p and q be as in (i) above, $c^* > 0$, $|\arg \omega| < c^*\pi$; $1 \leq \alpha \leq \delta$, $0 \leq \beta \leq \gamma < \delta$, $b^* > 0$, $|\arg \eta| = b^*\pi$ (or $b^* = 0$, $\arg \eta = 0$, $0 \leq \beta \leq \gamma \leq \delta-2$) and the conditions (3.2.4) to (3.2.9) are satisfied.

When $p \leq q$ and $\gamma \geq \delta$, then in view of (1.2.8) the integral can be written as

$$\int_0^\infty G^{m,n}_{p,q}\left(\omega x \,\Big|\, {a_p \atop b_q}\right) G^{\beta,\alpha}_{\delta,\gamma}\left(\frac{1}{\eta\, x^{\frac{k}{\rho}}} \,\Big|\, {1 - d_\delta \atop 1 - c_\gamma}\right) dx \,.$$

The conditions of validity in this situation can be obtained in a similar fashion. Also the cases (i) $p \geq q$ or $\gamma \leq \delta$, (ii) $p \geq q$ and $\gamma \geq \delta$ can be treated in the same way. In the latter case the conditions of validity can be written in exactly a similar manner as in the case $p \leq q$ and $\gamma \leq \delta$.

For $k = \rho = \sigma = 1$, (3.2.2) gives rise to Meijer's formula [205], namely.

$$\int_0^\infty G_{p,q}^{m,n}\left(\omega x \,\middle|\, {a_p \atop b_q}\right) G_{\gamma,\delta}^{\alpha,\beta}\left(\eta x \,\middle|\, {c_\gamma \atop d_\delta}\right) dx\ .$$

$$= \omega^{-1} G_{\gamma+q,\delta+p}^{\alpha+n,\beta+m}\left(\frac{\eta}{\omega} \,\middle|\, {c_1,\ldots,c_\beta,\ -b_q,\ c_{\beta+1},\ldots,c_\gamma \atop d_1,\ldots,d_\alpha,\ -a_p, d_{\alpha+1},\ldots,d_\delta}\right), \qquad (3.2.11)$$

valid under the same conditions with $k = \rho = \sigma = 1$.
The following integral involving products of G-functions is also given by Saxena [268].

$$\int_0^\infty t^{\sigma-1} G_{p,q}^{m,n}\left(\lambda t \,\middle|\, {a_p \atop b_q}\right) G_{\gamma,\delta}^{\alpha,\beta}\left(b+t \,\middle|\, {c_\gamma \atop d_\delta}\right) dt = \sum_{k=0}^{\infty} \frac{(-b)^k}{k!}$$

$$\times\, G_{\delta+p+1,\ \gamma+q+1}^{\beta+m,\ \alpha+n+1}\left(\lambda \,\middle|\, {1-\sigma, a_p\ \ 1+k-\sigma-d_\delta \atop b_q, 1+k-\sigma-c_\gamma, 1+k-\sigma}\right), \qquad (3.2.12)$$

where $R(\sigma + \min b_j) > 0$ $j = 1,2,\ldots, m$; $R(\sigma + \max a_j + \max c_k) < 2$, $j = 1,\ldots,n$; $k = 1,\ldots,\beta$; $|\arg \lambda| < c^* \pi$, $c^* = m+n- \frac{p}{2} - \frac{q}{2} > 0$, $|\arg b| < \frac{\pi}{2}$, $\alpha + \beta > \frac{\gamma}{2} + \frac{\delta}{2} + \frac{1}{2}$.

For the proof of this result, see the work of Saxena cited above.

Remark: It is interesting to note that a result due to Sundararajan [324] follows as a particular case of (3.2.12)

3.3. INTEGRAL TRANSFORMS OF THE G-FUNCTION.

By specializing the parameters in (3.2.2) and using the tables of the particular cases of the G-function given in Chapter II, we obtain many important integrals of the Special Functions occurring in Mathematical Physics and Statistics. However, for the sake of brevity, some interesting and useful special cases of (3.2.2) are given in the following sections which supply us with some of the integral transforms associated with a G-function. We have given one condition of validity with each result, since other conditions of validity of these results can be readily deduced from the cases associated with (3.2.2).

Another set of integral transforms of the G-function can be deduced from the result (3.2.12). This is left as an exercise for the reader.

3.3.1. Hankel transform of the G-function

$$\int_0^\infty x^{\sigma-1} J_\nu(\omega x^{\frac{1}{2}})\, G_{p,q}^{m,n}\left(\eta x^{\frac{k}{\rho}} \,\middle|\, {a_p \atop b_q}\right)dx$$

$$= \left(\frac{2}{\omega}\right)^{2\sigma} (2\pi)^{c^*(1-\rho)}\, k^{2\sigma-1}\, \rho^{U}$$

$$\times\, G_{\rho p+2k,\,\rho q}^{\rho m,\,\rho n+k}\left(\frac{\eta^\rho \rho^{\rho(p-q)}}{\omega^{2k}}(2k)^{2k} \,\middle|\, {\Delta(k,1-\sigma-\frac{\nu}{2}),\ \Delta(\rho,a_p),\ \Delta(k,1-\sigma+\frac{\nu}{2}) \atop \Delta(\rho,b_q)}\right), \tag{3.3.1}$$

where k and ρ are positive integers,

$$c^* = m+n-\frac{p}{2}-\frac{q}{2}\ ;\quad U = \sum_{j=1}^{q} b_j - \sum_{j=1}^{p} a_j + \frac{p}{2} - \frac{q}{2} + 1,$$

$$R(\sigma + \frac{\nu}{2} + \frac{k}{\rho}\min b_j) > 0 \quad (j = 1,\ldots,m),$$

$$R(\sigma\rho + k\max a_j) < k + \frac{3\rho}{4} \quad (j = 1,\ldots,n),$$

$$\omega > 0\ ,\ |\arg\eta| < c^*\pi,\quad c^* > 0.$$

$$\int_0^\infty x^{\sigma-1} J_\nu(\omega x) G_{p,q}^{m,n}\left(\omega x \,\middle|\, {a_p \atop b_q}\right)dx = \left(\frac{2}{\omega}\right)^{\sigma} (2\pi)^{-c^*}\, 2^{U-1}$$

$$\times G^{2m,2n+1}_{2p+2,2q}\left[\frac{\eta^2\, 2^{2p-2q+2}}{\omega^2}\,\middle|\,\begin{matrix}1-\frac{\sigma}{2}-\frac{\nu}{2},\ \Delta(2,a_p),\ 1-\frac{\sigma}{2}+\frac{\nu}{2}\\ \Delta(2,b_q)\end{matrix}\right],$$

$R(\sigma+\nu+\min b_j) > 0$ $(j = 1,\dots,m)$, $R(\sigma + \max a_j) < \frac{5}{2}$ $(j = 1,\dots,n)$, $\omega > 0$,
$c^* = m + n - \frac{p}{2} - \frac{q}{2} > 0$, $|\arg \eta| < c^*\pi$. Here $U = \sum_{j=1}^{q} b_j - \sum_{j=1}^{p} a_j + \frac{p}{2} - \frac{q}{2} + 1$.

(3.3.2)

3.3.2. Meijer Transform of the G-function.

$$\int_0^\infty x^{\sigma-1} K_\nu(\omega x^{\frac{1}{2}})\, G^{m,n}_{p,q}\left(\eta x^{\frac{k}{\rho}}\,\middle|\,\begin{matrix}a_p\\ b_q\end{matrix}\right)dx = \frac{1}{2}\left(\frac{2}{\omega}\right)^{2\sigma}(2\pi)^{c^*(1-\rho)+1-k}\rho^U k^{2\sigma-1}$$

$$\times G^{\rho m,\rho n+2k}_{\rho p+2k,\rho q}\left[\frac{\eta^\rho \rho^{\rho(p-q)}(2k)^{2k}}{\omega^{2k}}\,\middle|\,\begin{matrix}\Delta(k,1-\sigma\pm\frac{\nu}{2}),\ \Delta(\rho,a_p)\\ \Delta(\rho,b_q)\end{matrix}\right],$$

$R(\sigma \pm \frac{\nu}{2} + \frac{k}{\rho}\min b_j) > 0$ $(j = 1,\dots,m)$, $R(\omega) > 0$, $c^* > 0$, $|\arg \eta| < c^*\pi$.

(3.3.3)

$$\int_0^\infty x^{\sigma-1} K_\nu(\omega x)\, G^{m,n}_{p,q}\left(\eta x\,\middle|\,\begin{matrix}a_p\\ b_q\end{matrix}\right)dx = \left(\frac{2}{\omega}\right)^{\sigma}(2\pi)^{-c^*}\, 2^{U-2}$$

$$\times G^{2m,2n+2}_{2p+2,2n}\left[\frac{\eta^2\, 2^{2(p-q+1)}}{\omega^2}\,\middle|\,\begin{matrix}1\pm\frac{\nu}{2}-\frac{\sigma}{2},\ \Delta(2,a_p)\\ \Delta(2,b_q)\end{matrix}\right],$$

$R(\sigma \pm \nu + \min b_j) > 0$ $(j = 1,\dots,m)$, $R(\omega) > 0$, $c^* > 0$, $|\arg \eta| < c^*\pi$.

(3.3.4)

3.3.3. Y-Transform of the G-function .

$$\int_0^\infty x^{\sigma-1} Y_\nu(\omega x^{\frac{1}{2}})\, G^{m,n}_{p,q}\left(\eta x^{\frac{k}{\rho}}\,\middle|\,\begin{matrix}a_p\\ b_q\end{matrix}\right)dx = \left(\frac{2}{\omega}\right)^{2\sigma}(2\pi)^{c^*(1-\rho)} k^{2\sigma-1}\rho^U$$

$$\times G^{\rho m,\rho n+2k}_{\rho p+3k,\rho q+k}\left[\frac{\eta^\rho \rho^{\rho(p-q)}(2k)^{2k}}{\omega^{2k}}\,\middle|\,\begin{matrix}\Delta(k,1-\sigma\pm\frac{\nu}{2}),\ \Delta(\rho,a_p),\ \Delta(k,\frac{3}{2}-\sigma+\frac{\nu}{2})\\ \Delta(\rho,b_q),\ \Delta(k,\frac{3}{2}-\sigma+\frac{\nu}{2})\end{matrix}\right],$$

$R(\sigma \pm \frac{\nu}{2} + \frac{k}{\rho}\min b_j) > 0$ $(j = 1,\dots,m)$, $R(\rho\sigma + k\max a_j) < k + \frac{3\rho}{4}$
$(j = 1,\dots,n)$; $\omega > 0$, $c^* > 0$, $|\arg \eta| < c^*\pi$. (3.3.5)

$$\int_0^\infty x^{\sigma-1} Y_\nu(\omega x)\, G^{m,n}_{p,q}\left(\eta x \,\middle|\, {a_p \atop b_q}\right) dx = \left(\frac{2}{\omega}\right)^\sigma (2\pi)^{-a}\, 2^{U-1}$$

$$\times\, G^{2m,2n+2}_{2p+3,2q+1}\left[\frac{\eta^2\, 2^{2p-2q+2}}{\omega^2} \,\middle|\, {1-\frac{\sigma}{2} \pm \frac{\nu}{2},\ \Delta(2,a_p),\ \frac{3}{2}-\frac{\sigma}{2}+\frac{\nu}{2} \atop \Delta(2,b_q),\ \frac{3}{2}-\frac{\sigma}{2}+\frac{\nu}{2}}\right],$$

$R(\sigma \pm \nu + \min b_j) > 0 \quad (j = 1,\ldots,m), \quad R(\sigma + \max a_j) < \frac{5}{2} \ (j = 1,\ldots,n),$

$\omega > 0, \ c^* > 0, \quad |\arg \eta| < c^*\pi.$ (3.3.6)

3.3.4. H-Transform of the G-function .

$$\int_0^\infty x^{\sigma-1} H_\nu(\omega x^{\frac{1}{2}})\, G^{m,n}_{p,q}\left(\eta x^{\frac{k}{\rho}} \,\middle|\, {a_p \atop b_q}\right) dx = \left(\frac{2}{\omega}\right)^{2\sigma} (2\pi)^{c^*(1-\rho)}\, k^{2\sigma-1}\, \rho^U$$

$$\times\, G^{\rho m+k,\rho n+k}_{\rho p+3k,\rho q+k}\left[\frac{\eta^\rho \rho^{\rho(p-q)} (2k)^{2k}}{\omega^{2k}} \,\middle|\, {\Delta(k,\frac{1}{2}-\sigma-\frac{\nu}{2}),\ \Delta(\rho,a_p),\ \Delta(k,1-\sigma\pm\frac{\nu}{2}) \atop \Delta(k,\frac{1}{2}-\sigma-\frac{\nu}{2}),\ \Delta(\rho,b_q)}\right],$$ (3.3.7)

$R(\sigma + \frac{\nu}{2} + \frac{k}{\rho}\min b_j) > -\frac{1}{2} \ (j = 1,\ldots,m), \ R(\sigma + \frac{\nu}{2} + \frac{k}{\rho}\max a_j) < \frac{k}{\rho} + \frac{1}{2},$

$R(\sigma + \frac{k}{\rho}\max a_j) < \frac{k}{\rho} + \frac{3}{4}, \quad j = 1,\ldots,n; \quad \omega > 0, \ c^* > 0, \quad |\arg \eta| < c^*\pi.$

$$\int_0^\infty x^{\sigma-1} H_\nu(\omega x)\, G^{m,n}_{p,q}\left(\eta x \,\middle|\, {a_p \atop b_q}\right) dx = \left(\frac{2}{\omega}\right)^\sigma (2\pi)^{-c^*}\, 2^{U-1}$$

$$\times\, G^{2m+1,2n+1}_{2p+3,2q+1}\left[\frac{\eta^2\, 2^{2p-2q+2}}{\omega^2} \,\middle|\, {\frac{1}{2}-\frac{\sigma}{2}-\frac{\nu}{2},\ \Delta(2,a_p),\ 1-\frac{\sigma}{2}\pm\frac{\nu}{2} \atop \frac{1}{2}-\frac{\sigma}{2}-\frac{\nu}{2},\ \Delta(2,b_q)}\right],$$ (3.3.8)

$R(\sigma + \nu + \min b_j) > -1 \quad (j = 1,\ldots, m); \quad R(\sigma + \max a_j) < \frac{5}{2},$

$R(\sigma + \nu + \max a_j) < 2 \ (j = 1,\ldots,n); \ \omega > 0, \quad c^* > 0, \quad |\arg \eta| < c^*\pi.$

3.3.5. Gauss' Hypergeometric Transform of the G-Function.

$$\int_0^\infty x^{\sigma-1}\, {}_2F_1(\lambda,\mu;\nu;-\omega x)\, G^{m,n}_{p,q}\left(\eta x^{\frac{k}{\rho}} \,\middle|\, \begin{matrix} a_p \\ b_q \end{matrix}\right) dx$$

$$= \omega^{-\sigma}\,(2\pi)^{1-k+c^*(1-\rho)}\, k^{\lambda+\mu-\nu-1}\rho^{U}\, \frac{\Gamma(\nu)}{\Gamma(\lambda)\Gamma(\mu)}$$

$$\times\, G^{m\rho+2k,\,\rho n+k}_{p\rho+2k,\,q\rho+2k}\left[\frac{\eta^{\rho}\rho^{\rho(p-q)}}{\omega^{k}} \,\middle|\, \begin{matrix} \Delta(k,1-\sigma),\ \Delta(\rho,a_p),\ \Delta(\rho,\nu-\sigma) \\ \Delta(k,\lambda-\sigma),\ \Delta(k,\mu-\sigma),\ \Delta(\rho,b_q) \end{matrix}\right], \qquad (3.3.9)$$

$R(\sigma + \frac{k}{\rho}\min b_j) > 0 \quad (j = 1,\dots,m), \quad R(\sigma-\lambda + \frac{k}{\rho}\max a_j) < \frac{k}{\rho} \quad (j = 1,\dots,n),$

$R(\sigma-\mu + \frac{k}{\rho}\max a_j) < \frac{k}{\rho} \quad (j = 1,\dots,n), \quad c^* > 0, \quad |\arg \eta| < c^*\pi, \quad |\arg \omega| < \pi \,;$

$$U = \sum_{j=1}^{q} b_j - \sum_{j=1}^{p} a_j + \frac{p}{2} - \frac{q}{2} + 1 .$$

For a generalization of this result see Golas [104].

3.3.5a. Generalized Stieltjes Transform of the G-Function.

$$\int_0^\infty x^{\sigma-1}(\omega+x)^{\lambda-1}\, G^{m,n}_{p,q}\left(\eta x^{\frac{k}{\rho}} \,\middle|\, \begin{matrix} a_p \\ b_q \end{matrix}\right) dx = \frac{\omega^{\lambda+\sigma-1}}{\Gamma(1-\lambda)}(2\pi)^{1-k+c^*(1-\rho)}\, \rho^{U}\, k^{-\lambda}$$

$$\times\, G^{m\rho+k,\,n\rho+k}_{p\rho+k,\,q\rho+k}\left[\frac{\eta^{\rho}\rho^{\rho(p-q)}}{\omega^{k}} \,\middle|\, \begin{matrix} \Delta(k,1-\sigma),\ \Delta(\rho,a_p) \\ \Delta(k,1-\lambda-\sigma),\ \Delta(\rho,b_q) \end{matrix}\right], \qquad (3.3.10)$$

$|\arg \omega| < \pi, \quad R(\sigma+\min b_j) > 0 \quad (j = 1,\dots,m)\,,$

$R(\sigma + \lambda + \frac{k}{\rho}\max a_j) < \frac{k}{\rho} + 1 \quad (j = 1,\dots n), \quad c^* > 0, \quad |\arg \eta| < c^*\pi.$

3.3.6 Whittaker Transform of the G-function.

$$\int_0^\infty x^{\sigma-1}\exp\left(-\frac{\omega x}{2}\right) W_{\lambda,\mu}(\omega x)\, G^{m,n}_{p,q}\left(\eta x^{\frac{k}{\rho}} \,\middle|\, \begin{matrix} a_p \\ b_q \end{matrix}\right) dx$$

$$= \omega^{-\sigma} (2\pi)^{c^*(1-\rho)+\frac{(1-k)}{2}} k^{\lambda+\sigma-\frac{1}{2}} \rho^{U}$$

$$\times G^{\rho m, \rho n+2k}_{\rho p+2k, \rho q+k} \left[\frac{k^k \eta^\rho \rho^{\rho(p-q)}}{\omega^k} \middle| \begin{matrix} \Delta(k,\frac{1}{2}-\sigma \pm \mu), \Delta(\rho, a_p) \\ \Delta(\rho, b_q), \Delta(k, \lambda-\sigma) \end{matrix} \right],$$

$$R(\sigma \pm \mu + \frac{k}{\rho} \min b_j) > -\frac{1}{2} \; (j = 1,\ldots,m), \; R(\omega) > 0, \; c^* > 0, \; |\arg \eta| < c^* \pi. \quad (3.3.11)$$

3.3.6a. Laplace Transform of the G-function

$$\int_0^\infty x^{\sigma-1} e^{-\omega x} G^{m,n}_{p,q}\left(\eta x^{\frac{k}{\rho}} \middle| \begin{matrix} a_p \\ b_q \end{matrix}\right) dx = \omega^{-\sigma} (2\pi)^{\frac{1-k}{2}+c^*(1-\rho)} k^{\sigma-\frac{1}{2}} \rho^{U}$$

$$\times G^{\rho m, \rho n+k}_{\rho p+k, \rho q} \left[\frac{k^k \eta^\rho \rho^{\rho(p-q)}}{\omega^k} \middle| \begin{matrix} \Delta(k, 1-\sigma), \Delta(\rho, a_p) \\ \Delta(\rho, b_q) \end{matrix} \right], \quad (3.3.12)$$

$$R(\sigma + \frac{k}{\rho} \min b_j) > 0 \; (j = 1,\ldots,m), \; R(\omega) > 0, \; c^* > 0, \; |\arg \eta| < c^* \pi.$$

3.3.6b Laguerre Transform of the G-function.

$$\int_0^\infty x^{\sigma+\frac{\alpha-1}{2}} e^{-\omega x} L_u^\alpha(x\omega) G^{m,n}_{p,q}\left(\eta x^{\frac{k}{\rho}} \middle| \begin{matrix} a_p \\ b_q \end{matrix}\right) dx$$

$$= \frac{(-1)^u}{u!} \omega^{-\sigma} (2\pi)^{c^*(1-\rho)+\frac{1-k}{2}} k^{\sigma+u+\frac{\alpha}{2}} \rho^{U}$$

$$\times G^{\rho m, \rho n+2k}_{\rho p+2k, \rho q+k} \left[\frac{k^k \eta^\rho \rho^{\rho(p-q)}}{\omega^k} \middle| \begin{matrix} \Delta(k, \frac{1}{2} \pm \frac{\alpha}{2} - \sigma), \Delta(\rho, a_p) \\ \Delta(\rho, b_q), \Delta(k, \frac{1+\alpha}{2} + u - \sigma) \end{matrix} \right], \quad (3.3.13)$$

$$R(\sigma \pm \frac{\alpha}{2} + \frac{k}{\rho} \min b_j) > -\frac{1}{2} \; (j = 1,\ldots,m);$$

$$R(\omega) > 0, \; c^* > 0, \; |\arg \eta| < c^* \pi.$$

Here $L_u^\alpha(x)$ is the generalized Laguerre polynomial.

3.3.7. An Integral Involving Hypergeometric Function.

$$\int_0^\infty x^{\sigma-1} \, {}_2F_1(\lambda,\mu;\nu;\, 1-\omega x) \, G^{m,n}_{p,q}\left(\eta x^{\frac{k}{\rho}} \,\middle|\, {}^{a_p}_{b_q}\right) dx$$

$$= \omega^{-\sigma} (2\pi)^{c^*(1-\rho) + 2(1-k)} \, k^{\nu-2} \, \rho^{U}$$

$$\times \frac{\Gamma(\nu)}{\Gamma(\lambda)\,\Gamma(\mu)\,\Gamma(\nu-\lambda)\,\Gamma(\nu-\mu)}$$

$$\times G^{\rho m+2k,\rho n+2k}_{\rho p+2k,\rho q+2k}\left[\frac{\eta^{\rho}\,\rho^{\rho(p-q)}}{\omega^{k}} \,\middle|\, {}^{\Delta(k,1-\sigma),\ \Delta(1,1+\alpha+\beta-\gamma-\sigma),\ \Delta(\rho,a_p)}_{\Delta(k,\alpha-\sigma),\ \Delta(k,\beta-\sigma),\ \Delta(\rho,b_q)}\right], \tag{3.3.14}$$

$R(\sigma + \frac{k}{\rho} b_j) > 0$, $R(\sigma + -\lambda-\mu+ \frac{k}{\rho} b_j)>0$ $(j = 1,\dots,m)$;

$R(\sigma+ \frac{k}{\rho} \max a_j-\lambda) < \frac{k}{\rho}$, $R(\sigma+ \frac{k}{\rho} \max a_j-\mu) < \frac{k}{\rho}$ $(j = 1,\dots,n)$, $c^*> 0$,

$|\arg \omega| < 2\pi$, $|\arg \eta|< c^*\pi$. For $k = \rho = 1$, (3.3.14) reduces to one given by Sharma, K.C. [289].

<u>Remark:</u> Integral transforms associated with a Meijer's G-function are studied by Kapoor [131], Kapoor and Masood [132], Sharma, K.C. [295], Wimp[357] and Kalia [128].

3.4. INTEGRALS INVOLVING PRODUCTS OF GAUSS'S HYPERGEOMETRIC FUNCTION AND THE G-FUNCTION.

The integrals given in this section are due to Saxena [264] and Sharma, B.L. ([282],[284]) and can be readily established with the help of the integrals given in section 3.1.

In all the results of this section k and ρ are assumed to be positive integers and $c^* = m + n - \frac{p}{2} - \frac{q}{2}$.

$$\int_0^1 t^{\nu-1}(1-t)^{\nu-1} \, {}_2F_1(\alpha,\beta;\tfrac{\alpha+\beta+1}{2};\, t) \, G^{m,n}_{p,q}\left[z t^k (1-t)^k \,\middle|\, {}^{a_p}_{b_q}\right] dt$$

$$= \frac{\pi k^{-\frac{1}{2}}\, 2^{1-2\nu}\, \Gamma(\frac{\alpha+\beta+1}{2})}{\Gamma(\frac{\alpha+1}{2})\, \Gamma(\frac{\beta+1}{2})}$$

$$\times G^{m,n+2k}_{p+2k,q+2k}\left[\frac{z}{2^{2k}} \,\middle|\, \begin{matrix} \Delta(k,1-\nu),\ \Delta(k, \frac{\alpha+\beta+1}{2}-\nu),\ a_p \\ b_q,\ \Delta(k,\frac{\alpha+1}{2}-\nu),\ \Delta(k, \frac{\beta+1}{2}-\nu) \end{matrix} \right]. \quad (3.4.1)$$

The result is valid under the following conditions:

(i) $0 \leq n \leq p < q,\ \ 1 \leq m \leq q,\ \ R(\nu+kb_j) > 0,$

$R(\frac{1}{2} + \nu + kb_j - \frac{\alpha}{2} - \frac{\beta}{2}) > 0,\ (j = 1,\ldots,m);$

(ii) $0 \leq m \leq q < p,\ \ 1 \leq n \leq p,\ \ c^* > 0,\ \ |\arg z| = c^*\pi$ or

$0 \leq m \leq q \leq p-2,\ \ c^* = 0,\ |\arg z| = 0$; and

$R(\nu+kb_j) > 0,\ \ R(\frac{1}{2} + \nu + kb_j - \frac{\alpha}{2} - \frac{\beta}{2}) > 0\ \ (j = 1,\ldots,m),$

$R[k(\sum_{j=1}^{p} a_j - \sum_{j=1}^{p} b_j - \frac{1}{2}) + (p-q)(\nu - \frac{k}{2})] > -1,$

$R[k(\sum_{j=1}^{p} a_j - \sum_{j=1}^{q} b_j - \frac{1}{2}) + (p-q)(\nu + \frac{1}{2} - \frac{\alpha+\beta+k}{2})] > -1.$

$$\int_0^1 t^{\beta-1}(1-t)^{\beta-\nu}\, {}_2F_1(\alpha,1-\alpha,\nu;\ t)\, G^{m,n}_{p,q}\left(2t^k(1-t)^k \,\middle|\, \begin{matrix} a_p \\ b_q \end{matrix}\right)dt$$

$$= \frac{\pi k^{-\frac{1}{2}}\, 2^{1-2\beta}\, \Gamma(\nu)}{\Gamma(\frac{\alpha+\nu}{2})\, \Gamma(\frac{1+\nu-\alpha}{2})}\, G^{m,n+2k}_{p+2k,q+2k}\left(\frac{z}{2^{2k}} \,\middle|\, \begin{matrix} \Delta(k,1-\beta),\ \Delta(k,\beta-\nu),\ a_p \\ b_q,\ \Delta(k,\frac{1+\nu-\alpha}{2}-\beta),\ \Delta(k,\frac{\alpha+\nu}{2}-\beta) \end{matrix} \right).$$

(3.4.2)

The result is valid under the following conditions.

(i) $0 \leq n \leq p < q,\ 1 \leq m \leq q,\ R(\beta + kb_j) > 0,$

$R(1 + \beta - \nu + kb_j) > 0,\ \ j = 1,\ldots,m;$

(ii) $0 \le m \le q < p$, $1 \le n \le p$, $c^* > 0$, $|\arg z| = c^* \pi$ or

$0 \le m \le q \le p-2$, $c^* = 0$ $\arg z = 0$ and

$R(\beta + kb_j) > 0$, $R(1+\beta-\nu+kb_j) > 0$ $(j = 1,\dots,m)$,

$$R\,[k(\sum_{j=1}^{p} a_j - \sum_{j=1}^{q} b_j - \tfrac{1}{2}) + (p-q)(\beta - \tfrac{k}{2})] > -1 ,$$

$$R\,[k(\sum_{j=1}^{p} a_j - \sum_{j=1}^{q} b_j - \tfrac{1}{2}) + (p-q)(\beta-\nu-\tfrac{k}{2}+1)] > -1 .$$

$$\int_0^{\pi/2} \exp\{i(\alpha+\beta)\theta\}(\sin\theta)^{\alpha-1}(\cos\theta)^{\beta-1}\, G^{m,n}_{p,q}\left[ze^{ik\theta}\sin^k\theta \,\middle|\, \begin{matrix} a_p \\ b_q \end{matrix}\right] d\theta$$

$$= e^{i\frac{1}{2}\pi\alpha}\, k^{-\beta}\Gamma(\beta)\, G^{m,n+k}_{p+k,q+k}\left[ze^{\frac{i\pi k}{2}} \,\middle|\, \begin{matrix} \Delta(k,1-\alpha),\ a_p \\ b_q,\ \Delta(k,1-\alpha-\beta) \end{matrix}\right]. \quad (3.4.3)$$

The result in (3.4.3) is valid under the following conditions:

(i) $0 \le n \le p < q$, $1 \le m \le q$, $R(\beta) > 0$, $R(\alpha+kb_j) > 0$ for $j = 1,\dots,m$;

(ii) $0 \le m \le q < p$, $1 \le n \le p$, $c^* > 0$, $|\arg z| = c^* \pi$ or

$0 \le m \le q \le p-2$, $c^* = 0$, $\arg z = 0$; and $R(\beta) > 0$, $R(\alpha+kb_j) > 0$

for $j = 1,2,\dots,m$, $R\,[k(\sum_{j=1}^{p} a_j - \sum_{j=1}^{q} b_j - \frac{1}{2}) + (p-q)(\alpha - \frac{k}{2})] > -1.$

$$\int_0^{\pi/2} \exp\{i(\alpha+\beta)\theta\}(\sin\theta)^{\alpha-1}(\cos\theta)^{\beta-1}$$

$$\times G^{m,n}_{p,q}\left[ze^{i(k+\rho)\theta}\sin^k\theta\cos^\rho\theta \,\middle|\, \begin{matrix} a_p \\ b_q \end{matrix}\right] d\theta = e^{i\frac{\pi\alpha}{2}} (2\pi)^{\frac{1}{2}} \frac{k^{\alpha-\frac{1}{2}}\rho^{\beta-\frac{1}{2}}}{(k+\rho)^{\alpha+\beta-\frac{1}{2}}}$$

$$\times G^{m,n+k+\rho}_{p+k+\rho,q+k+\rho}\left[\frac{zk^k\rho^\rho e^{\frac{i\pi k}{2}}}{(k+\rho)^{k+\rho}} \,\middle|\, \begin{matrix} \Delta(k,1-\alpha),\ \Delta(\rho,1-\beta),\ a_p \\ b_q,\ \Delta(k+\rho,1-\alpha-\beta) \end{matrix}\right]. \quad (3.4.4)$$

$$\int_0^1 \frac{t^{\alpha-1}(1-t)^{\beta-1}}{\{1+ct+d(1-t)\}^{\alpha+\beta}}\, G^{m,n}_{p,q}\left[\frac{zt^k(1-t)^\rho}{\{1+ct+d(1-t)\}^{k+\rho}} \,\middle|\, \begin{matrix} a_p \\ b_q \end{matrix}\right] dt$$

$$= \frac{(2\pi)^{\frac{1}{2}} k^{\alpha-\frac{1}{2}} \rho^{\beta-\frac{1}{2}}}{(1+c)^{\alpha} (1+d)^{\beta} (k+\rho)^{\alpha+\beta-\frac{1}{2}}}$$

$$\times G^{m,n+k+\rho}_{p+k+\rho,q+k+\rho} \left[S \,\middle|\, \begin{matrix} \Delta(k,1-\alpha),\ \Delta(\rho,1-\beta),\ a_p \\ b_q,\ \Delta(k+\rho,1-\alpha-\beta) \end{matrix} \right], \qquad (3.4.5)$$

where
$$S = \frac{z k^k \rho^\rho}{(1+c)^k (1+d)^\rho (k+\rho)^{k+\rho}} .$$

Here (3.4.4) and (3.4.5) are valid under the following conditions:

(i) $0 \le n \le p < q$, $1 \le m \le q$, $R(\alpha+kb_j) > 0$, $R(\beta+\rho b_j) > 0$ for $j = 1,\ldots,m$;

(ii) $1 \le n \le p$, $0 \le m \le q < p$, $c^* > 0$, $|\arg z| = c^*\pi$ or $0 \le m \le q \le p-2$, $c^* = 0$, $\arg z = 0$; and $R(\alpha+kb_j) > 0$, $R(\beta+\rho b_j) > 0$ for $j = 1,\ldots,m$,

$R\,[k\,(\sum_{j=1}^{p} a_j - \sum_{j=1}^{q} b_j - \frac{1}{2}) + (p-q)\,(\alpha - \frac{k}{2})] > -1$, and

$R\,[\rho(\sum_{j=1}^{p} a_j - \sum_{j=1}^{q} b_j - \frac{1}{2}) + (p-q)(\beta - \frac{\rho}{2})] > -1$.

$$\int_0^1 t^{\lambda-1}(1-t)^{\mu-1}[1 + ct + d(1-t)]^{\lambda-\mu}$$

$$\times\ {}_2F_1\left(\begin{matrix} \alpha,\beta; \\ \mu; \end{matrix} \frac{(1-t)\ (1+d)}{1 + ct + d(1-t)} \right) G^{m,n}_{p,q} \left[z \left\{ \frac{t(1+c)}{1+ct+d(1-t)} \right\}^k \,\middle|\, \begin{matrix} a_p \\ b_q \end{matrix} \right] dt$$

$$= \frac{\Gamma(\mu)\ k^{-\mu}}{(1 + c)^{\lambda}(1+d)^{\mu}} G^{m,n+2k}_{p+2k,q+2k} \left(z \,\middle|\, \begin{matrix} \Delta(k,1-\lambda),\Delta(k,1-\lambda-\mu+\alpha+\beta,\alpha_p \\ \beta_q,\Delta(k,1-\lambda-\mu+\alpha),\ \Delta(k,1-\lambda-\mu+\beta) \end{matrix} \right),$$

(3.4.6)

where the constants c and d are such that noneof the expressions 1+c, 1+d and 1+ct + d(1-t), where $0 \le t \le 1$, is zero.

The result is valid under the following conditions.

(i) $1 \le m \le q$, $0 \le n \le p < q$, $R(\mu) > 0$, $R(\mu - \alpha - \beta) > 0$, $R(\lambda+k\beta_j) > 0$, $j = 1,\ldots,m$;

(ii) $1 \le n \le p$, $0 \le m \le q < p$, $c^* > 0$, $|\arg z| = c^*\pi$ or $0 \le m \le q \le p-2$,

$c^* = 0$, $\arg z = 0$, and $R(\mu) > 0$, $R(\mu-\alpha-\beta) > 0$, $R(\lambda+k\beta_j) > 0$, $j = 1,\ldots,m$;

$R[\, k\, (\sum_{j=1}^{p} a_j - \sum_{j=1}^{q} b_j - \frac{1}{2}) + (p-q)(\lambda - \frac{k}{2})] > -1.$

$$\int_0^{\pi/2} e^{i(\alpha+\beta)\theta} (\sin\theta)^{\alpha-1} (\cos\theta)^{\beta-1} {}_2F_1(c,d;\alpha;e^{-i(\frac{\pi}{2}-\theta)}\sin\theta)$$

$$\times G^{m,n}_{p,q}[\, ze^{-ik\theta} (\cos\theta)^k \mid {}^{a_p}_{b_q}]\, d\theta$$

$$= \frac{e^{i\frac{\pi}{2}\alpha}\,\Gamma(\alpha)}{k^{\alpha}} G^{m,n+2k}_{p+2k,q+2k}(z \mid {}^{\lambda,\,\mu,\,a_p}_{b_q,\,\nu,\,\tau}), \tag{3.4.7}$$

where $\lambda = \Delta(k, \frac{1-\beta}{2})$, $\mu = \Delta(k,1+c+d-\alpha-\beta)$, $\nu = \Delta(k,1+c-\alpha-\beta)$, $\tau = \Delta(k,1+d-\alpha-\beta)$.

The result is valid under the following conditions:

(i) $1 \le m \le q$, $0 \le n \le p < q$, $R(\alpha) > 0$, $R(\alpha-c-d) > 0$, $R(\beta+kb_j) > 0$

for $j = 1,\ldots,m$;

(ii) $0 \le m \le q < p$, $1 \le n \le p$, $c^* > 0$, $|\arg z| = c^*\pi$, or

$0 \le m \le q \le p-2$, $c^* = 0$, $\arg z = 0$; and $R(\alpha) > 0$, $R(\alpha-c-d) > 0$,

$R(\beta+kb_j) > 0$, $j = 1,\ldots,m$, $R[k\, (\sum_{j=1}^{p} a_j - \sum_{j=1}^{q} b_j - \frac{1}{2}) + (p-q)(\beta - \frac{k}{2})] > -1.$

$$\int_0^{\pi/2} e^{i(\alpha+\beta)\theta} (\sin\theta)^{\alpha-1} (\cos\theta)^{\beta-1} {}_2F_1(c,d;\beta;e^{i\theta}\cos\theta)$$

$$\times G^{m,n}_{p,q}[ze^{ik(\theta-\frac{\pi}{2})} \sin^k\theta \mid {}^{a_p}_{b_q}]\, d\theta$$

$$= \frac{e^{\frac{i\pi\alpha}{2}}\,\Gamma(\beta)}{k^{\beta}} G^{m,n+2k}_{p+2k,q+2k}(z \mid {}^{\lambda,\mu,\,a_p}_{b_q,\nu,\,\tau}), \tag{3.4.8}$$

where $\lambda = \Delta(k,1-\alpha)$, $\mu = \Delta(k,1+c+d-\alpha-\beta)$,

$\nu = \Delta(k,1+c-\alpha-\beta)$, $\tau = \Delta(k,1+d-\alpha-\beta)$.

The result (3.4.8) is valid under the same conditions given with (3.4.7) such that α and β are interchanged.

Remark 1: For further integrals of this type see Sharma, K.C. [292], Bajpai [21], Sharma, B.L. [282], Verma [347] and for the generalization of (3.4.2) and (3.4.8), see the work of Mathur [203].

Remark 2: Integrals involving associated Legendre functions and the G-function can be found in the work of Saxena ([261],[267]).

3.5. INTEGRALS OF G-FUNCTION WITH ARGUMENT CONTAINING THE FACTOR $x^k\{x^{\frac{1}{2}}+(1+x)^{\frac{1}{2}}\}^{2\rho}$.

In this section we give three integrals involving G-function in which the argument of the G-function contains a factor $x^k\{x^{\frac{1}{2}} + (1+x)^{\frac{1}{2}}\}^{2\rho}$, where k and ρ are positive integers and x is the variable of integration. Three different forms of the general result are given according as $k > \rho$, $k < \rho$, and $k = \rho$. The discussion of this section is based on the work of Saxena [263].

First Formula:

$$\int_0^\infty x^{\lambda-1}(1+x)^{-\frac{1}{2}}[x^{\frac{1}{2}} + (1+x)^{\frac{1}{2}}]^{2\mu} \quad G^{m,n}_{p,q}[ax^k\{x^{\frac{1}{2}} + (1+x)^{\frac{1}{2}}\}^{2\rho}\Big|^{a_p}_{b_q}]dx$$

$$= \frac{(2\pi)^{1-k-\rho} \quad k^{2\lambda-\frac{1}{2}}}{\Gamma(\frac{1}{2})(k+\rho)^{\lambda+\mu} \quad (k-\rho)^{\lambda-\mu}}$$

$$\times G^{m+k+\rho,n+2k}_{p+2k,q+2k}\left[\frac{ak^{2k}}{(k+\rho)^{k+\rho}\quad(k-\rho)^{k-\rho}}\Big|^{d,\ a_p}_{e,b_q,g}\right], \tag{3.5.1}$$

where k and ρ are positive and non-negative integers respectively and $k > \rho$.

$$d = \Delta(2k,1-2\lambda),\quad e = \Delta(k+\rho,\ \tfrac{1}{2} - \lambda - \mu),\quad g = \Delta(k-\rho,\ \tfrac{1}{2} - \lambda+\mu).$$

Here (3.5.1) readily follows from the integral (3.1.17).

The following are the conditions which are needed in the discussion of the conditions of the validity of (3.5.1)

$$R(\lambda+kb_j) > 0 \quad (j = 1,\dots,m), \quad R[\lambda+\mu-\tfrac{1}{2}+(k+\rho)(a_h-1)] < 0, \ (h = 1,\dots,n). \tag{3.5.2}$$

$$R[(k+\rho)(\sum_{h=1}^{p} a_h - \sum_{h=1}^{q} b_h - \tfrac{1}{2}) + (q-p)(\lambda+\mu- \frac{k+\rho+1}{2})] > -1 . \tag{3.5.3}$$

$$R[k(\sum_{h=1}^{p} a_h - \sum_{h=1}^{q} b_h - \tfrac{1}{2}) + (p-q)(\lambda-\tfrac{k}{2})] > -1 . \tag{3.5.4}$$

The result (3.5.1) is valid under the following conditions.

(i) $1 \le m \le q$, $1 \le n \le p < q$, the condition (3.5.2) holds $|\arg a| < c^*\pi$, $c^* = m+n- \frac{p}{2} - \frac{q}{2} > 0$. The result is also valid if (3.5.2) holds, $c^* > 0$, $|\arg a| < c^*\pi$, $p \ge 1$, $0 \le n \le p$, $1 \le m \le q + p+1$ (excluding $m = p+1$ and $n = 0$) or $p \ge 1$, $0 \le n \le p$, and $0 \le m \le q = p$, provided that in the last case $|\arg a| = (c^*-2j)\pi$, $j = 0,1,\dots, [\frac{c^*}{2}]$ is excluded.

(ii) $1 \le m \le q$, $0 \le n \le p < q$, (3.5.2) and (3.5.3) hold, $|\arg a| = c^*\pi$ $c^* > 0$ or (3.5.2) and (3.5.3) hold, $0 \le n \le p \le q-2$, $c^* = 0$ and $\arg a = 0$.

(iii) $n = 0$, $1 \le p+1 \le m \le q$, $c^* > 0$, $|\arg a| < c^*\pi$ and (3.5.2) is satisfied.

(iv) $1 \le n \le p$, $0 \le m \le q < p$, $c^* > 0$, $|\arg a| < c^*\pi$, (3.5.2) and (3.5.4) hold or $c^* = 0$ and $\arg z = 0$, $0 \le m \le q \le p-2$ and the conditions (3.5.2) and (3.5.4) are satisfied.

(v) $1 \le n \le p$, $1 \le m \le p < q$ or $q \ge 1$, $0 \le m \le q$, $1 \le n \le p = q+1$, $c^* > 0$ and $|\arg a| < c^*\pi$.

The following two integrals can be proved in a similar manner and are valid under the conditions given above.

Second Formula:

$$\int_0^\infty x^{\lambda-1}(1+x)^{-\frac{1}{2}} [x^{\frac{1}{2}} + (1+x)^{\frac{1}{2}}]^{2\mu} \, G^{m,n}_{p,q}[ax^k\{x^{\frac{1}{2}} + (1+x)^{\frac{1}{2}}\}^{2\rho} \,|\, {}^{a_p}_{b_q}]dx$$

$$= \frac{(2\pi)^{1-2k} \quad k^{2\lambda-\frac{1}{2}}}{\Gamma(\frac{1}{2})(k+\rho)^{\lambda+\mu} \, (\rho-k)^{\lambda-\mu}}$$

$$\times\, G^{m+k+\rho,\,n+2k}_{p+k+\rho,\,q+k+\rho}\left[\frac{2k^{2k}\,(\rho-k)^{\rho-k}}{(\rho+k)^{\rho+k}}\,\middle|\,\begin{matrix} d,\ a_p,\ h \\ e,\ b_q\end{matrix}\right],$$

where k and ρ are positive integers, d and e are defined in (3.5.1) and

$$h = \Delta(\rho-k,\ \tfrac{1}{2}+\lambda-\mu)\ .$$

Third Formula:

$$\int_0^\infty x^{\lambda-1}\,(1+x)^{-\frac{1}{2}}\,[x^{\frac{1}{2}}+(1+x)^{\frac{1}{2}}]^{2\mu}$$

$$\times\, G^{m,n}_{p,q}\left[ax^k\{x^{\frac{1}{2}}+(1+x)^{\frac{1}{2}}\}^{2k}\,\middle|\,\begin{matrix} a_p \\ b_q\end{matrix}\right]dx$$

$$= \frac{\pi^{1-2k}\,k^{\lambda-\mu-\frac{1}{2}}}{2^{\lambda+\mu+2k-\frac{3}{2}}\,\Gamma(\frac{1}{2}+\lambda-\mu)}\; G^{m+2k,\,n+2k}_{p+2k,\,q+2k}\left[\frac{a}{2^{2k}}\,\middle|\,\begin{matrix} d,\ a_p \\ e,\ b_q\end{matrix}\right], \qquad (3.5.3)$$

where k is a positive integer and d and e are defined in (3.5.1) such that ρ = k. The result is valid under the same conditionsas in (3.5.1) with ρ = k.

Taking m = 2, n = p = 1, q = 2, k = 1, ρ = 0, $a_1 = 1+\sigma$, $b_1 = \frac{1}{2}-\nu$, $b_2 = \frac{1}{2}+\nu$, in (3.6.1) it yields a known integral ([86] Vol. II, p.406), namely

$$\pi^{\frac{1}{2}}\int_0^\infty x^{\lambda-1}\,(1+x)^{-\frac{1}{2}}\,[x^{\frac{1}{2}}+(1+x)^{\frac{1}{2}}]^{2\mu}\; e^{\frac{ax}{2}}\; W_{\sigma,\nu}(ax)dx$$

$$= G^{3,3}_{3,4}\left(a\,\middle|\,\begin{matrix} 1+\sigma,\ \frac{1}{2}-2,1-\lambda \\ \frac{1}{2}-\nu\ ,\ \frac{1}{2}+\nu, \frac{1}{2}-\lambda-\mu,\ \frac{1}{2}-\lambda+\mu\end{matrix}\right), \qquad (3.5.4)$$

where $R(\lambda) > |R(\nu)| - \frac{1}{2}$, $R(\sigma+\lambda+\ ) < \frac{1}{2}$, $|\arg a| < \frac{3\pi}{2}$.

3.6. AN INTEGRAL INVOLVING G-FUNCTION AND JACOBI POLYNOMIALS.

The integral to be proved is

$$\int_{-1}^{1}(1+x)^{\mu-1}\,(1-x)^{\lambda-1}\,P^{(\alpha,\beta)}_{\nu}[1-\frac{\gamma y}{2}(1-x]\;G^{m,n}_{p,q}\left[z(1-x)^h\,\middle|\,\begin{matrix} a_p \\ b_q\end{matrix}\right]dx$$

$$= \frac{2^{\lambda+\mu+1}\ (\alpha+1)_\nu\ \Gamma(\mu)}{h^{\mu}\ \ \nu!} \sum_{r=o}^{\nu} \frac{(-\nu)_r (1+\alpha+\beta+\nu)_r (\frac{\gamma y}{2})^r}{r!\ \ (\alpha+1)_r}$$

$$\times G^{m,n+h}_{p+h,q+h}\left[2^h z \,\middle|\, {\Delta(h,1-\lambda-r),\ a_p \atop b_q,\ \Delta(h,1-\lambda-\mu-r)}\right], \tag{3.6.1}$$

where h is a positive integer, $R(\mu) > 0$, $p \leq q$ $R(\lambda+hb_j) > 0$ $(j = 1,\ldots,m)$, $c^* = m+n- \frac{p}{2} - \frac{q}{2} > 0$ and $|\arg z| < c^*\pi$. This result is due to Bajpai [25].

Result (3.6.1) can be established with the help of the modified form of the formula (Erdélyi, A. et. al, [86], Vol. I, p.192(46)), namely,

$$\int_{-1}^{1} (1+x)^{\mu-1}\ (1-x)^{\lambda-1}\ P^{(\alpha,\beta)}_{\nu}\ [1 - \frac{\gamma y(1-x)}{2}]\ dx$$

$$= \frac{2^{\lambda+\mu-1}\ \ (\alpha+1)_\nu\ \Gamma(\mu)}{\nu\ !} \sum_{r=o}^{\infty} \frac{(-\nu)_r (1+\alpha+\beta+\nu)_r\ \Gamma(\lambda+r)}{r!(\alpha+1)_r\ \Gamma(\lambda+\mu+r)} (\frac{\gamma y}{2})^r ,$$

where $R(\lambda) > 0$, $R(\mu) > 0$ and in this form it is given by Bajpai [25,p.669].

In particular if $\gamma = 2$, $y = 1$, $\lambda = \sigma+1$, $\mu = \beta+1$ then expressing the G-function on the right hand side of (3.6.1) in terms of the equivalent integral (1.1.1), interchanging the order of summation and integration and making use of Saalchūtz's theorem (4.1.12), we find that

$$\int_{-1}^{1} (1+x)^{\beta} (1-x)^{\sigma}\ P^{(\alpha,\beta)}_{\nu}(x)\ \ G^{m,n}_{p,q}[z(1-x)^h| {a_p \atop b_q}]dx$$

$$= \frac{(-1)^{\nu}\ 2^{\sigma+\beta+1}\ \ (\beta+1)_\nu \Gamma(1+\beta)}{2^{\beta+1}\ \ \ \nu\ !}$$

$$\times G^{m,n+2h}_{p+2h,q+2h}\left[\ 2^h z \,\middle|\, {\Delta(h,-\sigma),\ \Delta(h,\alpha-\sigma),\ \ a_p \atop b_q,\ \Delta(h,-1-\sigma-\beta-\nu),\ \Delta(h,\alpha-\sigma+\nu)}\right], \tag{3.6.2}$$

where h is a positive integer, $p \leq q$, $c^* = m+n-\frac{p}{2}-\frac{q}{2} > 0$, $|\arg z| < c^*\pi$, $R(\beta) > 0$, $R(\sigma+hb_j) > -1$ for $j = 1,\ldots,m$.

On account of the following identity, which readily follows from the definition of the G-function,

$$G^{m,n+2}_{p+2,q+2}\left[z\,\middle|\,\begin{matrix}\Delta(h,-\sigma),\ \Delta(h,\alpha-\sigma),\ a_p\\ b_q, \Delta(h,\alpha-\sigma-\nu),\ \Delta(h,-1-\beta-\sigma-\nu)\end{matrix}\right]$$

$$= (-1)^{\nu}\ G^{m+1,n+1}_{p+2,q+2}\left[z\,\middle|\,\begin{matrix}\Delta(h,-\sigma),\ a_p,\ \Delta(h,\alpha-\sigma)\\ \Delta(h,\alpha-\sigma+\nu),\ b_q,\ \Delta(h,-1-\beta-\sigma-\nu)\end{matrix}\right], \qquad (3.6.3)$$

the results (3.6.3) reduces to the form given by Saxena [267], namely,

$$\int_{-1}^{1} (1+x)^{\beta}(1-x)^{\sigma}\ P_{\nu}^{(\alpha,\beta)}(x)\ G^{m,n}_{p,q}\left[z(1-x)^{h}\,\middle|\,\begin{matrix}a_p\\ b_q\end{matrix}\right]dx = \frac{2^{\beta+\sigma+1}\ \Gamma(1+\nu+\beta)}{h^{\beta+1}\nu\,!}$$

$$\times\, G^{m+1,n+1}_{p+2,q+2}\left[2^{h}z\,\middle|\,\begin{matrix}\Delta(h,-\sigma),\ a_p,\ \Delta(h,\alpha-\sigma)\\ \Delta(h,\alpha-\sigma+\nu),\ b_q,\ \Delta(h,-1-\beta-\sigma-\nu)\end{matrix}\right], \qquad (3.6.4)$$

which holds under the same conditions as given with (3.6.3), which itself is a generalization of a result due to Bhonsle [40].

EXERCISES

3.1. Deduce the following integrals as particular cases of (3.2.2) or prove directly from the integrals given in Section 3.1.

(i) $$\int_{0}^{\infty} x^{2\lambda-1}\ K_{2\mu}(zx)K_{2\nu}(zx)\ G^{m,n}_{p,q}\left[yx^{2\rho}\,\middle|\,\begin{matrix}a_p\\ b_q\end{matrix}\right]dx = \frac{\pi^{\frac{3}{2}-\rho}\ \rho^{2\lambda-\frac{3}{2}}}{z^{2\lambda}\ \ 2^{\rho+1}}$$

$$\times\, G^{m,n+4\rho}_{p+4\rho,q+2\rho}\left[\frac{y\rho^{2\rho}}{z^{2\rho}}\,\middle|\,\begin{matrix}\Delta(\rho,1-\lambda\pm\mu\pm\nu),\ a_p\\ b_q,\ \Delta(2\rho,1-2\lambda)\end{matrix}\right],$$

provided that ρ is a positive integer, $c^* = m+n-\frac{p}{2}-\frac{q}{2} > 0$, $|\arg y| < c^*\pi$, $|\arg z| < \frac{\pi}{2}$, $R(z) > 0$, and $R(\lambda+\rho b_j) > |R(\mu)| + |R(\nu)|$, $(j = 1,2,\ldots,m)$ (Srivastava, H.M. and Joshi, 1969, [314]).

$$\text{(ii)} \int_1^\infty x^{-\rho}(x-1)^{\sigma-1} {}_2F_1(\alpha+\sigma-\rho, \beta+\sigma-\rho;\sigma;1-x)\, G^{m,n}_{p,q}\left[\omega x\,\Big|\,{a_p \atop b_q}\right] dx$$

$$= \Gamma(\sigma)\, G^{m+2,n}_{p+2,q+2}\left[\omega\,\Big|\,{a_p \quad \alpha+\beta+\sigma-\rho,\ \rho \atop \alpha,\beta,\quad b_q}\right];$$

The conditions of the validity are $c^* > 0$, $|\arg\omega| < c^*\pi$, $R(\sigma) > 0$, $R(\alpha) \geq R(\beta) > R(a_j) - 1$, $j = 1,2,\ldots,n$; $c^* = m+n-\frac{p}{2}-\frac{q}{2}$.

$$\text{(iii)} \int_0^\infty x^{2\sigma-1} W_{\lambda,\mu}(zx)\, W_{-\lambda,\mu}(zx)\, G^{m,n}_{p,q}\left[yx^{2\rho}\,\Big|\,{a_p \atop b_q}\right] dx = \frac{(2\pi)^{\frac{1}{2}-\sigma}\rho^{2\lambda-\frac{3}{2}}}{z^{2\lambda}}$$

$$\times\, G^{m,n+4\rho}_{p+4\rho,q+2\rho}\left[\frac{y(2\rho)^{2\rho}}{z^{2\rho}}\,\Big|\,{\Delta(2\rho,-2\sigma),\ \Delta(\rho,\frac{1}{2}\pm\mu-\sigma),\ a_p \atop b_q,\ \Delta(\rho,-\sigma\pm\lambda)}\right],$$

where $c^* > 0$, $|\arg y| < c^*\pi$, $|\arg z| < \frac{\pi}{2}$, $R(z) > 0$, $R(\sigma + \rho b_j) > |R(\mu)| - \frac{1}{2}$ $(j = 1,2,\ldots,m)$, ρ being a positive integer, $c^* = m+n-\frac{p}{2}-\frac{q}{2}$.

(Srivastava, H.M. and Joshi, 1969,[314]).

$$\text{(iv)} \int_0^1 x^{\sigma-1}(1-x)^{\gamma-1} {}_2F_1(\alpha,\beta;\gamma;1-x)\, G^{m,n}_{p,q}\left[\omega x^\rho\,\Big|\,{a_p \atop b_q}\right] dx$$

$$= \Gamma(\gamma)\, G^{m,n+2\rho}_{p+2\rho,q+2\rho}\left[\omega\,\Big|\,{\Delta(\rho,1-\sigma),\ \Delta(\rho,1+\alpha+\beta-\gamma-\sigma),\ a_p \atop b_q,\ \Delta(\rho,1+\alpha-\sigma-\gamma),\ \Delta(\rho,1+\beta-\sigma-\gamma)}\right],$$

where ρ is a positive integer, $R(\sigma + \min \rho\, b_j) > 0$ $j = 1,\ldots,m$; $R(\gamma) > 0$, $c^* > 0$, $|\arg\omega| < c^*\pi$, $c^* = m+n-\frac{p}{2}-\frac{q}{2}$.

From (iv) above deduce the following integrals:

$$\text{(v)} \int_0^1 x^{\rho-1}(1-x)^{\sigma-1} G^{m,n}_{p,q}\left[\omega x\,\Big|\,{a_p \atop b_q}\right] dx = \Gamma(\sigma)\, G^{m,n+1}_{p+1,q+1}\left[\omega\,\Big|\,{1-\rho,\ a_p \atop b_q,\ 1-\rho-\sigma}\right],$$

where $c^* > 0$ $|\arg\omega| < c^*\pi$, $R(\rho+b_j) > 0$, $j = 1,2,\ldots,m$ and $R(\sigma) > 0$;

(vi) $$\int_1^\infty x^{-\rho}(x-1)^{\sigma-1} G^{m,n}_{p,q}\left[\omega x \middle| \begin{matrix} a_p \\ b_q \end{matrix}\right] dx = \Gamma(\sigma)\, G^{m+1,n}_{p+1,q+1}\left[\omega \middle| \begin{matrix} a_p, \rho \\ \rho-\sigma, b_q \end{matrix}\right],$$

where $c^* > 0$, $|\arg \omega| < c^*\pi$, $R(\sigma) > 0$, $R[\sigma-\rho+a_j-1] < 0$, $j = 1,2,\dots,n)$;

(vii) $$\int_0^1 x^{\gamma-1}(1-x)^{\rho-1}\ {}_2F_1(\alpha,\beta;\gamma;x)\ G^{m,n}_{p,q}\left[\omega(1-x) \middle| \begin{matrix} a_p \\ b_q \end{matrix}\right] dx$$

$$= \Gamma(\gamma)\, G^{m,n+2}_{p+2,q+2}\left[\omega \middle| \begin{matrix} 1-\rho, 1-\gamma-\rho+\alpha+\beta, a_p \\ b_q, 1-\gamma-\rho+\alpha, 1-\gamma-\rho+\beta \end{matrix}\right] = \Gamma(\gamma)\,(-1)^{\beta-\gamma}$$

$$\times G^{m+1,n+1}_{p+2,q+2}\left[\omega \middle| \begin{matrix} 1-\gamma-\rho+\alpha+\beta, a_p, 1-\rho \\ 1-\gamma-\rho+\beta, b_q, 1-\gamma-\rho-\alpha \end{matrix}\right],$$

$R(\gamma) > 0$, $R(\rho+b_j) > 0$, $R(\rho+\gamma-\alpha-\beta+b_j) > 0$, $(j = 1,\dots,m)$, $|\arg \omega| < c^*\pi$, $c^* = m+n-\frac{p}{2} - \frac{q}{2} > 0$.

3.2. Establish the following loop integrals giving the conditions of their validity.

(i) $$\int_\infty^{(+1)} x^{-\rho}(1-x)^{\sigma-1} G^{m,n}_{p,q}\left[zx \middle| \begin{matrix} a_p \\ b_q \end{matrix}\right] dx$$

$$= -\frac{2\pi i}{\Gamma(1-\sigma)} G^{m+1,n}_{p+q,q+1}\left[z \middle| \begin{matrix} a_p, \rho \\ \rho-\sigma, b_q \end{matrix}\right].$$

(ii) $$\int_0^{(+1)} x^{-\rho}(1-x)^{\sigma-1} G^{m,n}_{p,q}\left[zx \middle| \begin{matrix} a_p \\ b_q \end{matrix}\right] dx = \frac{2\pi i}{\Gamma(1-\sigma)} G^{m,n+1}_{p+1,q+1}\left[z \middle| \begin{matrix} \rho, a_p \\ b_q, \rho-\sigma \end{matrix}\right].$$

3.3. Prove the following results:

(i) $$\int_0^1 x^{\rho-1}(1-x)^{\beta-\gamma-r}\ {}_2F_1(-r,\beta;\gamma;x)\ G^{m,n}_{p,q}\left[zx^k \middle| \begin{matrix} a_p \\ b_q \end{matrix}\right] dx$$

$$= \frac{\Gamma(\gamma)\,\Gamma(\beta-\gamma+1)}{\Gamma(\gamma+r)}\, k^{\gamma-\beta+r-1}$$

$$\times G^{m+k,n+k}_{p+2k,q+2k}\left[z \middle| \begin{matrix} \Delta(k,1-\rho), a_p, \Delta(k,\gamma-\rho) \\ \Delta(k,\gamma-\rho+r), b_q, \Delta(k,\gamma-\beta-\rho) \end{matrix}\right],$$

where k and r are both positive integers. Show that the integral exists if $R(\rho+k\, b_h) > 0$, for $h = 1,\dots,m$; $R(\beta-\gamma) > r-1$ and one of the following conditions is fulfilled: -

(a) $c^* > 0$, $|\arg z| < c^*\pi$; (b) $c^* \geq 0$, $|\arg z| \leq c^*\pi$,

$$R(\Sigma a_j;-\Sigma b_j + \frac{q-p}{2} + \beta-\gamma - r) > 1.$$

(ii) $$\int_0^1 x^{\rho-1}(1-x)^{\beta-\rho-1}\ {}_2F_1(\alpha,\beta;\gamma;x)\ G^{m,n}_{p,q}\left[zx^r(1-x)^{-r}\ \Big|\ {a_p \atop b_q}\right] dx$$

$$= \frac{\Gamma(\gamma)}{\Gamma(\beta)\Gamma(\gamma-\alpha)} (2\pi)^{1-r}\ r^{\beta-\alpha-1}$$

$$\times\ G^{m+2r,n+r}_{p+2r,q+2r}\left[z\ \Big|\ {\Delta(r,1-\rho),\ a_p,\ \Delta(r,\gamma-\rho \atop \Delta(r,\beta-\rho),\ b_q,\ \Delta(r,\gamma-\alpha-\rho)}\right],$$

where r is a positive integer $R(\rho+rb_h) > 0$, $h = 1,\dots,m$; $R(\gamma-\alpha-\rho-ra_k + r) > 0$, $R(\beta-\rho-ra_k + r) > 0$, $k = 1,\dots,n$; $c^* > 0$, $|\arg z| < c^*\pi$, (Sharma, K.C. 1964, [292] p.539-540).

From 3.3(i) deduce the following integrals:

$$\int_0^\pi \sin(2\rho+1)\theta\ (\sin\theta)^{1-2\xi}\ G^{m,n}_{p,q}\left[x\sin^{2k}\theta\ \Big|\ {a_p \atop b_q}\right] d\theta$$

$$= \frac{\Gamma(\frac{1}{2})}{k^{\frac{1}{2}}}\ G^{m+k,n+k}_{p+2k,q+2k}\left[x\ \Big|\ {\Delta(k,\ \xi-\frac{1}{2}),\ a_p,\ \Delta(k,\xi) \atop \Delta(k,\xi+\rho),\ b_q,\ \Delta(k,\ \xi-\rho-1)}\right]$$

where k is a positive integer ;

$$\int_0^\pi \cos(\rho\theta)\ (\sin\tfrac{\theta}{2})^{-2\xi}\ G^{m,n}_{p,q}\left[x\sin^{2k}(\tfrac{\theta}{2})\ \Big|\ {a_p \atop b_q}\right] d\theta$$

$$= \frac{\Gamma(\frac{1}{2})}{k^{\frac{1}{2}}}\ G^{m+k,n+k}_{p+2k,q+2k}\left[x\ \Big|\ {\Delta(k,\xi+\frac{1}{2}),\ a_p,\ \Delta(k,\xi) \atop \Delta(k,\xi+\rho),\ b_q,\ \Delta(k,\xi-\rho)}\right]$$ (Saxena, 1962, [262]).

3.4. Prove that

$$\int_0^\infty x^{\sigma-\nu-\frac{3}{2}} (\alpha+\beta x+\gamma x^2)^{\frac{1}{2}-\sigma} G^{m,n}_{p,q}\left[z \left(\frac{\alpha+\beta x+\gamma x^2}{x}\right)^k \Big| {a_p \atop b_q}\right]dx$$

$$= \frac{1}{\pi^{\frac{1}{2}}} \left(\frac{\gamma}{\alpha}\right)^{\frac{\nu}{2}} (2\pi)^{\frac{3}{2}-c^*} \, 2^{\sum_{j=1}^{q} b_j - \sum_{j=1}^{p} a_j + \frac{p}{2} - \frac{q}{2} - \frac{1}{2}}$$

$$\times \sum_{\nu,-\nu} \sum_{r=0}^{\infty} \frac{\beta^{\frac{1}{2}-\sigma-\nu-2r} (2k\alpha^{\frac{1}{2}}\gamma^{\frac{1}{2}})^{\nu+2r}}{r! \, \sin(-\pi\nu) \, \Gamma(\nu+r+1)}$$

$$\times G^{2m+2k,\ 2n}_{2p+2k,\ 2q+2k}\left[\frac{z^2\beta^{2k}}{2^{2q-2p}} \Big| {\Delta(2,a_p),\ \Delta(2k,\sigma-\frac{1}{2}) \atop \Delta[k,\ \frac{\sigma+\nu+2r\pm\frac{1}{2}}{2}],\ \Delta(2,b_q)}\right],$$

where k is a positive integer, $R(\sigma \pm \nu - k - ka_j) > \frac{1}{2}$, $j = 1,\dots,n$, $R(\alpha) > 0$, $R(\gamma) > 0$, $c^* > 0$, $|\arg z| < c^*\pi$, $c^* = m+n-\frac{p}{2}-\frac{q}{2}$.

Show that this integral can be put in an alternative form as

$$\int_0^\infty \cosh \nu\theta \, (\alpha+\beta \cosh \theta)^{\frac{1}{2}-\sigma} G^{m,n}_{p,q}\left[z (\alpha+\beta \cosh \theta)^k \Big| {a_p \atop b_q}\right]d\theta$$

$$= \frac{\pi \, 2^{\sum_{j=1}^{q} b_j - \sum_{j=1}^{p} a_j + \frac{p}{2} - \frac{q}{2}}}{(2\pi)^{c^*}} \sum_{\nu,-\nu} \sum_{r=0}^{\infty} \frac{\beta^{\nu+2r} \alpha^{\frac{1}{2}-m-\nu-2r}}{r! \, \sin(-\nu\pi)}$$

$$\times \frac{k^{\nu+2r}}{\Gamma(\nu+r+1)} G^{2m+2k,2n}_{2p+2k,2q+2k}\left[\frac{z^2\alpha^{2k}}{2^{2q-2p}} \Big| {\Delta(2,a_p),\ \Delta(2k,\sigma-\frac{1}{2}) \atop \Delta(k,\ \frac{\sigma+\nu+2r\pm\frac{1}{2}}{2}),\ \Delta(2,b_q)}\right],$$

where $R(\sigma \pm \nu + k - ka_j) > \frac{1}{2}$, $j = 1,\dots,n$; $R(\beta) > 0$, $c^* > 0$, $|\arg z| < c^*\pi$, $c^* = m+n-\frac{p}{2}-\frac{q}{2}$.

The symbol $\sum\limits_{\nu,-\nu}$ in the above expressions indicates that to the expression following it, a similar expression, in which ν has been replaced by $-\nu$, is to added.

Also deduce the following integrals.

$$\int_0^\infty x^{1-k\sigma}(\alpha+\beta x+\gamma x^2)^{k\sigma-\frac{3}{2}} G^{m,n}_{p,q}\left[\left(\frac{x}{\alpha+\beta x+\gamma x^2}\right)^k \Big| \begin{matrix} a_p \\ b_q \end{matrix}\right] dx = \sqrt{\frac{\pi}{k\gamma}}(\beta+2\sqrt{\gamma\alpha})^{k\sigma-1}$$

$$\times G^{m,n+k}_{p+k,q+k}\left[(\beta+2\sqrt{\gamma\alpha})^{-k} \Big| \begin{matrix} \Delta(k,\sigma k),\ a_p \\ b_q,\ \Delta(k,\sigma k-\frac{1}{2}) \end{matrix}\right]. \quad \text{(Saxena, 1960,[259])}$$

$$\int_0^\infty x^{\frac{1}{2}}\left(\frac{\alpha+\beta x+\gamma x^2}{x}\right)^{\sigma} G^{m,n}_{p,q}\left[\left(\frac{\alpha+\beta x+\gamma x^2}{x}\right)^k \Big| \begin{matrix} a_p \\ b_q \end{matrix}\right] dx$$

$$= \frac{\pi^{\frac{1}{2}}}{2(k\gamma)^{3/2}}\left[G^{m+k,n}_{p+k,q+k}\left((\beta+2(\alpha\gamma)^{\frac{1}{2}}\Big| \begin{matrix} (a_p)+\frac{2\sigma+3}{2k},\Delta(k,\frac{3}{2}) \\ \Delta(k,0),\ (b_q)+\frac{2\sigma+3}{2k} \end{matrix}\right)\right.$$

$$\left. + 2k(\gamma\alpha)^{\frac{1}{2}} G^{m+k,n}_{p+k,q+k}\left((\beta+2(\alpha\gamma)^{\frac{1}{2}}\Big| \begin{matrix} (a_p)+\frac{2\sigma+1}{2k},\ \Delta(k,\frac{1}{2}) \\ \Delta(k,0),\ (b_q)+\frac{2\sigma+1}{2k} \end{matrix}\right)\right).$$

(Saxena, 1964, [265]).

3.5. Evaluate the following integrals, giving their conditions of validity.

(i) $\int_0^\infty t^{\lambda-1} J_\nu(at^{\frac{1}{2}}) J_\nu(bt^{\frac{1}{2}}) G^{m,n}_{p,q}\left[\frac{16}{c^2t^2}\Big| \begin{matrix} \alpha_p \\ \beta_q \end{matrix}\right] dt$.

(ii) $\int_0^\infty t^{\frac{\lambda}{2}-1} J_\nu(at^{\frac{1}{2}}) J_\nu(bt^{\frac{1}{2}}) G^{m,n}_{p,q}\left[\frac{c^2t}{4}\Big| \begin{matrix} \alpha_p \\ \beta_q \end{matrix}\right] dt$.

(iii) $\int_0^\infty t^{\frac{\lambda}{2}-1} \exp[-t(\rho^2+\delta^2)]\, I_\nu(2\rho\delta t)\, G^{m,n}_{p,q}\left[\alpha t\Big| \begin{matrix} a_p \\ b_q \end{matrix}\right] dt$.

(iv) $\int_0^\infty t^{2\lambda-1} I_\nu(\delta t) K_\nu(\rho t)\, G^{m,n}_{p,q}\left[\frac{c^2 t^4}{16} \middle| \begin{matrix}\alpha_p \\ \beta_q\end{matrix}\right] dt.$ (Maloo, 1966, [170]).

(v) $\int_0^\infty y^{\rho+\nu} G^{m,n}_{p,q}\left[p^2 y^2 \middle| \begin{matrix}\alpha_p \\ \beta_q\end{matrix}\right] F_4\left[\alpha,\beta;\, 1+\nu,\, 1+\sigma;\, -\frac{y^2}{b^2}, \frac{x^2}{b^2}\right] dy$

where F_4 denotes the Appell's hypergeometric function of two variables of the fourth type. (Rathie, P.N. 1965, [247]).

3.6 Establish the following lemma.

$$\int_0^\infty t^\lambda \; {}_pF_q\left(\begin{matrix}a_i \\ b_j\end{matrix}; -x^2 t^2\right) {}_PF_Q\left(\begin{matrix}A_I \\ B_J\end{matrix}; -y^2 t^2\right) f(t)dt$$

$$= \sum_{n=0}^\infty \frac{(\lambda+2n)\,\Gamma(\lambda+n)}{n!} \; F\left[\begin{matrix}-n, \lambda+n;\, a_i;\, A_I \\ b_j;\, B_J\end{matrix} ; x^2, y^2\right]$$

$$\times \int_0^\infty J_{\lambda+2}(2t)\, f(t)dt,$$

provided that $R(\lambda+\zeta+1-2a_j-2A_J) < 0$ for $j = 1,2,\ldots,p$, and $J = 1,2,\ldots,P$ and $R(\lambda+\xi+1) > 0$ where $f(t) = O(t^\zeta)$ for large t and $f(t) = O(t^\xi)$ for small t. F denotes Kampé de Fériet function of two variables.

By the application of above lemma, show that

$$\int_0^\infty t^\lambda \; {}_pF_q\left(\begin{matrix}a_i \\ b_j\end{matrix}; -x^2 t^2\right) {}_PF_Q\left(\begin{matrix}A_I \\ B_J\end{matrix}; -y^2 t^2\right) G^{\alpha,\beta}_{\gamma,\delta}\left(z^2 \middle| \begin{matrix}a_\gamma \\ b_\delta\end{matrix}\right) dt$$

$$= \sum_{n=o}^{\infty} \frac{(\lambda+2n)\,\Gamma(\lambda+n)}{2\;n!}\; F\left(\begin{matrix} -n,\lambda+n;\; a_i; A_I \\ b_j;\; B_J \end{matrix};\; x^2, y^2\right)$$

$$\times\, G^{\alpha,\beta+1}_{\gamma,\delta}\left(z^2 \middle| \begin{matrix} \frac{(1-\lambda-2n)}{2},\; a_\gamma,\; \frac{(1+\lambda+2n)}{2} \\ b_\delta \end{matrix}\right),$$

provided $\alpha+\beta > \frac{1}{2}(\gamma+\delta)$, $|\arg z^2| < (\alpha+\beta - \frac{\gamma}{2} - \frac{\delta}{2})\pi$, $R(2a_j - \frac{3}{2}) < 0$,

$R(2\beta_h + \lambda+1) > 0$ where $j = 1,2,\ldots,\beta$ and $h = 1,2,\ldots,\alpha$.

From the above integral deduce the Hankel, Meijer and Laplace transforms of the product

$$t^{\lambda}\; {}_pF_q\left(\begin{matrix} a_i \\ b_j \end{matrix}; -x^2t^2\right)\; {}_PF_Q\left(\begin{matrix} A_I \\ B_J \end{matrix}; -y^2t^2\right).$$ (Bora and Saxena, 1971, [49]).

3.7. Establish the formula

$$\int_o^t x^{\rho-1}(1-x)^{\beta-1}\; {}_uF_v\;[\alpha_u;\, \beta_v;\, cx^{\nu}(1-x)^{\mu}]\;\; G^{m,n}_{p,q}\left[zx^{\frac{k}{h}}\middle| \begin{matrix} a_p \\ b_q \end{matrix}\right]dx$$

$$= (2\pi)^{c^*(1-h)}\; h^{\overset{*}{B}}\; k^{-\beta}\;\; t^{\rho+\beta-1}$$

$$\times \sum_{r=0}^{\infty} \frac{\prod_{j=o}^{u} (\alpha_j;r)\, \Gamma(\beta+r\mu)\, c^r t^{r(\nu+\mu)}}{\prod_{j=1}^{v} (\beta_j;r)\; r!\; m^{r\mu}}$$

$$\times G^{mh,nh+k}_{ph+k,qh+k} \left[z^h t^k h^{(p-q)} \middle| \begin{matrix} \Delta(k,1-\rho-rv),\ \Delta(h,a_p) \\ \Delta(h,b_q),\ \Delta(k,1-\rho-\beta-rv-r\mu) \end{matrix} \right],$$

where $c^* = m+n-\frac{p}{2}-\frac{q}{2}$, $B^* = \sum_{j=1}^{q} b_j - \sum_{j=1}^{p} a_j + \frac{p}{2} - \frac{q}{2} + 1$; α_u and β_v represent the set of parameters $\alpha_1,\ldots,\alpha_u$ and $\beta_1,\ldots,\beta_v$ respectively.

The above formula holds provided that $u \leq v$ (or $u \leq v+1$ and $|c| < 1$), and none of the parameters $\beta_1,\ldots,\beta_v$ is zero or a negative integer, $R(\beta) > 0$, $R[\rho + \frac{kb_j}{h}] > 0$, $j = 1,\ldots,k$; $|\arg z| < c^*\pi$; $c^* = m+n-\frac{p}{2}-\frac{q}{2} > 0$. (Chhabra and Singh, 1969, [64]). For similar types of results see the work of Gupta and Olkha (1969, [114]).

3.8 Show that

$$\int_0^\infty x^{\beta-1} (x+y)^{-\alpha-\beta}\, G^{m,n}_{p,q} \left[z\, x^{\rho-k} (x+y)^k \middle| \begin{matrix} a_p \\ b_q \end{matrix} \right] dx$$

$$= (2\pi)^{\frac{1}{2}}\, \rho^{\alpha-\frac{1}{2}}\, k^{\frac{1}{2}-\alpha-\beta}\, (k-\rho)^{\beta-\frac{1}{2}}\, y^{-\alpha}$$

$$\times G^{m+k,n}_{p+k,q+k} \left[\frac{k^k (k-\rho)^{\rho-k} z\, y^\rho}{\rho^\rho} \middle| \begin{matrix} a_p,\ (k,\alpha+\beta) \\ (k-\rho,\beta),\ (\rho,\alpha),\ b_q \end{matrix} \right],$$

where k and ρ are positive integers, $\rho < k$ and $c^* > 0$, $|\arg z\, y^\rho| < a\pi$, $R(1-\alpha_j - \frac{\alpha}{\rho}) > 0$, $R(1-\alpha_j + \frac{\rho}{(k-\rho)}) > 0$, $j = 1,\ldots,n$ and

$$\int_0^\infty x^{\beta-1}(x+y)^{-\alpha-\beta}\, G^{m,n}_{p,q}\left[zx^{\rho-k}(x+y)^k \Big| {a_p \atop b_q}\right] dx$$

$$= (2\pi)^{\frac{1}{2}}\, \rho^{\alpha-\frac{1}{2}}\, k^{\frac{1}{2}-\alpha-\beta}\, (k-\rho)^{\beta-\frac{1}{2}}\, y^{-\alpha}$$

$$\times G^{m+k,n}_{p+k,q+k}\left[\frac{k^k(k-\rho)^{\rho-k}\, zy^\rho}{\rho^\rho} \Big| {a_p\,,\ \Delta(k,\alpha+\beta) \atop \Delta(k-\rho,\beta),\ \Delta(\rho,\alpha) b_q}\right]$$

where $c^* > 0$, $|\arg zy^\rho| < c^*\pi$, $R(b_h + \frac{\beta}{\rho-k}) > 0$, $R(1-a_j + \frac{\alpha}{\rho}) > 0$, $h = 1,\ldots,m$, $j = 1,\ldots,n$. (Sharma, K.C., 1964, [290]).

3.9 Show that

$$\int_{-\infty}^{\infty} G^{m,n}_{p,q}\left(x \Big| {a_1,\ldots,a_{p-2},\ a_{p-1}+\epsilon,\ a_p+\epsilon \atop b_1,\ldots,b_{q-2},\ b_{q-1}+\epsilon,\ b_q+\epsilon}\right) d\epsilon$$

$$= \frac{\Gamma(a_{p-1}+a_p-b_{q-i}-b_q-1)}{\Gamma(a_{p-1}-b_{q-1})\,\Gamma(a_p-b_{q-1})\,\Gamma(a_p-b_q)\,\Gamma(a_{p-1}-b_q)}$$

$$\times G^{m,n}_{p-2,q-2}\left(x \Big| {a_1,\ldots,a_{p-2} \atop b_1,\ldots,b_{q-2}}\right)$$

where $\sum_{j=1}^{p} a_j - \sum_{j=1}^{q} b_j > 1$, $|\arg x| < c^*\pi$, $c^* = m+n - \frac{p}{2} - \frac{q}{2} > 0$.

Hint: Apply the integral

$$\int_{-\infty}^{\infty} \frac{dx}{\Gamma(\alpha + x)\,\Gamma(\beta - x)\,\Gamma(\gamma + x)\,\Gamma(\delta - x)}$$

$$= \frac{\Gamma(\alpha + \beta + \gamma + \delta - 3)}{\Gamma(\alpha + \beta - 1)\,\Gamma(\beta + \gamma - 1)\,\Gamma(\gamma + \delta - 1)\,\Gamma(\delta + \alpha - 1)}$$

(Maheshwari, 1969, [168]).

CHAPTER IV

FINITE AND INFINITE SERIES OF G-FUNCTIONS

In this present chapter we discuss the expansion formulas of G-functions expressed in terms of the related G-functions. In order to present the material of this chapter in a proper form we have classified the expansion formulas under the following four categories:

Category 1: Expansions of G-functions in series of G-functions whose coefficients are gamma functions;

Category 2: Infinite series of G-functions whose sums are constants;

Category 3: Expansions of G-functions in series of products of two G-functions:

Category 4: Expansions of G-functions in series of orthogonal polynomials and the G-function.

The work presented in this chapter and these classifications are not avaiable in any book. As far as possible an exhaustive list of results, falling under these categories either in the text or in the form of exercises, is given.

The results are useful and will find applications in Pure as well as Applied Mathematics, Mathematical Physics, Statistics and other related fields.

It is worthwhile to mention that the work of Meijer [209],[210], Bhise [38], Swaroop [327], Chhabra [63], Anandani [12],[13], Olkha [225] and others fall in category 1.

The second category involves mainly the work of Sharma, B.L. [287], Saxena and Bhagchandani [273], Abiodun and Sharma, B.L. [1], and Srivastava, H.M. and Daoust [312].

The third category is characterized by the work of Verma, A. [345], Abiodun and Sharma, B.L. [1].

The work of Bajpai ([22],[25],[30],[32]), Anandani ([12],[13],[14]), Parashar [226], Shah ([279],[280],[281]) and others come in the category 4.

Because of the importance of the elementary Special Functions such as Bessel Functions, Whittaker Functions and Hypergeometric Functions in various problems arising in Physical and Biological Sciences, it should be remarked that the results associated with these functions and their related functions can be derived from the general results,by giving suitable values to the parameters and using the results of Chapter II. A more detailed discussion of the various particular cases of the formulas of this chapter as well as an exhaustive Bibliography concerning them can be found in the works of Meijer ([204],[209],[210]), Knotternus [137], Erdélyi, A et al [85], Sharma, B.L. [287], Abiodun and Sharma, B.L. [1], Srivastava, H.M. and Daoust [312], Srivastava, H.M. [311], Magnus, Oberhettinger and Soni [167], and Slater ([301],[302]).

Throughout this chapter it is assumed that the poles, of the Mellin-Barnes integral representation for the G-function,are simple. The logarithmic cases of the G-function are discussed in Chapter V.

4.1. SUMMATION FORMULAS FOR HYPERGEOMETRIC FUNCTIONS WITH SPECIALIZED ARGUMENTS.

Following are some preliminary summation theorems for the hypergeometric functions which are needed in the analysis that follows. These results follow easily from the definitions but for the details of the proofs the reader is referred to Appell and Kampe de Fériet [17] and Slater ([301],(302]). It should be remarked here that whenever the Special Functions reduce to gamma products the results are very important from the point of view of applications. For example see the applications of the following results to Optimization Problems in Management Sciences from Mathai and Sancho [193].

Gauss's theorem:

$$ {}_2F_1(a,b;c;1) = \frac{\Gamma(c)\,\Gamma(c-a-b)}{\Gamma(c-a)\,\Gamma(c-b)} . \tag{4.1.1} $$

provided $c \neq 0, -1, -2, \ldots$, and $R(c-a-b) > 0$.

Vandermonde's theorem:

$$ {}_2F_1\,(-n,b;c;1) = \frac{(c-b)_n}{(c)_n} \quad c \neq 0,\,-1,\,-2,\dots\,. \tag{4.1.2}$$

A deduction from Gauss's theorem:

$$ {}_2F_1\left(\begin{matrix} -\frac{n}{2},\ -\frac{n}{2}+\frac{1}{2} \\ b+\frac{1}{2} \end{matrix};\ 1\right) = \frac{\Gamma(b+\frac{1}{2})\,\Gamma(b+n)}{\Gamma(b+\frac{n}{2})\,\Gamma(b+\frac{n}{2}+\frac{1}{2})}\,. \tag{4.1.3}$$

Kummer's theorem:

$$ {}_2F_1(a,b;1+a-b;\,-1) = \frac{\Gamma(1+a-b)\,\Gamma(1+\frac{a}{2})}{\Gamma(1+\frac{a}{2}-b)\,\Gamma(1+a)}\,, \tag{4.1.4}$$

provided $1+a-b$ is neither zero nor a negative integer.

$$ {}_2F_1\,(a,b;\ \frac{a+b+1}{2}\ ;\ \frac{1}{2}\ ) = \frac{\Gamma(\frac{1}{2})\ \Gamma(\frac{1+a+b}{2})}{\Gamma(\frac{1+a}{2})\ \Gamma(\frac{1+b}{2})}\,, \tag{4.1.5}$$

provided $a+b+1 \neq 0,\,-2,\,-4,\,-6,\dots\,$.

$$ {}_2F_1\,(a,1-a;\ c;\ \frac{1}{2}) = \frac{2^{1-c}\ \Gamma(c)\ \Gamma(\frac{1}{2})}{\Gamma(\frac{c+a}{2}\ )\ \Gamma(\frac{c-a+1}{2})}\,, \tag{4.1.6}$$

provided $c \neq 0,\,-1,\,-2,\,\dots\,$.

$$ {}_2F_1\,(a,b;\ a-b+2;\ -1) = 2^{-a}\ \pi^{\frac{1}{2}}\ (b-1)^{-1}\ \Gamma(a-b+2)$$

$$ \times\ \left[\frac{1}{\Gamma(\frac{a}{2})\ \Gamma(\frac{3+a-2b}{2})} - \frac{1}{\Gamma(\frac{1+a}{2}\ )\ \Gamma(\frac{2+a-2b}{2})}\right], \tag{4.1.7}$$

provided $a-b+2 \neq 0,\,-1,\,-2,\dots\,$.

$$ {}_2F_1\,(a,b;\ \frac{a+b+2}{2};\ \frac{1}{2}\ ) = 2\pi^{\frac{1}{2}}\ (a-b)^{-1}\ \Gamma(\frac{a+b}{2}+1)$$

$$\times \left[\frac{1}{\Gamma(\frac{a}{2})\,\Gamma(\frac{b+1}{2})} - \frac{1}{\Gamma(\frac{b}{2})\,\Gamma(\frac{a+1}{2})} \right], \tag{4.1.8}$$

provided $a+b \neq -2, -4, -6, \ldots$.

$${}_2F_1\left(a, a+\tfrac{1}{2}\,;\,\tfrac{3}{2}-2a\,;\,-\tfrac{1}{3}\right) = \left(\tfrac{8}{9}\right)^{-2a} \frac{\Gamma(\frac{4}{3})\,\Gamma(\frac{3}{2}-2a)}{\Gamma(\frac{3}{2})\,\Gamma(\frac{4}{3}-2a)}, \tag{4.1.9}$$

provided $\frac{3}{2} - 2a \neq 0, -1, -2, \ldots$.

$${}_2F_1\left(a, a+\tfrac{1}{2}\,;\,a+\tfrac{5}{6}, \tfrac{1}{9}\right) = \left(\tfrac{3}{4}\right)^{a} \frac{\pi^{\frac{1}{2}}\,\Gamma(\frac{5}{6}+\frac{2a}{3})}{\Gamma(\frac{1}{2}+\frac{a}{3})\,\Gamma(\frac{5}{6}+\frac{a}{3})}, \tag{4.1.10}$$

provided $\frac{2a}{3} + \frac{5}{6} \neq 0, -1, -2, \ldots$.

$${}_2F_1\left(a, \tfrac{a}{3}+\tfrac{1}{3}\,;\,\tfrac{2(a+1)}{3}; e^{\frac{i\pi}{3}}\right) = 2^{\frac{2a}{3}+\frac{2}{3}}\,\pi^{\frac{1}{2}}\,3^{-\frac{(a+1)}{2}}$$

$$\times\, e^{\frac{i\pi a}{6}} \frac{\Gamma(\frac{a}{3}+\frac{5}{6})}{\Gamma(\frac{2}{3})\,\Gamma(\frac{a+2}{3})}, \tag{4.1.11}$$

provided $\frac{5}{6} + \frac{a}{3} \neq 0, -1, -2, \ldots$.

<u>Saalschütz's theorem:</u>

$${}_3F_2\left(\begin{matrix} a, b, -n \\ a, 1+a+b-c-n \end{matrix}; 1\right) = \frac{(c-a)_n (c-b)_n}{(c)_n\,(c-a-b)_n}, \tag{4.1.12}$$

where $n = 0, 1, 2, \ldots$.

We also have

$${}_3F_2\left(\begin{matrix} -n, n+a, b \\ c, a+b+1-c \end{matrix}; 1\right) = \frac{(c-b)_n\,(a-c+1)_n}{(c)_n\,(a+b+1-c)_n} .$$

Dixon's theorem:

$$ {}_3F_2\left(\begin{matrix} a,b,c;1 \\ 1+a-b,\ 1+a-c \end{matrix}\right) $$

$$ = \frac{\Gamma(1+\frac{a}{2})\,\Gamma(1+a-b)\,\Gamma(1+a-c)\,\Gamma(1+\frac{a}{2}-b-c)}{\Gamma(1+a)\,\Gamma(1+\frac{a}{2}-b)\,\Gamma(1+\frac{a}{2}-c)\,\Gamma(1+a-b-c)}, \quad R(a-2b-2c) > -2\ . \qquad (4.1.13) $$

Watson's theorem:

$$ {}_3F_2\left(\begin{matrix} a,b,c;1 \\ \frac{1}{2}(a+b+1),\ 2c \end{matrix}\right) $$

$$ = \frac{\Gamma(\frac{1}{2})\,\Gamma(\frac{1}{2}+\frac{a}{2}+\frac{b}{2})\,\Gamma(\frac{1}{2}+c)\,\Gamma(\frac{1}{2}-\frac{a}{2}-\frac{b}{2}+c)}{\Gamma(\frac{1}{2}+\frac{a}{2})\,\Gamma(\frac{1}{2}+\frac{b}{2})\,\Gamma(\frac{1}{2}-\frac{a}{2}+c)\,\Gamma(\frac{1}{2}-\frac{b}{2}+c)}, \quad R(2c-a-b) > -1. \qquad (4.1.14) $$

Another form of Dixon's theorem:

$$ {}_3F_2\left(\begin{matrix} a,b,c;\ 1 \\ 1+a-b,\ 1+a-c \end{matrix}\right) $$

$$ = \frac{\cos(\frac{\pi a}{2})\,\sin((b-\frac{a}{2})\pi)\,\Gamma(1-a)\,\Gamma(b-\frac{a}{2})\,\Gamma(1+a-c)\,\Gamma(1+\frac{a}{2}-b-c)}{\sin((b-a)\pi)\,\Gamma(1-\frac{a}{2})\,\Gamma(b-a)\,\Gamma(1+\frac{a}{2}-c)\,\Gamma(1+a-b-c)} \qquad (4.1.15) $$

$R(a-2b-2c) > -2$.

Whipple's theorem 1.

$$ {}_3F_2\left(\begin{matrix} a,b,c;1 \\ e,f \end{matrix}\right) = \frac{\pi\,\Gamma(e)\,\Gamma(f)}{2^{2c-1}\,\Gamma(\frac{a}{2}+\frac{e}{2})\,\Gamma(\frac{a}{2}+\frac{f}{2})\,\Gamma(\frac{b}{2}+\frac{e}{2})\,\Gamma(\frac{b}{2}+\frac{f}{2})}, \qquad (4.1.16) $$

provided $a+b = 1$, $e+f = 2c + 1$, $R(c) > 0$.

Modified form of Whipple's theorem due to Dzrbasjan.

$$ {}_3F_2\left(\begin{matrix}-n, a, 1-a \\ f, -2n-f+1\end{matrix}; 1\right) = \frac{2^{2n}\left(\frac{a}{2}+\frac{f}{2}\right)_n\left(\frac{a}{2}-\frac{f}{2}+\frac{1}{2}-n\right)_n}{(f)_n\ (1-2n-f)_n}. \qquad (4.1.17) $$

Whipple's theorem 2.

$$ {}_4F_3\left(\begin{matrix}a, 1+\frac{a}{2}, b, c; -1 \\ \frac{a}{2}, 1+a-b, 1+a-c\end{matrix}\right) = \frac{\Gamma(1+a-b)\ \Gamma(1+a-c)}{\Gamma(1+a)\ \Gamma(1+a-b-c)}, \ R(a-2b-2c) > -2. \qquad (4.1.18) $$

When $a = h$, $b = 1$, $c = 1-a_1 + s$, (4.1.18) reduces to

$$ {}_3F_2\left(\begin{matrix}1-a_1+s, \frac{h}{2}+1, 1; -1 \\ h+a_1-s, \frac{h}{2}\end{matrix}\right) = \frac{\Gamma(h)\ \Gamma(h+a_1-s)}{\Gamma(h+1)\ \Gamma(h+a_1+1-s)}, $$

$R(h + 2a_1 - 2 - 2s) > 0$.

Dougall's first theorem:

$$ {}_7F_6\left(\begin{matrix}a, 1+\frac{a}{2}, b, c, d, e, -n; 1 \\ \frac{a}{2}, 1+a-b, 1+a-c, 1+a-d, 1+a-e, 1+a+n\end{matrix}\right) $$

$$ = \frac{(1+a)_n(1+a-b-c)_n(1+a-b-d)_n(1+a-c-d)_n}{(1+a-b)_n(1+a-c)_n(1+a-d)_n(1+a-b-c-d)_n}, \qquad (4.1.19) $$

where $n = 0,1,2,\ldots$ and $1+a = b+c+d+e-n$.

$$ {}_5F_4\left(\begin{matrix}a, 1+\frac{a}{2}, c, d, e; 1 \\ \frac{a}{2}, 1+a-c, 1+a-d, 1+a-e\end{matrix}\right) $$

$$ = \frac{\Gamma(1+a-c)\ \Gamma(1+a-d)\ \Gamma(1+a-e)\ \Gamma(1+a-c-d-e)}{\Gamma(1+a)\ \Gamma(1+a-d-e)\ \Gamma(1+a-c-e)\ \Gamma(1+a-c-d)}, \qquad (4.1.20) $$

provided $R(a-c-d-e) > -1$.

For the source of the following results see Luke [152]

$${}_3F_2\left(\begin{matrix}-n,\ n+2\nu,\mu+\nu+1;\\ 2\nu+1,\mu+\nu+2\end{matrix};1\right)=\frac{(2\nu)\ n!\ (\nu-\mu\ \ 1)_n}{(\nu-\mu-1)\ (2\nu)_n(\mu+\nu+2)_n},\ n>0,\quad (4.1.21)$$

$${}_3F_2\left(\begin{matrix}a,b,1\\ c,2\end{matrix};1\right)=\frac{(c-1)}{(a-1)(b-1)}\left[\frac{\Gamma(c-1)\ \Gamma(c-a-b+1)}{\Gamma(c-a)\ \Gamma(c-b)}-1\right],\quad (4.1.22)$$

$a\neq 1,\ b\neq 1,\ R(c-a-b)>-1$.

$${}_3F_2\left(\begin{matrix}a,b,1;\ 1\\ c,3\end{matrix}\right)=\frac{2(c-2)_2}{(a-2)_2(b-2)_2}\left[\frac{\Gamma(c-2)\ \Gamma(c-a-b+2)}{\Gamma(c-a)\ \Gamma(c-b)}-1\right]-\frac{2(c-1)}{(a-1(b-1)}$$

(4.1.23)

$a\neq 1,2;\quad b\neq 1,2\ ;\ R(c-a-b)>-2$.

$${}_4F_3\left(\begin{matrix}-n,\ f+1,\ 1,\ \frac{f(n+e-1)+z(e-f-1)}{n+f}\\ e,f,1-z\end{matrix};1\right)=\frac{z(n+f)\ (e-1)}{f(z-n)(n+e-1)},\quad (4.1.24)$$

where n is zero or a positive integer and $f(z-n)(n+e-1)\neq 0$.

$${}_4F_3\left(\begin{matrix}-n,\ \beta+1,\ 1,\ z+2\beta\\ n+2\beta+1,\ \beta,\ 1-z\end{matrix};1\right)=\frac{z(n+2\beta)}{2\beta(z-n)},\quad n=0,1,\ldots,\ ;\ \beta(z-n)\neq 0.$$

(4.1.25)

$${}_3F_2\left(\begin{matrix}-n,\ f+1,1;\ \frac{z+f}{n+f}\\ f,1-z\end{matrix}\right)=\frac{z(n+f)}{f(z-n)},\quad n=0,1,\ldots;f(z-n)\neq 0\quad (4.1.26)$$

$${}_3F_2\left(\begin{matrix}f+1,\ f(\lambda+1)+\lambda z\\ f,1-z\end{matrix};-\frac{1}{\lambda}\right)=-\frac{\lambda z}{f(\lambda+1)},\quad (4.1.27)$$

$f(\lambda+1)\neq 0,\quad |\lambda|>1\quad$ or $\quad|\lambda|=1$ and $R[\{(\lambda+1)(f+z)\}]<-1$.

Carlitz [58] has shown that

$${}_4F_3\left(\begin{matrix}-n,\ \frac{a+1}{2},\ \frac{a}{2}+1,\ b+n;\ 1\\ a+1,\ \frac{b+1}{2},\ \frac{b}{2}+1\end{matrix}\right)=\frac{b(b-a)}{(b+2n)(b)_n}.\quad (4.1.28)$$

Finally we record a result due to Kalla and Saxena [130] which is a generalization of Saalchütz's theorem (4.1.12):

$$ {}_3F_2\left(\begin{matrix}\alpha,\beta+m,\gamma\\ \beta,\gamma+1\end{matrix};1\right) = \frac{\Gamma(1-\alpha)\,\Gamma(1+\gamma)\,\Gamma(\beta-\gamma)_m}{\Gamma(\beta)_m\,\Gamma(1+\gamma-\alpha)}, \tag{4.1.29} $$

$R(1-\alpha) > m;\ m = 0,1,2,\ldots$.

The following four transformation formulas can be readily deduced from the multiplication formula for gamma functions (1.2.6).

$$ \prod_{k=o}^{m-1} \Gamma\left(\frac{\alpha+r+k}{m}\right) = m^{-r}\,(\alpha)_r \prod_{k=o}^{m-1} \Gamma\left(\frac{\alpha+k}{m}\right) . \tag{4.1.30} $$

$$ \prod_{k=o}^{m-1} \Gamma\left(\frac{\alpha+r-k}{m}\right) = m^{-r}\,(\alpha-m+1)_r \prod_{k=o}^{m-1} \Gamma\left(\frac{\alpha-k}{m}\right) \tag{4.1.31} $$

$$ \prod_{k=o}^{m-1} \Gamma\left(\frac{\alpha-r+k}{m}\right) = \frac{(-m)^r}{(1-\alpha)_r} \prod_{k=o}^{m-1} \Gamma\left(\frac{\alpha+k}{m}\right) . \tag{4.1.32} $$

$$ \prod_{k=o}^{m-1} \Gamma\left(\frac{\alpha-r-k}{m}\right) = \frac{(-m)^r}{(m-\alpha)_r} \prod_{k=o}^{m-1} \Gamma\left(\frac{\alpha-k}{m}\right) . \tag{4.1.33} $$

4.1.1. Finite Summation Formulas for Hypergeometric and Whittaker Functions.

The results presented here are due to Pathan ([229], p.1045) and MacRobert ([158],[162]) where

$$ \binom{k}{r} = \frac{k(k-1)\ldots(k-r+1)}{r\,!} . $$

$$ \sum_{r=o}^{s} (-1)^{s-r}\binom{s}{r} x^{\frac{s-r}{2}}\, W_{k+\frac{r}{2},\,\mu+\frac{r}{2}}(x) $$

$$ = \frac{\Gamma(\mu+s-k+\frac{1}{2})}{\Gamma(\mu-k+\frac{1}{2})}\, W_{k-\frac{s}{2},\,\mu+\frac{s}{2}}(x) . \tag{4.1.34} $$

$$\sum_{r=0}^{s} (-1)^{s-r}\binom{s}{r} \frac{\Gamma(\mu-k-r+\frac{1}{2})}{\Gamma(\mu-k-s+\frac{1}{2})} W_{k+r,\mu}(x) = x^{\frac{s}{2}} W_{k+\frac{s}{2},\, \mu-\frac{s}{2}}(x). \tag{4.1.35}$$

$$\sum_{r=0}^{s} \frac{(-1)^{s-r}\binom{s}{r}\Gamma(\gamma-\beta-r+1)}{\Gamma(\gamma-r+1)} \; {}_2F_1(\alpha,\beta;\gamma-r+1;x)$$

$$= \frac{\Gamma(s+\beta)\,\Gamma(\gamma-\beta-s+1)}{\Gamma(\beta)\;\Gamma(\gamma+1)} \; {}_2F_1(\alpha,\beta+s;\gamma+1;x)\,. \tag{4.1.36}$$

$$\sum_{r=0}^{s} (-1)^{s-r}\binom{s}{r} \frac{\Gamma(\beta-r)}{\Gamma(\gamma-r)} \; {}_2F_1(\alpha,\beta-r;\gamma-r;x)$$

$$= \frac{\Gamma(\beta-s)\,\Gamma(\gamma-\beta+s)}{\Gamma(\gamma)\,\Gamma(\gamma-\beta)} \; {}_2F_1(\alpha,\beta-s;\gamma;x)\,. \tag{4.1.37}$$

$$(\sin\tfrac{\theta}{2})^{-2\rho} = \frac{\Gamma(\frac{1}{2}-\rho)}{\Gamma(\frac{1}{2})\,\Gamma(1-\rho)} \left[\, 1+2\sum_{r=1}^{\infty} \frac{(\rho)_r}{(1-\rho)_r} \cos(r\theta)\right], \tag{4.1.38}$$

where $0 \le \theta \le \pi$, $R(1-2\rho) > 0$.

$$(\sin\theta)^{1-2\rho} = \frac{\Gamma(\frac{3}{2}-\rho)}{\Gamma(\frac{3}{2})\,\Gamma(2-\rho)} \sum_{r=0}^{\infty} \frac{(\rho)_r}{(2-\rho)_r} \sin(2r+1)\theta\,, \tag{4.1.39}$$

where $0 \le \theta \le \pi$, $R(1-2\rho) \ge 0$.

4.2 SUMMATION FORMULAS FOR THE G-FUNCTION

In the present chapter we shall use the following notations freely.

$$(a_{R,S}) \equiv (a_R, a_{R+1}, \ldots, a_S) .$$

$$(a_{1,S}) \equiv (a_S) \equiv (a_1, a_2, \ldots, a_S) .$$

$$\Delta(\lambda, a + \left| \begin{matrix} r_1 \\ r_2 \\ \vdots \\ r_n \end{matrix} \right.) \equiv \Delta(\lambda, a + r_1), \ldots, \Delta(\lambda, a + r_n) .$$

$$\Gamma[(a_n) + k] \equiv \prod_{j=1}^{n} \Gamma(a_j + k).$$

$$(\alpha_n)_r \equiv (\alpha_1)_r \ldots (\alpha_n)_r \equiv \prod_{j=1}^{n} [\alpha_j(\alpha_j + 1) \ldots (\alpha_j + r-1)] \cdot$$

$$\Delta[k, (\alpha_n) \pm r] \equiv \Delta(k, \alpha_1 \pm r), \Delta(k, \alpha_2 \pm r), \ldots, \Delta(k, \alpha_n \pm r).$$

The value zero of suffix S in (a_S) will mean the total absence of these parameters. For other notations appearing in the following sections see the list of symbols at the end of this book. $\Delta(\cdot)$ is discussed in Section 3.1 Chapter III.

In this section we give ten finite summation formulas and four infinite summation formulas for the G-function. Further results of this type are given in the excercises at the end of the chapter.

4.2.1. Finite Summation Formulas.

The discussion of this section is based on the work of Anandani [13], Bhise [38] and Chhabra [63]. The series are summed up by expressing the G-function as a Mellin-Barnes integral (1.1.1) and interchanging the order of summation and integration and making use of the summation formulas of the hypergeometric functions given in Section 4.1.

In this section λ, k and r are taken to be positive integers.

First Summation Formula:

$$\sum_{r=o}^{k} (-1)^{k+r} \binom{k}{r} G^{m,n+\lambda}_{p+\lambda,q+\lambda} \left[x \middle| \begin{matrix} \Delta(\lambda,\alpha-r), a_p \\ b_q, \Delta(\lambda,\beta-r) \end{matrix} \right]$$

$$= \frac{\Gamma(1-\alpha+\beta)}{\lambda^k \Gamma(1-\alpha+\beta-k)} G^{m,n+\lambda}_{p+\lambda,q+\lambda} \left[x \middle| \begin{matrix} \Delta(\lambda,\alpha), a_p \\ b_q, \Delta(\lambda,\beta-k) \end{matrix} \right], \qquad (4.2.1)$$

$R(\alpha-\beta+k) > 0$, $c^* = m+n - \frac{p}{2} - \frac{q}{2} > 0$.

Proof: If we substitute the value of the G-function on the L.H.S. of (4.2.1) from (1.1.1), interchange the order of summation and integration and make use of (4.1.31) then the series under consideration transforms into

$$\frac{1}{2\pi i} \int_L \frac{\prod_{j=1}^{m} \Gamma(b_j-s) \prod_{j=1}^{n} \Gamma(1-a_j+s) \prod_{j=o}^{\lambda-1} \Gamma(1-\frac{\alpha+j}{\lambda}-s)}{\prod_{j=m+1}^{q} \Gamma(1-b_j+s) \prod_{j=n+1}^{p} \Gamma(a_j-s) \prod_{j=o}^{\lambda-1} \Gamma[1-\frac{\beta+j}{\lambda}+s]}$$

$$\times (-1)^k x^s {}_2F_1(-k, 1-\alpha+\lambda s ; 1-\beta+\lambda s; 1) ds .$$

By the application of (4.1.1), the interesting formula

$$\frac{(-1)^k}{(\alpha)_k} = (1-\alpha)_{-k}$$

and (1.1.1), the result follows immediately.

Second Summation Formula:

$$\sum_{r=o}^{k} (-1)^{k+r} \lambda^{-r} \binom{k}{r} \Gamma(\alpha-\beta+k+r) G^{m+\lambda,n}_{p,q+2\lambda} \left(x \middle| \begin{matrix} a_1,\dots,a_p \\ \Delta(\lambda,\alpha), b_1,\dots,b_q, \Delta(\lambda,\beta-r) \end{matrix} \right)$$

$$= \Gamma(\alpha-\beta+k) G^{m+\lambda,n}_{p,q+2\lambda} \left(x \middle| \begin{matrix} a_1,\dots,a_p \\ \Delta(\lambda,\alpha+k), b_1,\dots,b_q, \Delta(\lambda,\beta-k) \end{matrix} \right), \qquad (4.2.2)$$

where $R(\alpha) < 1$, $m+n > \frac{p}{2} + \frac{q}{2}$.

Third Summation Formula:

$$\sum_{r=o}^{k} \frac{(-1)^{k+r}\,\lambda^{r-k}\,\binom{k}{r}}{\Gamma(\beta+r)}\, G^{m,n+\lambda}_{p+\lambda,q}\left(x \,\middle|\, \begin{matrix} \Delta(\lambda,\ \alpha-r),\ a_1,\dots,a_p \\ b_1,\dots,b_q \end{matrix}\right)$$

$$= \frac{1}{\Gamma(\beta+k)}\, G^{m,n+2\lambda}_{p+2\lambda,q+\lambda}\left(x \,\middle|\, \begin{matrix} \Delta(\lambda,\alpha),\ \Delta(\lambda,\alpha+\beta-1),\ a_1,\dots,a_p \\ b_1,\dots,b_q,\ \Delta(\lambda,\ \alpha+\beta+k-1) \end{matrix}\right), \quad (4.2.3)$$

where $R(\alpha+\beta+k) > 1$ and $m+n > \frac{p}{2} + \frac{q}{2}$.

Fourth Summation Formula:

$$\sum_{r=o}^{k} \binom{k}{r} \lambda^{k}\, G^{m,n}_{p+\lambda-1,q+\lambda-1}\left(x \,\middle|\, \begin{matrix} a_1,\dots,a_{p-1},\ \Delta(\lambda,\alpha+r) \\ b_1,\dots,b_{q-1},\ \Delta(\lambda,\beta+r) \end{matrix}\right)$$

$$= \frac{\Gamma(\alpha-\beta+k+\lambda-1)}{\Gamma(\alpha-\beta+\lambda-1)}\, G^{m,n}_{b+\lambda-1,q+\lambda-1}\left(x \,\middle|\, \begin{matrix} a_1,\dots,a_{p-1},\ \Delta(\lambda,\alpha+k) \\ b_1,\dots,b_{q-1}\ \ \Delta(\lambda,\beta) \end{matrix}\right),$$

where $R(\alpha-\beta+k) > 0$, $c^{*} = m+n- \frac{p}{2} - \frac{q}{2} > \lambda-1$. (4.2.4)

Fifth Summation Formula:

$$\sum_{r=o}^{k} \frac{(-1)^{k+r}\binom{k}{r}\lambda^{r}}{\Gamma(1+b-\alpha_p+r)}\, G^{m+\lambda,n}_{p,q+\lambda}\left(x \,\middle|\, \begin{matrix} a_1,\dots,a_p \\ \Delta(\lambda,b+r),\ b_1,\dots,b_q \end{matrix}\right)$$

$$= \frac{\lambda^{k}}{\Gamma(1+b-a_p+k)}\, G^{m+2\lambda,n}_{p+\lambda,q+2\lambda}\left(x \,\middle|\, \begin{matrix} a_1,\dots,a_p,\ \Delta(\lambda,a_p-k) \\ \Delta(\lambda,a_p),\ \Delta(\lambda,b),b_1,\dots,b_q \end{matrix}\right), \quad (4.2.5)$$

where $R(a_p) < k+1$ and $c^{*} + \frac{\lambda}{2} > 0$; $c^{*} = m+n - \frac{p}{2} - \frac{q}{2}$.

Sixth Summation Formula:

$$\sum_{r=o}^{k} (-1)^{k+r}\,\lambda^{-r}\binom{k}{r}\,\Gamma(\beta-\alpha+k+r)\; G^{m,n+\lambda}_{p+2\lambda,q}\left(x \,\middle|\, \begin{matrix} (\lambda,\alpha),\ a_1,\dots,a_p,\ \Delta(\lambda,\beta+r) \\ b_1,\dots,b_q \end{matrix}\right)$$

$$= \Gamma(\beta-\alpha+k)\, G^{m,n+\lambda}_{p+2\lambda,q}\left(x \,\Big|\, \begin{matrix} \Delta(\lambda,\alpha-k),\ a_1,\dots,a_p,\ \Delta(\lambda,\beta+k) \\ b_1,\dots,b_q \end{matrix}\right),$$

where $R(\alpha) > 0$, $c^* > 0$; $c^* = m+n - \frac{p}{2} - \frac{q}{2}$. (4.2.6)

Seventh Summation Formula:

$$\sum_{r=o}^{k} \frac{\binom{k}{r}\lambda^{2r}}{\Gamma(1+c+r)\,\Gamma(1-c-k+r)}\, G^{m,n+\lambda}_{p+\lambda,q+\lambda}\left(x \,\Big|\, \begin{matrix} \Delta(\lambda,\alpha-r),\ a_1,\dots,a_p \\ b_1,\dots,b_q,\ \Delta(\lambda,\alpha+r) \end{matrix}\right)$$

$$= \frac{\lambda^{2k}}{\Gamma(1-c)\,\Gamma(1+c+k)}$$

$$\times\, G^{m,n+2\lambda}_{p+2\lambda,q+2\lambda}\left(x \,\Big|\, \begin{matrix} \Delta(\lambda,\alpha+c),\ \Delta(\lambda,\alpha-k-c),\ a_1,\dots,a_p \\ b_1,\dots,b_q,\ \Delta(\lambda,\alpha-c),\ \Delta(\lambda,\alpha+k+c) \end{matrix}\right), \qquad (4.2.7)$$

where $m+n > \frac{p}{2} + \frac{q}{2}$.

The result in (4.2.7) can be established in the same way by using the Saalschütz's theorem (4.1.12).

In a similar manner the following formulas can be established.

Eighth Summation Formula:

$$\sum_{r=o}^{k} \frac{\binom{k}{r}\lambda^{2r}}{\Gamma(1+c+r)\,\Gamma(1-k-c+r)}\, G^{m+\lambda,n}_{p+q,q+\lambda}\left(x \,\Big|\, \begin{matrix} a_1,\dots,a_p,\ \Delta(\lambda,\alpha-r) \\ \Delta(\lambda,\alpha+r),\ b_1,\dots,b_q \end{matrix}\right)$$

$$= \frac{\lambda^{2k}}{\Gamma(1-c)\Gamma(1+c+k)}\, G^{m+2\lambda,n}_{p+2\lambda,q+2\lambda}\left(x \,\Big|\, \begin{matrix} a_1,\dots,a_p,\ \Delta(\lambda,\alpha-c-k),\ \Delta(\lambda,\alpha+c) \\ \Delta(\lambda,\alpha-c),\ \Delta(\lambda,\alpha+c+k),\ b_1,\dots,b_q \end{matrix}\right),$$

where $m+n > \frac{p}{2} + \frac{q}{2}$. (4.2.8)

Ninth Summation Formula:

$$\sum_{r=o}^{k} \frac{\binom{k}{r}(2\lambda)^{2r}}{\Gamma(2\beta+r)\,\Gamma(1-2\beta-2k+r)}\, G^{m+2\lambda,n}_{p+2\lambda,q+2\lambda}\left(x\,\middle|\,\begin{matrix} a_1,\ldots,a_p,\ \Delta(2\lambda,\lambda\alpha-r) \\ \Delta(2\lambda,2\alpha+r),\ b_1,\ldots,b_q \end{matrix}\right)$$

$$= \frac{(2\lambda)^{2k}}{\Gamma(2\beta+k)\,\Gamma(1-2\beta-k)}\, G^{m+2\lambda,n}_{p+2\lambda,q+2\lambda}\left(x\,\middle|\,\begin{matrix} a_1,\ldots,a_p,\ \Delta(\lambda,\alpha+\beta),\Delta(\lambda,\alpha-\beta-k+\frac{1}{2}) \\ \Delta(\lambda,\alpha+\beta+k),\ (\lambda,\alpha-\beta+\frac{1}{2}),b_1,\ldots,b_q \end{matrix}\right)$$

where $m+n > \frac{p}{2} + \frac{q}{2}$. (4.2.9)

By adopting a similar procedure and using (4.1.17) the above formula can be proved.

Finally if we proceed in the same way and use Carlitz's theorem [58] the following interesting result is obtained.

Tenth Summation Formula.

$$\sum_{r=o}^{k} \frac{\binom{k}{r}\Gamma(\beta+k+r)\,(-\lambda)^{r}}{\Gamma(1+\beta+2r)}\, G^{m+\lambda,n}_{p+\lambda,q+\lambda}\left(x\,\middle|\,\begin{matrix} a_1,\ldots,a_p,\ \Delta(\lambda,a+r) \\ \Delta(\lambda,a+2r),\ b_1,\ldots,b_q \end{matrix}\right)$$

$$= \frac{(-1)^{k}}{(\beta+2k)}\, G^{m+\lambda,n}_{p+\lambda,q+\lambda}\left(x\,\middle|\,\begin{matrix} a_1,\ldots,a_p,\ \Delta(\lambda,\alpha-\beta-k) \\ \Delta(\lambda,\alpha-\beta),\ b_1,\ldots,b_q \end{matrix}\right), \qquad (4.2.10)$$

where $m+n > \frac{p}{2} + \frac{q}{2}$.

For a general expansion theorem as well as the generalizations of the results of this section the reader is referred to the work of Swaroop [327]. The exercises given at the end of this chapter also include some of his results.

4.2.2. Infinite Summation Formulas.

Here we give four summation formulas for the G-function due to Jain, R.N. [121] and Narain [218] which are obtained in a manner similar to that employed in the preceding section.

Eleventh Summation Formula.

$$\sum_{r=o}^{\infty} \frac{z^{2a_1 - 2+r} t^r}{r!} G^{m,n}_{p,q}\left(z^{-2} \middle| \begin{matrix} a_1, a_2+r, \ldots, a_p+r \\ b_1+r, \ldots, b_q+r \end{matrix}\right)$$

$$= (z^2-zt)^{a_1-1} G^{m,n}_{p,q}\left[(z^2-zt)^{-1} \middle| \begin{matrix} a_1, \ldots, a_p \\ b_1, \ldots, b_q \end{matrix}\right], \tag{4.2.11}$$

where $|\arg z| < \frac{c^* \pi}{2}$; $c^* = m+n - \frac{p}{2} - \frac{q}{2} > 0$, $|z-t| < c^* \frac{\pi}{2}$ and $|\frac{t}{z}| < 1$.

The formula in (4.2.11) can be proved with the help of (1.1.1) and the formula

$$z^{2s} \; {}_1F_o\,(-s; - ; \tfrac{t}{z}) = (z^2 - zt)^s \; . \tag{4.2.12}$$

The following sums can be obtained in the same way from the results (4.1.20), (4.1.38) and (4.1.39) respectively.

Twelth Summation Formula.

$$\sum_{r=o}^{\infty} \frac{(h+2r)\,\Gamma(h+r)\,(h-b+1)_r\,\Gamma(b-k)\,(k)_r}{r!\,\Gamma(b+r)\,\Gamma(h-k+1+r)\,z^r}$$

$$\times G^{m+1,n}_{p+2,q+1}\left(z \middle| \begin{matrix} a_1, a_2+r, \ldots, a_p+r, \; h+a+2r, b+a-k+r-1 \\ h+a-k+r, \; b_1+r, \ldots, b_q+r \end{matrix}\right)$$

$$= G^{m,n}_{p+1,q}\left(z \middle| \begin{matrix} a_1, \ldots, a_p, \; b+a-1 \\ b_1, \ldots, b_q \end{matrix}\right), \tag{4.2.13}$$

$|\arg z| < c^* \pi$; $c^* = m+n - \frac{p}{2} - \frac{q}{2} > 0$; $R(b-k) > 0$.

Thirteenth Summation Formula.

$$G^{m+1,n}_{p+1,q+1}\left(z\,\middle|\,\begin{matrix}a_1,\ldots,a_p,a_1\\ a_1-\frac{1}{2},\,b_1,\ldots,b_q\end{matrix}\right)$$

$$+\;2\sum_{r=1}^{\infty}\frac{\cos r\theta}{z^r}\,G^{m+1,n}_{p+1,q+1}\left(z\,\middle|\,\begin{matrix}a_1,a_2+r,\ldots,a_p+r,a_1+2r\\ a_1+r-\frac{1}{2},\,b_1+r,\ldots,b_q+r\end{matrix}\right)$$

$$=\;(\sin\tfrac{\theta}{2})^{2a_1-2}\,\Gamma(\tfrac{1}{2})\,G^{m,n}_{p,q}\left(\frac{z}{\sin^2\frac{\theta}{2}}\,\middle|\,\begin{matrix}a_1,\ldots,a_p\\ b_1,\ldots,b_q\end{matrix}\right), \qquad (4.2.14)$$

$|\arg z| < c^*\pi$; $c^* = m+n-\frac{p}{2}-\frac{q}{2} > 0$; $0 \le \theta \le \pi$.

Fourteenth Summation Formula:

$$\sum_{r=o}^{\infty} z^{-r}\,G^{m+1,n}_{p+1,q+1}\left(z\,\middle|\,\begin{matrix}a_1,a_2+r,\ldots,a_p+r,\ a_1+2r+1\\ a_1+r+\frac{1}{2}\ ,\ b_1+r,\ldots,b_q+r\end{matrix}\right)$$

$$\times\sin(2r+1)\theta = \frac{1}{2}\,\Gamma(\tfrac{1}{2})(\sin\theta)^{2a_1-1}\,G^{m,n}_{p,q}\left(\frac{z}{\sin^2\theta}\,\middle|\,\begin{matrix}a_p\\ b_q\end{matrix}\right), \qquad (4.2.15)$$

$|\arg z| < c^*\pi$; $c^* = m+n-\frac{p}{2}-\frac{q}{2} > 0$, $0 \le \theta \le \pi$.

From (4.2.14) and (4.2.15) it readily follows that

$$\int_0^{\pi}(\sin\tfrac{\theta}{2})^{2a_1-2}\cos r\theta\; G^{m,n}_{p,q}\left(\frac{z}{\sin^2\frac{\theta}{2})}\,\middle|\,\begin{matrix}a_1,\ldots,a_p\\ b_1,\ldots,b_q\end{matrix}\right)d\theta$$

$$=\Gamma(\tfrac{1}{2})z^{-r}\;G^{m+1,n}_{p+1,q+1}\left(z\,\middle|\,\begin{matrix}a_1,a_2+r,\ldots,a_p+r,\ a_1+2r\\ a_1+r-\frac{1}{2},\ b_1+r,\ldots,b_q+r\end{matrix}\right). \qquad (4.2.16)$$

$$\int_0^{\pi}\sin(2r+1)\theta\,\sin^{2a_1-1}\theta\; G^{m,n}_{p,q}\left(\frac{z}{\sin^2\theta}\,\middle|\,\begin{matrix}a_p\\ b_q\end{matrix}\right)d\theta$$

$$=\Gamma(\tfrac{1}{2})z^{-r}\,G^{m+1,n}_{p+1,q+1}\left(z\,\middle|\,\begin{matrix}a_1,a_2+r,\ldots,a_p+r,\ a_1+2r+1\\ a_1+r+\frac{1}{2},\ b_1+r,\ldots,b_q+r\end{matrix}\right). \qquad (4.2.17)$$

It is interesting to observe that if a_1 is a positive integer, (4.1.16) can be written as,

$$\int_0^{\pi} (\sin \phi)^{2a_1-2} \cos 2r\phi \, G^{m,n}_{p,q}\left(\frac{z}{\sin^2\phi}\middle| \begin{matrix} a_p \\ b_q \end{matrix}\right) d\phi$$

$$= \Gamma(\tfrac{1}{2}) z^{-r} \; G^{m+1,n}_{p+1,q+1}\left(z \middle| \begin{matrix} a_1, a_2+r, \ldots, a_p+r, \; a_1+2r \\ a_1+r-\frac{1}{2}, \; b_1+r, \ldots, b_q+r \end{matrix}\right). \qquad (4.2.18)$$

Hence on applying the formula

$$\sin(2r+1)\theta \sin\theta = \frac{1}{2}[\cos 2r\theta - \cos(2r+2)\theta]$$

to (4.2.17) and using (4.2.18), we find that

$$G^{m+1,m}_{p+1,q+1}\left(z \middle| \begin{matrix} a_1, a_2+r, \ldots, a_p+r, \; a_1+2r \\ a_1+r-\frac{1}{2}, b_1+r, \ldots, b_q+r \end{matrix}\right)$$

$$-z^{-1} G^{m+1,n}_{p+1,q+1}\left(z \middle| \begin{matrix} a_1, a_1+r+1, \ldots, a_p+r+1, a_1+2r+2 \\ a_1+r+\frac{1}{2}, \; b_1+r+1, \ldots, b_q+r+1 \end{matrix}\right)$$

$$= 2 \; G^{m+1,n}_{p+1,q+1}\left(z \middle| \begin{matrix} a_1, a_2+r, \ldots, a_p+r, \; a_1+2r+1 \\ a_1+r-\frac{1}{2}, \; b_1+r, \ldots, b_q+r \end{matrix}\right). \qquad (4.2.19)$$

Remark: The results of this section reduce to the corresponding results for the E-function given by MacRobert ([158],[162]) on reducing the G-function to E-function by virtue of the formula (1.1.13).

4.3. G-FUNCTION SERIES WHOSE SUMS ARE CONSTANTS.

The following theorems will be established here.

Theorem 4.3.1. If $c^* = m+n-\frac{p}{2}-\frac{q}{2} > 0$, $|\arg z| < c^*\pi$, $0 \le n \le p$, $0 \le m \le q$, then

$$\sum_{r=0}^{\infty} \frac{(2\lambda+\lambda r)\,\Gamma(2\lambda+r)}{r!} \; G^{m,n+1}_{p+2,q}\left(\frac{1}{z} \middle| \begin{matrix} 1-\lambda-r, \; a_p, \; 1+\lambda+r \\ b_q \end{matrix}\right)$$

$$= z^{\lambda} \frac{\prod_{j=1}^{m} \Gamma(b_j+\lambda) \prod_{j=1}^{n} \Gamma(1-a_j-\lambda)}{\prod_{j=m+1}^{q} \Gamma(1-b_j-\lambda) \prod_{j=n+1}^{p} \Gamma(a_j+\lambda)} . \tag{4.3.1}$$

Theorem 4.3.2. If $c^* + \frac{1}{2} > 0$; $c^* = m+n - \frac{p}{2} - \frac{q}{2}$,

$|\arg z| < (c^* + \frac{1}{2})\pi$, $0 \leq m \leq q$ and $0 \leq n \leq p$, then

$$\sum_{r=o}^{\infty} \frac{1}{r!} G^{m,n+1}_{p+1,q} \left(\frac{1}{z} \middle| \begin{matrix} 1-\lambda-r, a_p \\ b_q \end{matrix} \right)$$

$$= z^{\lambda} \frac{\prod_{j=1}^{m} \Gamma(b_j+\lambda) \prod_{j=1}^{n} \Gamma(1-a_j-\lambda)}{\prod_{j=m+1}^{q} \Gamma(1-b_j-\lambda) \prod_{j=n+1}^{p} \Gamma(a_j+\lambda)} . \tag{4.3.2}$$

For similar results see exercise 4.16.

Remark: Theorem 4.3.1 and Theorem 4.3.2 are respectively given by Sharma, B.L. [287] and Saxena and Bhagchandani [273]. It goes without saying that theorems similar to these can be obtained for other Special Functions by specializing the parameters in (4.3.1) and (4.3.2).

The following integrals and series are required in the sequel.

$$\int_{o}^{\infty} z^{\lambda-1} G^{m,n}_{p,q} \left(zx^2 \middle| \begin{matrix} a_1,\dots,a_p \\ b_1,\dots,b_q \end{matrix} \right) dx$$

$$= \frac{1}{2} z^{-\frac{\lambda}{2}} \frac{\prod_{j=1}^{m} \Gamma(b_j + \frac{\lambda}{2}) \prod_{j=1}^{n} \Gamma(1-a_j - \frac{\lambda}{2})}{\prod_{j=m+1}^{q} \Gamma(1-b_j - \frac{\lambda}{2}) \prod_{j=n+1}^{p} \Gamma(a_j + \frac{\lambda}{2})} , \tag{4.3.3}$$

$0 \leq m \leq q$, $0 \leq n \leq p$, $c^* = m+n- \frac{p}{2} - \frac{q}{2} > 0$,

$|\arg z| < c^*\pi$, $- \min_{1 \leq j \leq m} 2R(b_j) < R(\lambda) < 1 - \max_{1 \leq j \leq n} 2R(a_j)$.

$$\int_0^\infty x^{\lambda-1} J_\nu(ax)\, G^{m,n}_{p,q}\left(zx^2 \left| \begin{matrix} a_1,\ldots,a_p \\ b_1,\ldots,b_q \end{matrix}\right.\right) dx$$

$$= \frac{2^{\lambda-1}}{a^\lambda} G^{m,n+1}_{p+2,q}\left(\frac{4z}{a^2} \left| \begin{matrix} 1-\frac{\nu}{2}-\frac{\lambda}{2},\ a_p,\ 1+\frac{\nu}{2}-\frac{\lambda}{2} \\ b_q \end{matrix}\right.\right), \tag{4.3.4}$$

$0 \leq m \leq q$, $0 \leq n \leq p$, $c^* = m+n-\frac{p}{2}-\frac{q}{2} > 0$, $|\arg z| < c^*\pi$, $a > 0$, $R(\lambda+\nu+2b_j) > 0$, $j = 1,\ldots,m$; $R(\lambda+2a_h) < \frac{3}{2}$, $h = 1,\ldots,n$.

$$x^\lambda = \sum_{n=o}^{\infty} \frac{(\lambda+2n)\,\Gamma(\lambda+n)}{n!} J_{\lambda+2n}(2x) \tag{4.3.5}$$

and

$$x^\lambda = 2^\lambda\, \Gamma(1+\tfrac{\lambda}{2}) \sum_{n=o}^{\infty} \frac{(\frac{x}{2})^{\frac{\lambda}{2}+n}}{n\,!} J_{\frac{\lambda}{2}+n}(x). \tag{4.3.6}$$

Remark 2: It is easy to see that (4.3.3) and (4.3.4) follow from (3.3.1) and (3.2.1) respectively.

Remark 3: (4.3.5) is the Gegenbauer's formula [100], whereas (4.3.6) represents the modified Lommel's expansion, see Watson ([352], p.140(5)).

Proof of Theorem 4.3.1: It can be readily established on using (4.3.3),(4.3.4) and (4.3.5).

Theorem 4.3.2 can be proved in the same way such that instead of (4.3.5) we employ the formula (4.3.6).

4.3.1. Expansions of Kampé de Fériet Function in Series of G-Functions.

Here we derive two expansion formulas for Kampé de Fériet function in a series of products of Kampé de Fériet function and a G-function by the application of the theorems of the previous section. The discussion of this article is based on the work of Sharma, B.L. [287] and Saxena and Bhagchandani [273].

In what follows we have used the following notation due to Chaundy [62] to represent the Kampé de Fériet function.

$$F\left[\begin{matrix}(a_P);\ (b_Q);\ (c_R);\ x,y\\ (d_S);\ (e_H);\ (f_L)\end{matrix}\right]$$

$$= \sum_{r=o}^{\infty}\sum_{s=o}^{\infty} \frac{[(a_P)]_{r+s}[(b_Q)]_r\ [(c_R)]_s\ x^r y^s}{[(d_S)]_{r+s}\ [(e_H)]_r\ (f_L)_s\ r!\ s!} \tag{4.3.7}$$

where (a_P) and $[(a_P)]_{r+s}$ denote the set of parameters

$$a_1,\ldots,a_P \quad\text{and}\quad (a_1)_{r+s},\ldots,(a_P)_{r+s}\ ,$$

respectively; and

$$[a]_n = a(a+1)\ \ldots\ (a+n-1);\ [a]_o = 1\ .$$

In terms of this notation the generalized hypergeometric series ${}_rF_s$ is written as follows.

$${}_rF_s\left(\begin{matrix}a_r\\ b_s\end{matrix};\ z\right) = \sum_{n=o}^{\infty} \frac{[(a_r)]_n}{[1]_n[b_s]_n}\ z^n\ , \tag{4.3.8}$$

$r < s + 1$ or $r = s + 1$ and $|z| < 1$.

The contracted notation

$$\Gamma[(a_p),(b_q);(c_r),(d_s)] \equiv \left[\begin{matrix}(a_p),(b_q)\\ (c_r),(d_s)\end{matrix}\right] = \frac{\Gamma(a_1)\ldots\ \Gamma(a_p)\ \Gamma(b_1)\ldots\ \Gamma(b_q)}{\Gamma(c_1)\ldots\Gamma(c_r)\ \Gamma(d_1)\ldots\Gamma(d_s)}$$

will also be used throughout this section.

First expansion:

The first formula to be proved is

$$x^\lambda F\left[\begin{matrix}(A_L);\ (C_{P_1});\ (E_{P_2});\ -ax,\ -bx\\ (B_M);\ (D_{Q_1});\ (H_{Q_2})\end{matrix}\right]$$

$$= \frac{\prod_{j=m+1}^{q} \Gamma(1-b_j-\lambda) \prod_{j=n+1}^{p} \Gamma(a_j+\lambda)}{\prod_{j=1}^{m} \Gamma(b_j+\lambda) \prod_{j=1}^{n} \Gamma(1-a_j-\lambda)}$$

$$\times \sum_{r=o}^{\infty} \frac{(2\lambda+2r)\,\Gamma(2\lambda+r)}{r!} G^{m,n+1}_{p+2,q}\left(\frac{1}{x}\,\middle|\,\begin{matrix}1-\lambda-r,\ a_p,\ 1+\lambda+r\\ b_q\end{matrix}\right)$$

$$\times F\left[\begin{matrix}-r,2\lambda+r,\ (a_p)+\lambda,\ (A_L);\ (C_{P_1});\ (E_{P_2})\\ (b_q)+\lambda,\ (B_M);\ (D_{Q_1});\ (H_{Q_2});\end{matrix};\ (-1)^{m+n-q}\,a,\ (-1)^{m+n-q}\,b\right]$$

(4.3.9)

provided that $L + F_1 = M + Q_1 + 1$ or $L + P_1 < M + Q_1$, $L + P_2 = 1 + M + Q_2$ or $L + P_2 < M + Q_2$, $m + n > \frac{p}{2} + \frac{q}{2}$ and $|\arg x| < (m+n - \frac{p}{2} - \frac{q}{2})\pi$.

<u>Proof:</u> If we substitute the expansion formula (4.3.1) in the power series expansion of the hypergeometric series of two variables on the L.H.S. of (4.3.9) and collect the terms associated with the same G-functions, we arrive at the result (4.3.9).

This result (4.3.9) is given by Sharma, B.L. [287].

Let A = B and L = M in (4.3.9), then it reduces to the interesting expansion,

$$\frac{\prod_{j=m+1}^{q} \Gamma(1-b_j-\lambda) \prod_{j=n+1}^{p} \Gamma(a_j+\lambda)}{\prod_{j=1}^{m} \Gamma(b_j+\lambda) \prod_{j=1}^{n} \Gamma(1-a_j-\lambda)} \sum_{r=o}^{\infty} \frac{(2\lambda+2r)\,\Gamma(2\lambda+r)}{r!}$$

$$\times G^{m,n+1}_{p+2,q}\left[\frac{1}{x}\,\middle|\,\begin{matrix}1-\lambda-r,\ a_p,\ 1+\lambda+r\\ b_q\end{matrix}\right]$$

$$\times F\left[\begin{matrix}-r,2\lambda+r,\ (a_p)+\lambda\ ;\ (C_{P_1});\ (E_{P_2});(-1)^{m+n-q}a,(-1)^{m+n-q}b\\ (b_q)+\lambda;\ (D_{Q_1});\ (H_{Q_2})\end{matrix}\right]$$

$$= x^{\lambda}\ {}_{P_1}F_{Q_1}\,[(C_{P_1});\ (D_{P_1});\ -ax]\ {}_{P_2}F_{Q_2}\,[(E_{P_2});\ (H_{Q_2});\ -bx]. \qquad (4.3.10)$$

Remark: If we take $m = n = p = q = 0$ then (4.3.10) reduces to a result due to Srivastava H.M. ([311], p.246). Further on setting $\lambda = \frac{(\lambda_1+\lambda_2)}{2}$, $n = 0, p = 2$, $m = q = 2$, it gives another result due to Srivastava, H. M. ([311], p.246).

A formula due to Verma, A. [343] can be derived from (4.3.10) on taking $n = 0$, $m = q$.

Second expansion:

$$x^{\lambda}\ F\left[\begin{matrix}(A_L);\ (C_{P_1});\ (E_{P_2});\ -ax,\ -bx\\ (B_M);\ (D_{Q_1});\ (H_{Q_2})\end{matrix}\right] = \frac{\prod_{j=m+1}^{q}\Gamma(1-b_j-\lambda)\ \prod_{j=n+1}^{p}\Gamma(a_j+\lambda)}{\prod_{j=1}^{m}\Gamma(b_j+\lambda)\ \prod_{j=1}^{n}\Gamma(1-a_j-\lambda)}$$

$$\times \sum_{r=0}^{\infty}\frac{1}{r!}\,G^{m,n+1}_{p+1,q}\left(\frac{1}{x}\,\middle|\,\begin{matrix}1-\lambda-r,\ a_p\\ b_q\end{matrix}\right)$$

$$\times F\left[\begin{matrix}-r,(a_p)+\lambda,\ (A_L);\ (C_{P_1});\ (E_{P_2});\ (-1)^{m+n-q}a,\ (-1)^{m+n-q}b\\ (b_q)+\lambda,\ (B_M);\ (D_{Q_1});\ (H_{Q_2})\end{matrix}\right], \qquad (4.3.11)$$

provided that $L + P_1 = M + Q_1 + 1$ or $L + P_1 < M + Q_1$, $L + P_2 = 1 + M + Q_2$ or $L + P_2 < M + Q_2$, $m + n > \frac{p}{2} + \frac{q}{2} - \frac{1}{2}$ and $|\arg x| < (m+n- \frac{p}{2} - \frac{q}{2} + \frac{1}{2})\pi$.

For $A = B$, $L = M$ (4.3.11) reduces to the following result.

$$x^{\lambda}\ {}_{P_1}F_{Q_1}\,[(C_{P_1});\ (D_{Q_1});\ -ax]\ {}_{P_2}F_{Q_2}\,[(E_{P_2});\ (H_{Q_2});\ -bx]$$

$$= \frac{\prod_{j=m+1}^{q} \Gamma(1-b_j-\lambda) \prod_{j=n+1}^{p} \Gamma(a_j+\lambda)}{\prod_{j=1}^{m} \Gamma(b_j+\lambda) \prod_{j=1}^{n} \Gamma(1-a_j-\lambda)} \sum_{r=o}^{\infty} \frac{1}{r!} G^{m,n+1}_{p+1,q} \left[\frac{1}{x}\Big|\begin{matrix} 1-\lambda-r, a_p \\ b_q \end{matrix}\right]$$

$$\times F\left[\begin{matrix} -r,(a_p)+\lambda;\ (C_{P_1});\ (E_{P_2});\ (-1)^{m+n-q} a,\ (-1)^{m+n-q} b \\ (b_q)+\lambda;\ (D_{Q_1});\ (H_{Q_2}) \end{matrix}\right]. \qquad (4.3.12)$$

Let $m = n = q = 0$, $p = 1$, $a_1 = 1$ in (4.3.12) then it reduces to one given by Srivastava, H.M.([311], p.246).

Some interesting particular cases of (4.3.9) are given below.

$$x^{2\lambda} F\left[\begin{matrix} (A_L);\ (C_{P_1});\ (E_{P_2});\ -ax^2,\ -bx^2 \\ (B_M);\ (D_{Q_1});\ (H_{Q_2}) \end{matrix}\right]$$

$$= \frac{\Gamma(\frac{1}{2}-\lambda)\ (-\Gamma(\frac{1}{2}))}{\Gamma(1+\lambda)} \sum_{r=o}^{\infty} \frac{(2r+2\lambda)\ \Gamma(2\lambda+r)}{r!}$$

$$\times J_{\lambda+r}(x)\ Y_{\lambda+r}(x)\ F\left[\begin{matrix} -r,2\lambda+r,1+\lambda,(A_L);(C_{P_1});\ (E_{P_2});\ a,b \\ \frac{1}{2}+\lambda,\ (B_M);\ (D_{Q_1});\ (H_{Q_2}) \end{matrix}\right]. \qquad (4.3.13)$$

If we employ the formula

$$F_4[-s,\ \beta+s;\ \gamma,\ \beta-\gamma+1;\ zx,\ (1-z)(1-x)]$$

$$= \frac{(-1)^s (\gamma)_s}{(\beta-\gamma+1)_s}\ {}_2F_1(-s,\beta+s;\gamma;z)\ \ {}_2F_1(-s,\beta+s;\gamma,x), \qquad (4.3.14)$$

then (4.3.13) takes the interesting form

$$x^{2\lambda} F\left[\begin{matrix} \frac{1}{2}+\lambda;\ -;-;-yzx^2,\ -(1-y)(1-s)\ x^2 \\ 1+\lambda;\ 2\lambda-\mu+1 \end{matrix}\right]$$

$$= \frac{\Gamma(\frac{1}{2}-\lambda)\,\Gamma(\frac{1}{2})}{\Gamma(1+\lambda)} \sum_{r=o}^{\infty} \frac{(-1)^{r+1}\,(2r+2\lambda)}{r!}$$

$$\times \frac{\Gamma(2\lambda+r)\,(\mu)_r}{(2\lambda-\mu+1)_r} J_{\lambda+r}(x)\, Y_{\lambda+r}(x)\; {}_2F_1(-r,2\lambda+r;\mu;y)\; {}_2F_1(-r,2\lambda+r;\mu;z).$$

Now by applying the formula

$$x^{\sigma}\, I_{\nu}(x) K_{\mu}(x) = \frac{1}{2\pi^{1/2}} G^{2,2}_{2,4}\left(x^2 \,\middle|\, \begin{matrix} \frac{\sigma}{2}, \frac{\sigma}{2}+\frac{1}{2} \\ \frac{\nu+\mu+\sigma}{2}, \frac{\nu-\mu+\sigma}{2}, \frac{\mu+\sigma+\nu}{2}, \frac{\sigma-\mu-\nu}{2} \end{matrix}\right), \quad (4.3.15)$$

we find,

$$x^{\mu+\nu}\, F\left[\begin{matrix} (A_L);\,(C_{P_1});\,(E_{P_2});\, ax^2, bx^2 \\ (B_M);\,(D_{Q_1});\,(H_{Q_2}) \end{matrix}\right]$$

$$= \frac{2^{\mu+\nu+1}\Gamma(1+\nu)}{\Gamma(-\mu)\,\Gamma(1+\mu+\nu)} \sum_{r=1}^{\infty} \frac{(\mu+\nu+2r)\,\Gamma(\mu+\nu+r)}{r!}$$

$$\times I_{\nu+r}(x)\, K_{\mu+r}(x)\; F\left[\begin{matrix} -r,\mu+\nu+r,\, 1+\mu, 1+\nu\;,\, (A_L);\,(C_{P_1});\,(E_{P_2});\, a,b \\ 1+\frac{\mu}{2}+\frac{\nu}{2}, \frac{1}{2}+\frac{\mu}{2}+\frac{\nu}{2}\;,\, (B_M);\,(D_{Q_1});\,(H_{Q_2}); \end{matrix}\right]$$

(4.3.16)

Again using (4.3.14), we obtain

$$x^{\mu+\nu}\, F\left[\begin{matrix} 1+\frac{\mu}{2}+\frac{\nu}{2}, \frac{1}{2}+\frac{\mu}{2}+\frac{\nu}{2}\;;\, -;\, -:\, yzx^2,\, (1-y)(1-z)x^2 \\ 1+\mu,\, 1+\nu\;,\, \rho;\, \mu+\nu-\rho+1\;; \end{matrix}\right]$$

$$= \frac{2^{\mu+\nu+1}\,\Gamma(1+\nu)}{\Gamma(-\mu)\Gamma(1+\mu+\nu)} \sum_{r=o}^{\infty} \frac{(\mu+\nu+2r)(-1)^r(\rho)_r\,\Gamma(\mu+\nu+r)}{r!\;(\mu+\nu-\rho+1)_r} I_{\nu+r}(x)\, K_{\mu+r}(x)$$

$$\times {}_2F_1(-r,\mu+\nu+r;\,\rho;\,y)\;\; {}_2F_1(-r,\mu+\nu+r;\rho;x). \quad (4.3.17)$$

Results similar to (4.3.13), (4.3.15), (4.3.16) and (4.3.17) can also be obtained from the formula (4.3.11) but for the sake of brevity they are not presented here.

4.4. EXPANSIONS OF G-FUNCTIONS BY LAPLACE TRANSFORM TECHNIQUES.

The following theorem is due to Verma [342]

Theorem 4.4.1.

$$
{}_{p+s}F_q\left[\begin{matrix}(a_p),(b_s);\ x\omega\\ (c_q)\end{matrix}\right] = h\sum_{n=o}^{\infty}\frac{(h-n\alpha+1)_{n-1}[(b_s)]_n[(e_u)]_n(-x)^n}{n!\,[(c_q)]_n}
$$

$$
\times\ {}_{p+2}F_{u+2}\left[\begin{matrix}(a_p),\ 1+h(1-\alpha)^{-1},\ -n;\ \omega\\ (e_u),\ h(1-\alpha)^{-1},\ h-n\alpha+1\end{matrix}\right]
$$

$$
\times\ {}_{s+u+1}F_q\left[\begin{matrix}(b_s)+n,\ (e_u)+n,\ h+n(1-\alpha);\ x\\ (c_q)+n\end{matrix}\right], \qquad (4.4.1)
$$

provided $p+s \leq q$ or $p+s = 1+q$, and $|x\omega| < 1$, $s+u+1 \leq q$, or $s+u = q$ and $|x| < 1$, and the series of the hypergeometric functions on the right hand side is absolutely convergent.

Proof: To establish (4.4.1), let us compare the coefficient of $[(a_p)]_N\,\frac{\omega^N}{N!}$ on both sides of (4.4.1). We obtain

$$
\frac{[(b_s)]_N\,x^N}{[(c_q)]_N} = \{h+N(1-\alpha)\}\sum_{n=N}^{\infty}(h-n\,\alpha+1)_{n-1}\frac{[(b_s)]_n[(e_u)]_n(-x)^n}{n!\,[(c_q)]_n}
$$

$$
\times\ \frac{(-n)_N}{(h-n\,\alpha+1)_N}\ {}_{s+u+1}F_q\left[\begin{matrix}(b_s)+n,(e_u)+n,\ h+n(1-\alpha);\ x\\ (c_q)+n\end{matrix}\right].
$$

Writing $n = N+r$, the above series reduces to

$$1 = \{h+N(1-\alpha)\} \sum_{r=0}^{\infty} \frac{[(h+N(1-\alpha)+1-r\alpha]_{r-1}[(b_s)+N]_r[(e_u)+N]_r\,(-1)^r}{r!\quad[(c_q)+N]_r}$$

$$\times {}_{s+u+1}F_q\left[\begin{matrix}(b_s)+N+r,\ (e_u)+N+r,\ h+(1-\alpha)(N+r);x\\ (c_q)+N+r\end{matrix}\right].$$

It can be easily seen that the term independent of x on the R.H.S. of the above expression is unity. It thus remains to prove that the coefficient of any power of x vanishes on the right. In other words, when $M > 0$,

$$\frac{[(b_s)+N]_M[(e_u)+N)]_M}{[(c_q)+N]_M} \sum_{r=0}^{M} \frac{(-1)^r[h+N(1-\alpha)+1-r\alpha)]_{M-1}}{r!\quad(M-r)!} = 0\,.$$

This, however, is the coefficient of x^{M-1} in the expression,

$$\frac{[(b_s)+N]_M\;[(e_u)+N]_M}{M[(c_q)+N]_M}\,(1-x)^{-h-N(1-\alpha)-1}\,[1-(1-x)^{\alpha}]^M,$$

which has x^M as its lowest term. This completes the proof of (4.4.1). The rearrangement of the infinite series is permissible due to absolute convergence.

Remark: It is interesting to observe that for $\omega = 1$ and $u = q$, (4.4.1) reduces to one given by Niblett [219]. On the other hand if we put $\alpha = 0$, it gives a result due to Meijer [209] on assuming $u = q$, $(e_u) = (c_q)$ and after slight adjustments of the parameters. For a generalization of (4.4.1) see Abiodun and Sharma, B.L. ([1], p.258 (18)).

Verma [342] has also extended the expansion formula (4.4.1) to G-functions in the following form:

Theorem 4.4.2.

$$G^{h+s+t,\ 1+v}_{1+\rho+q+w,h+s+m}\left(x\omega\,\middle|\,\begin{matrix}1,(k_\rho),\ (c_q),\ (\delta_w)\\ (a_p),\ (b_s),\ (g_m)\end{matrix}\right) = \Gamma[(a_p);(e_u),\ (\delta_w)]$$

$$\times \sum_{n=o}^{\infty} \frac{1}{n!\,\Gamma(1-\alpha n+h)} \; {}_{p+2}F_{w+u+2}\left[\begin{matrix}(a_p),\ 1+h(1-\alpha)^{-1},\ -n;\ \frac{1}{\omega} \\ (\delta_w),\ (e_u),\ h(1-\alpha)^{-1},\ h-n\alpha+1\end{matrix}\right]$$

$$\times\, G^{1+s+u+t,\ 1+v}_{1+\rho+q,1+s+u+m}\left(x\,\Big|\,\begin{matrix}1-n,\ (k_\rho),\ (c_q) \\ (b_s),(e_u),\ h-\alpha n,\ (g_m)\end{matrix}\right), \tag{4.4.2}$$

provided that $v \leq \rho$, $t \leq m$,

$$\rho + m + q + w < 1+h+s+2t+2v,$$

$$|\arg x\omega| < \frac{\pi}{2}\,(1+h+s+2t+2v-\rho-m-q-w),$$

$$t + v < 2 + 2\rho + 2m + 2q + s + u,$$

$$|\arg x\,| < \frac{\pi}{2}\,(2 + 2\rho + 2m + 2q + s + u - t - v),$$

and the series on the right hand side has a meaning. The abbreviated gamma function appearing in (4.4.2) is mentioned following (4.3.8) also.

Proof: Now (4.4.2) will be proved by Mathematical induction. Let us assume that (4.4.2) is true for some fixed values of ρ,m,p,q,s,t,u,v and w. To effect the induction with respect to v multiply both sides by $z^{-k_{v+1}}$, replace x by xz and take the Laplace transform of both sides with respect to z.

If we now use the known integral

$$\int_o^\infty e^{-y}\,y^{-\alpha}\,G^{m,n}_{p,q}\left(xy\,\Big|\,\begin{matrix}a_p \\ b_q\end{matrix}\right)dy = G^{m,n+1}_{p+1,q}\left(x\,\Big|\,\begin{matrix}\alpha,\ a_p \\ b_q\end{matrix}\right), \tag{4.4.3}$$

provided that $c^* = m+n-\frac{p}{2}-\frac{q}{2} > 0$, $R(\alpha) < R(b_h) + 1$, $h = 1,\ldots,m$, $|\arg x| < c^*\pi$, on both sides, we get a relation in which v has been replaced by $v+1$.

In order to effect the induction with respect to m, multiply both sides by $z^{g_{m+1}-1}$, replace x by $\frac{x}{z}$, and take the inverse Laplace transforms of both sides with respect to z. If we now apply the integral,

$$\frac{1}{2\pi i}\int_C e^t\, t^{\alpha-1}\, G^{m,n}_{p,q}\left(\frac{x}{t}\,\Big|\,\begin{matrix}a_p\\ b_q\end{matrix}\right)dt = G^{m,n}_{p,q+1}\left(x\,\Big|\,\begin{matrix}a_p\\ b_q,\alpha\end{matrix}\right), \qquad (4.4.4)$$

which is valid when $R(\alpha) > 0$, together with the conditions stated with (4.5.3), on both sides of (4.4.2), we get a result in which m has been replaced by m+1.

In a similar manner the induction with respect to ρ,N,p,q,s,t,u and w can be effected.

Since $\rho = m = t = v = w = 0$, (4.4.2) yields the relation (4.4.1) the result (4.4.2) is established completely.

Remark 1: In view of (1.2.2) and (1.2.8) we can obtain an expansion of $G(\lambda x)$ in a series of products of $G(x)$ and $F(\lambda)$, where G denotes a Meijer's G-function and F, the generalized hypergeometric function.

Remark 2: An interesting particular case of (4.4.2) when $\alpha = 0$, $g_m = 1$, $a_p = c_q = h$ gives rise to a result due to Wimp and Luke [359].

Remark 3: The conditions on the parameters arise due to the particular method followed and can be waived off by analytic continuation.

4.5. EXPANSION OF G-FUNCTION IN SERIES OF PRODUCTS OF G-FUNCTIONS

In this article we have recorded two expansions of a G-function in series of products of G-functions given by Verma ([345], pp.109,110).

Theorem 4.5.1. The first expansion to be proved is

$$2\sum_{k=1}^{\infty} G^{m,n+1}_{p+2,q}\left(\frac{x}{k^2}\,\Big|\,\begin{matrix}1,(a_p),\ 1+\mu\\ (b_q)\end{matrix}\right) G^{M,N+1}_{P+2,Q}\left[\frac{y}{k^2}\,\Big|\,\begin{matrix}(1,\alpha_P),\ 1+\nu\\ (\beta_Q)\end{matrix}\right].$$

$$= -\Gamma\left[\begin{matrix}(b_m),\ 1-(a_n),\ (\beta_M),\ 1-(\alpha_N);\\ 1+\nu,1+\mu,\ 1-(b_{m+1,q}),\ (a_{n+1,p}),\ 1-(\beta_{M+1,Q}),(\alpha_{N+1,P})\end{matrix}\right]$$

$$+ \frac{y^{\frac{1}{2}}}{\pi} \sin [\pi(\nu + \tfrac{1}{2})]\, G^{1+m+N,\ 2+n+M}_{2+p+Q,\ 2+q+P} \left(- \frac{x}{y} \,\middle|\, X\right), \tag{4.5.1}$$

where X stands for the set of parameters,

$$X = \left[\begin{matrix} 1,\ (a_n),\ \frac{3}{2} - (\beta_Q),\ (a_{n+1,p}),\ 1+\mu \\ \frac{1}{2} - \nu,\ \frac{1}{2},\ (b_m),\ \frac{3}{2} - (\alpha_P),\ (b_{m+1,q}) \end{matrix}\right],$$

provided that $\pi > y > x > 0$, $\mu, \nu > -\frac{1}{2}$, $0 \leq m \leq q$, $0 \leq n \leq p$, $0 \leq M \leq Q$, $0 \leq N \leq P$,

$$|\arg x| < (m+n- \frac{p}{2} - \frac{q}{2})\pi, \quad |\arg y| < (M+N- \frac{P}{2} - \frac{Q}{2})\pi ,$$

$$m+n > \frac{p}{2} + \frac{q}{2}, \quad M+N > \frac{P}{2} + \frac{Q}{2},$$

and the series of products of G-functions on the R.H.S. has a meaning. For the notation of a gamma product appearing in (4.5.1) see the discussion following (4.3.8).

Proof: To establish (4.5.1) let us start from the result, given by Cooke ([73], p.240) namely

$$\sum_{k=1}^{\infty} \left(\frac{kx}{2}\right)^{-\mu} J_\mu(kx) \left(\frac{kx}{2}\right)^{-\nu} J_\nu(ky) = \frac{1}{2\Gamma(\mu+1)\,\Gamma(\nu+1)} + \frac{\pi^{\frac{1}{2}}}{y\Gamma(\mu+1)\,\Gamma(\nu+\frac{1}{2})}$$

$$\cdot {}_2F_1 \left(\frac{1}{2} - \nu,\ \frac{1}{2};\ \mu+1;\ \frac{x^2}{y^2}\right) \tag{4.5.2}$$

where $\pi > y > x > 0$; $\mu, \nu > -\frac{1}{2}$. Now (4.5.2) can be rewritten in the convenient form.

$$2 \sum_{k=1}^{\infty} G^{0,1}_{2,0} \left(\frac{x}{k^2} \,\middle|\, 1, 1+\mu\right) G^{0,1}_{2,0} \left(\frac{y}{k^2} \,\middle|\, 1, 1+\nu\right)$$

$$= \Gamma[;\ 1+\nu,\ 1+\mu] + \frac{y^{\frac{1}{2}}}{\pi} \sin [\pi(\nu+\tfrac{1}{2})]\, G^{1,2}_{2,2}\left(- \frac{x}{y} \,\middle|\, \begin{matrix} 1, 1+\mu \\ \frac{1}{2} - \nu,\ \frac{1}{2} \end{matrix}\right),$$

$\mu,\nu > -\frac{1}{2}$. The result can now be proved by following the method of multi-dimensional Mathematical induction on employing the Laplace transform and its inverse.

Similarly, starting with a result given by Buchholz [52], namely,

$$\sum_{k=1}^{\infty} \frac{J_\nu(x\,\gamma_{\nu,k})\,J_\nu(x\,\gamma_{\nu,k})}{[J_{\nu+1}(\gamma_{\nu,k})]^2(z^2-\gamma_{\nu,k}^2)} = \frac{\pi\,J_\nu(xz)}{4\,J_\nu(z)}\,[J_\nu(z)Y_\nu(Xz)-J_\nu(Xz)Y_\nu(z)], \tag{4.5.3}$$

where $0 \le x \le X \le 1, \nu$ is not an integer, and the zeros of $z^{-\nu}\,J_\nu(z)$ being arranged in the ascending magnitude of $R(\gamma_{\nu,k}) > 0$, (the zeros of $J_\nu(x\,\gamma_{\nu,k})$ being $\gamma_{\nu,k}(k = 0,1,2,\ldots,))$, the following general result can be established.

Theorem 4.5.2.

$$\sum_{k=1}^{\infty} \gamma_{\nu,k}^2\, G^{m,n+1}_{p+2,q}\left(\frac{x}{\gamma_{\nu,k}^2}\,\middle|\,\begin{matrix}1,(a_p),\,1+\nu\\(b_q)\end{matrix}\right)$$

$$\times\frac{G^{M,N+1}_{P+2,Q}\left(\frac{x}{\gamma_{\nu,k}^2}\,\middle|\,\begin{matrix}1,\,(\alpha_P),\,1+\nu\\(\beta_Q)\end{matrix}\right)}{[J_{\nu+1}(\gamma_{\nu,k})]^2\ (z^2-\gamma_{\nu,k}^2)}$$

$$= \pi z^{2\nu}\, G^{m,n+1}_{p+2,q}\left(\frac{x}{z^2}\,\middle|\,\begin{matrix}1,(a_p),\,1+\nu\\(b_q)\end{matrix}\right)$$

$$\times\left[\{\cot(\pi\nu)\,J_\nu(z) - Y_\nu(z)\}\,G^{M,N+1}_{P+2,Q}\left(\frac{x}{z^2}\,\middle|\,\begin{matrix}1,(\alpha_P),\,1+\nu\\(\beta_Q)\end{matrix}\right)\right.$$

$$\left.-\operatorname{cosec}(\pi\nu)\,G^{M,N+1}_{P+2,Q}\left(\frac{x}{z^2}\,\middle|\,\begin{matrix}1+\nu,\,(\alpha_P),1\\(\beta_Q)\end{matrix}\right)\right], \tag{4.5.4}$$

where $0 \le x \le X \le 1$, $0 \le n \le p$, $0 \le M \le Q$, $0 \le N \le P$, $0 \le m \le q$, $p+q < 2(m+n)$, $P+Q < 2(M+N)$, $|\arg \frac{1}{z^2}| < \pi[p+q-\frac{m}{2}-\frac{n}{2}]$, $|\arg(\frac{1}{z^2})| < \pi[P+Q-\frac{M}{2}-\frac{N}{2}]$ and the series of products of G-functions on the left hand side has a meaning. Here $J_\nu(z)$ and $Y_\nu(z)$ are the Bessel functions.

4.6. EXPANSION OF A G-FUNCTION IN SERIES OF JACOBI POLYNOMIALS.

The expansion to be proved here are as follows:

Theorem 4.6.1.

$$(1-x)^\sigma \, G^{m,n}_{p,q}\left[z(1-x)^h \,\middle|\, {a_p \atop b_q}\right] = \frac{2^\sigma}{h^{\beta+1}} \sum_{r=0}^{\infty} (-1)^r \frac{(\alpha+\beta+2r+1)\,\Gamma(\alpha+\beta+r+1)}{\Gamma(\alpha+r+1)}$$

$$\times\, G^{m,n+2h}_{p+2h,q+2h}\left(2^h z \,\middle|\, {\Delta(h,-\alpha-\sigma),\ \Delta(h,-\sigma),\ a_p \atop b_q,\ \Delta(h,r-\sigma),\ \Delta(h,-1-\alpha-\beta-\sigma-r)}\right) P^{(\alpha,\beta)}_r(x) \tag{4.6.1}$$

$$= \frac{2^\sigma}{h^{\beta+1}} \sum_{r=0}^{\infty} \frac{(\alpha+\beta+2r+1)\,\Gamma(\alpha+\beta+r+1)}{\Gamma(\alpha+r+1)}$$

$$\times\, G^{m+h,n+h}_{p+2h,q+2h}\left(2^h z \,\middle|\, {\Delta(h,-\alpha-\sigma),\ a_p,\ \Delta(h,-\sigma) \atop \Delta(h,r-\sigma),\ b_q,\ \Delta(h,-1-\alpha-\beta-\sigma-r)}\right] P^{(\alpha,\beta)}_r(x) \tag{4.6.2}$$

where h is a positive integer $c^* = m+n-\frac{p}{2}-\frac{q}{2}$, $|\arg z| < c^*\pi$, $R(\alpha) > -1$, $R(\beta) > -1$, $-1 < x < 1$, $R(\sigma+\alpha+hb_j) > -1$ $(j = 1,\ldots,m)$.

If we assume $\frac{(1-x)}{2} = \omega$, $2^h z = \zeta$, $h = 1$, $m = q = 1$, $n = p = 0$, $a_1 = 0$ in (4.6.1) we arrive at a result given by Wimp and Luke ([359], p.352 (1.3)).

Proof: To establish (4.6.1), let

$$f(x) = (1-x)^\sigma \, G^{m,n}_{p,q}\left[z(1-x)^h \,\middle|\, {a_p \atop b_q}\right] = \sum_{r=0}^{\infty} a_r P^{(\alpha,\beta)}_r(x), \tag{4.6.3}$$

$-1 < x < 1$.

The above equation is valid since $f(x)$ is continuous and of bounded variation in the open interval $(-1,1)$, when $\sigma \geq 0$.

Multiplying both the sides of (4.6.3) by $(1-x)^{\alpha} (1+x)^{\beta} P_{\nu}^{(\alpha,\beta)}(x)$ and integrating with respect to x from -1 to +1, we obtain

$$\int_{-1}^{1} (1-x)^{\sigma+\alpha} (1+x)^{\beta} P_{\nu}^{(\alpha,\beta)}(x)\ G_{p,q}^{m,n}\left[z(1-x)^{h}\Big|\begin{matrix} a_p \\ b_q \end{matrix}\right] dx$$

$$= \sum_{r=0}^{\infty} a_r \int_{-1}^{1} (1-x)^{\alpha} (1+x)^{\beta} P_{\nu}^{(\alpha,\beta)}(x)\ P_{r}^{(\alpha,\beta)}(x) dx.$$

Using the orthogonality property of Jacobi polynomials (3.1.34) and (3.7.3) we obtain

$$a_{\nu} = \frac{(-1)^{\nu}\, 2^{\sigma} (\alpha+\beta+2\nu+1)\, \Gamma(\alpha+\beta+\nu+1)}{h^{1+\beta}\, \Gamma(1+\alpha+\nu)}$$

$$\times G_{p+2h,q+2h}^{m,n+2h}\left(2^{h} z\Big|\begin{matrix} \Delta(h,-\alpha-\sigma),\ \Delta(h,\ -\sigma),\ a_p \\ b_q,\ \Delta(h,\ \nu-\sigma),\ \Delta(h,-1-\alpha-\beta-\sigma-\nu) \end{matrix}\right). \qquad (4.6.4)$$

Now, (4.6.1) follows from (4.6.3) and (4.6.4).

The formula (4.6.2) can be established in a similar way on using the result (3.7.4) instead of (3.7.3).

<u>Remark:</u> Since Jacobi polynomials include, as special cases, all the known orthogonal polynomials such as Tchebichef polynomials, Legendre polynomials and Gegenbauer polynomials, the expansion formulae for the G-functions in series of products of these polynomials and the G-function, can be derived, as particular cases, of the results (4.6.1) and (4.6.2).

The results of this section are obtained by Bajpai [25].

4.7. EXPANSION FORMULAS OF G-FUNCTIONS IN SERIES OF BESSEL FUNCTIONS

From the work of Bajpai [22] we have the following infinite expansion formula of G-functions in series of products of a Bessel function and a G-function.

Theorem 4.7.1.

$$e^{ix}\, x^{\mu}\, G^{m,n}_{p,q}\left(zx^{\delta}\,\middle|\,{a_p \atop b_q}\right) = \frac{\delta^{\mu-1}(2\pi)^{\frac{1}{2}-\frac{\delta}{2}}}{\Gamma(\frac{1}{2})\,2^{\mu-1}} \sum_{s=0}^{\infty} r\,\exp[\tfrac{i}{2}(\mu+r)\pi]\, J_r(x)$$

$$\times\, G^{m+\delta,n+\delta}_{p+2\delta,q+\delta}\left[z\left(\frac{\delta e^{\frac{i\pi}{2}}}{2}\right)^{\delta}\,\middle|\,{\Delta(\delta,1-\mu-r),\ a_p,\ \Delta(\delta,1-\mu+r) \atop \Delta(\delta,\frac{1}{2}-\mu),\ b_q}\right], \qquad (4.7.1)$$

where $c^* = m+n-\frac{p}{2}-\frac{q}{2} > 0$ $|\arg z| < c^*\pi$, $R(\mu+\delta b_j) > 0$, $j = 1,\ldots,m$; $R(\mu) < \frac{1}{2}$, $r = u+2s+1$.

Theorem 4.7.2.

$$x^{\mu}\, G^{m,n}_{p,q}\left(zx^{2\delta}\,\middle|\,{a_p \atop b_q}\right) = 2(2\delta)^{\mu-1} \sum_{s=0}^{\infty} r\, J_r(x)$$

$$\times\, G^{m,n+2\delta}_{p+2\delta,q}\left[z(2\delta)^{2\delta}\,\middle|\,{\Delta(\delta,\frac{2-\mu-r}{2}),\ a_p,\ \Delta(\delta,\frac{2-\mu+r}{2}) \atop b_q}\right], \qquad (4.7.2)$$

where $|\arg z| < c^*\pi$; $c^* = m+n-\frac{p}{2}-\frac{q}{2} > 0$, $R(\mu+2\delta b_j) > 0$, $j = 1,2,\ldots, m$, $R(\mu+2\delta a_j) < 2\delta$, $j = 1,2,\ldots,n$, $r = u+2s+1$.

Theorem 4.7.3.

$$x^{-\rho}\, J_{\mu}(x)\, G^{m,n}_{p,q}\left(zx^{2\delta}\,\middle|\,{a_p \atop b_q}\right) = \frac{\delta^{-\rho-\frac{3}{2}}}{\Gamma(\frac{1}{2})} \sum_{s=0}^{\infty} r\, J_r(x)$$

$$\times\, G^{m+2\delta,n+\delta}_{p+4\delta,q+2\delta}\left[z\delta^{2\delta}\,\middle|\,\begin{matrix}\Delta(\delta,\frac{2+\rho-\mu+r}{2})\ a_p,\ \Delta(\delta,\frac{2+\rho\pm\mu+r}{2}),\ \Delta(\delta,\frac{2+\rho+\mu-r}{2})\\ \Delta(2\delta,\rho+1)\qquad b_q\end{matrix}\right]$$

(4.7.3)

where $c^* = m+n-\frac{p}{2}-\frac{q}{2} > 0$, $|\arg z| < c^*\pi$,

$R(\mu-\rho+2\delta b_j) > 0$, $j = 1,\ldots,m$, $R(2\delta a_j - \rho) < 2\delta+1$, $j = 1,\ldots,n$, $r = u+2s+1$.

<u>Proof</u>: Put

$$f(x) = e^{ix}\, x^{\mu}\, G^{m,n}_{p,q}\left(zx^{2\delta}\,\middle|\,\begin{matrix}a_p\\ b_q\end{matrix}\right) = \sum_{s=o}^{\infty} C_s\, J_{u+2s+1}(x),\quad (0 < x < \infty). \tag{4.7.4}$$

This equation is valid since $f(x)$ is continuous and of bounded variation in the open interval $(0,\infty)$, when $\mu \geq 0$. Now multiplying both sides of (4.7.4) by $x^{-1}J_{u+2t+1}(x)$ and integrating from 0 to ∞, with respect to x, we get

$$\int_0^{\infty} e^{ix}\, x^{\mu-1}\, J_{u+2t+1}(x)\, G^{m,n}_{p,q}\left(zx^{2\delta}\,\middle|\,\begin{matrix}a_p\\ b_q\end{matrix}\right)dx$$

$$= \sum_{s=o}^{\infty} C_s \int_0^{\infty} x^{-1}\, J_{u+2t+1}(x)\, J_{u+2s+1}(x)\, dx\,.$$

Now using the integral

$$\int_0^{\infty} e^{ix}\, x^{\mu}\, J_{\nu}(x)\, G^{m,n}_{p,q}\left(zx^{\delta}\,\middle|\,\begin{matrix}a_p\\ b_q\end{matrix}\right)dx = \frac{\exp[\frac{i}{2}(\mu+\nu+1)\pi]\,(2\pi)^{\frac{1}{2}-\frac{\delta}{2}}}{\Gamma(\frac{1}{2})\, 2^{\mu+1}\,\delta^{-\mu}}$$

$$\times\, G^{m+\delta,n+\delta}_{p+2\delta,q+2\delta}\left[z\left(\frac{\delta e^{\frac{i\pi}{2}}}{2}\right)^{\delta}\,\middle|\,\begin{matrix}\Delta(\delta,-\nu-\mu)\ a_p,\ \Delta(\delta,\nu-\mu)\\ \Delta(\delta,-\mu-\frac{1}{2}),\ b_q\end{matrix}\right], \tag{4.7.5}$$

where $c^* > 0$, $|\arg z| < c^*\pi$, $R(\mu+\nu+\delta b_j) > -1$, $j = 1,2,\ldots,m$, $R(\mu) < -\frac{1}{2}$, which readily follows from

$$\int_0^\infty e^{ix} x^{\mu} J_\nu(x)dx = \frac{\exp \frac{i}{2}[(\mu+\nu+1)\pi]\, \Gamma(\mu+\nu+1)\, \Gamma(-\mu-\frac{1}{2})}{2^{\mu+1}\, \Gamma(\frac{1}{2})\, \Gamma(\nu-\mu)},$$

$R(\mu+\nu) > -1$, $R(\mu) < -\frac{1}{2}$, and from the orthogonality property of Bessel functions (3.1.52), we obtain

$$C_t = (u+2t+1) \frac{\delta^{\mu-1} (2\pi)^{\frac{1}{2}-\frac{\delta}{2}} \exp[\frac{\pi}{2} i(\mu+\nu)]}{\Gamma(\frac{1}{2})2^{\mu-1}}$$

$$\times G^{m+\delta,n+\delta}_{p+2\delta,q+\delta}\left[z(\delta e^{\frac{i\pi}{2}})^{\delta} \,\middle|\, \begin{matrix} \Delta(\delta,1-\mu-\nu),\ a_p,\ \Delta(\delta,1-\mu+\nu) \\ \Delta(\delta,\frac{1}{2}-\mu),\ b_q \end{matrix}\right]. \qquad (4.7.6)$$

From (4.7.4) and (4.7.6) the result (4.7.1) follows:

The remaining two theorems can be established in a similar manner on using the integrals (3.3.1) and the exercise 3.1(iv) respectively.

Remark: Expansion formulas for the G-functions involving exponential functions can be seen from the work of Bajpai [33].

EXERCISES

4.1. Show that

$$G^{m,n}_{p,q}\left[\frac{x}{\lambda} \,\middle|\, \begin{matrix}(\alpha_p)\\(\beta_q)\end{matrix}\right] = \frac{\lambda^{1-\alpha_1}}{\Gamma(\alpha_p-\alpha_1+1)}$$

$$\times \sum_{r=0}^{\infty} \frac{(\nu+2r)\,\Gamma(\nu+r)}{r!}\, {}_2F_1\left(\begin{matrix}\nu+r,\ -r\ ;\\ \alpha_p-\alpha_1+1\ ;\end{matrix}\ \lambda\right)$$

$$\times G^{m,n}_{p,q}\left[x \,\middle|\, \begin{matrix}\alpha_1-r,\ \alpha_2,\ldots,\alpha_{p-1},\ \alpha_1+\nu+r\\ \beta_q\end{matrix}\right],$$ Rathie, P.N. [247] (1966-67))

where $\alpha_p-\alpha_1-\nu$, $\alpha_p-\alpha_1,\nu$ are not negative integers, $1 \leq n \leq p-1$, $R(2\alpha_j-\alpha_1-\alpha_p-\frac{3}{2}) < 0$, $j = 2,\ldots,n$ and $R(2\beta_j-\alpha_1-\alpha_p+\nu+2) > 0$, $j = 1,\ldots,m$. Hence deduce a result for MacRobert's E-function.

4.2. Show that

$$\Gamma(k)\, G^{m,n}_{p,q}\left[x \,\middle|\, \begin{matrix}\alpha_p\\ \beta_q\end{matrix}\right]$$

$$= \sum_{r=o}^{\infty} (-1)^r (\alpha_p-\alpha_1+k+2r)\,(r+1)_{k-1}(\alpha_p-\alpha_1+1+r)_{k-1}$$

$$\times G^{m,n}_{p,q}\left[x \,\middle|\, \begin{matrix}\alpha_1-r,\ \alpha_2,\ldots,\alpha_{p-1},\ \alpha_p+k+r\\ \beta_q\end{matrix}\right],$$

for $1 \leq n \leq p-1$, $|\arg x| < c^*\pi$; $c^* = m+n-\frac{p}{2}-\frac{q}{2} > 0$, $R(\beta_j-\alpha_1+1) > 0$, $j = 1,\ldots,m$, $R(2\alpha_i-\alpha_1-\alpha_p-\frac{3}{2}) < 0$, $i = 2,\ldots,n$ and k a non-negative integer. Deduce a result for MacRobert's E-function.

4.3. Show that

$$2^{2k} \sum_{s=1}^{[\frac{1}{2}(k+1)]} (-1)^s\,(2s-1)!\; C_{k,2s-1}$$

$$\times G^{m,n}_{p,q}\left[x \,\middle|\, \begin{matrix}\alpha_1-\frac{\nu}{2}-s,\ \alpha_2,\ldots,\alpha_{p-1},\ \alpha_1+\frac{\nu}{2}-s\\ \beta_q\end{matrix}\right]$$

$$+\ 2^{2k} \sum_{s=o}^{[\frac{k}{2}]} (-1)^s(2s)!\; C_{k,2s}$$

$$\times G^{m,n}_{p,q}\left[x \,\middle|\, \begin{matrix}\alpha_1-\frac{\nu}{2}-s,\ \alpha_2,\ldots,\alpha_{p-1},\ \alpha_1+\frac{\nu}{2}-s-1\\ \beta_q\end{matrix}\right]$$

$$= \sum_{r=o}^{\infty} (-1)^r (\nu+2r)^{2k+1} \; G^{m,n}_{p,q}\left[x \middle| \begin{matrix} \alpha_1 - \frac{\nu}{2} - r, \alpha_2, \ldots, \alpha_{p-1}, \alpha_1 + \frac{\nu}{2} + r \\ \beta_q \end{matrix}\right]$$

for $1 \leq n \leq p-1$, $|\arg x| < c^*\pi$; $c^* = m+n- \frac{p}{2} - \frac{q}{2} > 0$, k is a non-negative integer, is not equal to an integer and $c_{k,2s}$ and $c_{k,2s-1}$ stand for

$$\frac{2}{(2s)!} \sum_{j=-s}^{s} (-1)^{s-j} \binom{2s}{s+j} \left(\frac{\nu}{2} + j\right)^{2k+1} \prod_{i=-s}^{s-1} (\nu + i + j)^{-1}$$

and

$$\frac{2}{(2s-1)!} \sum_{j=-s}^{s-1} (-1)^{s-j-i} \binom{2s-1}{s+j} \left(\frac{\nu}{2} + j\right)^{2k+1} \prod_{i=-s}^{s-1} (\nu + i + j)^{-1}$$

respectively. Hence or otherwise deduce that

(i)

$$G^{m,n}_{p,q}\left[x \middle| \begin{matrix} \alpha_1 - \frac{\nu}{2}, \alpha_2, \ldots, \alpha_{p-1}, \alpha_1 + \frac{\nu}{2} - 1 \\ \beta_q \end{matrix}\right]$$

$$= \sum_{r=o}^{\infty} (-1)^r (\nu+2r) \; G^{m,n}_{p,q}\left[x \middle| \begin{matrix} \alpha_1 - \frac{\nu}{2} - r, \alpha_2, \ldots, \alpha_{p-1}, \alpha_1 + \frac{\nu}{2} + r \\ \beta_q \end{matrix}\right]$$

for $1 \leq n \leq p-1$, $|\arg x| < c^*\pi$; $c^* = m+n- \frac{p}{2} - \frac{q}{2} > 0$, and ν is not an integer;

(ii)

$$\nu^2 G^{m,n}_{p,q}\left[x \middle| \begin{matrix} \alpha_1 - \frac{\nu}{2}, \alpha_2, \ldots, \alpha_{p-1}, \alpha_1 + \frac{\nu}{2} - 1 \\ \beta_q \end{matrix}\right]$$

$$-4 \, G^{m,n}_{p,q}\left[x \middle| \begin{matrix} \alpha_1 - \frac{\nu}{2} - 1, \alpha_2, \ldots, \alpha_{p-1}, \alpha_2 + \frac{\nu}{2} - 1 \\ \beta_q \end{matrix}\right]$$

$$= \sum_{r=o}^{\infty} (-1)^r (\nu + \nu r)^3 \; G^{m,n}_{p,q}\left[x \middle| \begin{matrix} \alpha_1 - \frac{\nu}{2} - r, \alpha_2, \ldots, \alpha_{p-1}, \alpha_1 + \frac{\nu}{2} + r \\ \beta_q \end{matrix}\right]$$

for $1 \leq n \leq p-1$, $|\arg x| < c^*\pi$, $c^* = m+n- \frac{p}{2} - \frac{q}{2}$, and ν is not an integer.

(Rathie P,N., 1967, [248]).

Hint: [By putting k = 0 and 1 we obtain (i) and (ii) above].

4.4. Show that

$$G^{m,n}_{p,q}\left[x\left|\begin{matrix}\alpha_1-\frac{\nu}{2},\alpha_2,\ldots,\alpha_{p-1},\alpha_1+\frac{\nu}{2}\\ \beta_q\end{matrix}\right.\right]$$

$$-(-1)^{-\lambda}G^{m,n}_{p,q}\left[x\left|\begin{matrix}\alpha_1-\frac{\nu}{2}-\lambda,\alpha_2,\ldots,\alpha_{p-1},\alpha_1+\frac{\nu}{2}+\lambda\\ \beta_q\end{matrix}\right.\right]=\sum_{r=o}^{\lambda-1}(-1)^r(\nu+2r+1)$$

$$\times G^{m,n}_{p,q}\left[x\left|\begin{matrix}\alpha_1-\frac{\nu}{2}-r,\alpha_2,\ldots,\alpha_{p-1},\alpha_1+\frac{\nu}{2}+1+r\\ \beta_q\end{matrix}\right.\right]$$

for $1 \leq n \leq p-1$, $|\arg x| < c^*\pi$; $c^* = m+n-\frac{p}{2}-\frac{q}{2} > 0$ and ν is a positive integer. (Rathie, P.N., 1967, [248]).

4.5. Establish the integral

$$\int_o^\infty x^\gamma e^{-x} L_k^{(\sigma)}(x)\, G^{m,n}_{p,q}\left[zx\left|\begin{matrix}a_p\\ b_q\end{matrix}\right.\right]dx = \frac{(-1)^k}{k!}G^{m,n+2}_{p+2,q+1}\left[x\left|\begin{matrix}-\gamma,\sigma-\gamma,a_p\\ b_q,\sigma-\gamma+k\end{matrix}\right.\right].$$

Hence or otherwise deduce

$$x^\gamma G^{m,n}_{p,q}\left(zx\left|\begin{matrix}a_p\\ b_q\end{matrix}\right.\right)=\sum_{r=o}^{\infty}\frac{(-1)^r}{\Gamma(\sigma+r+1)}G^{m,n+2}_{p+2,q+1}\left[x\left|\begin{matrix}-\gamma-\sigma,-\gamma,a_p\\ b_q,-\gamma+r\end{matrix}\right.\right]L_r^{(\sigma)}(x),$$

where $c^* = m+n-\frac{p}{2}-\frac{q}{2}$, $|\arg z| < c^*\pi$ and $R(\gamma+1+b_j) > -1$, $(j = 1,\ldots,m)$.

(Shah, 1969 [281]).

4.6. Show that

$$\sum_{r=o}^{\infty}\frac{z^{-r}}{r!}E(p;\alpha_p+r:q;\beta_q+r:z)=\frac{\prod_{j=1}^{p}\Gamma(\alpha_j)}{\prod_{j=1}^{q}\Gamma(\beta_j)},$$

where E denotes MacRobert's E-function. (MacRobert and Ragab, 1962 [166]).

4.7. Show that

$$G^{m,n}_{p,q}\left[z\,(\sin\varphi)^{-2r}\,\middle|\,\begin{matrix}a_p\\ b_q\end{matrix}\right] = \frac{1}{\sqrt{\pi r}}\,G^{m+r,n}_{p+r,q+r}\left[z\,\middle|\,\begin{matrix}a_p,\ \Delta(r,1)\\ \Delta(r,\frac{1}{2}),\ b_q\end{matrix}\right]$$

$$+\ \frac{2}{\sqrt{\pi r}}\sum_{h=1}^{\infty} G^{m+r,n+r}_{p+2r,q+2r}\left[z\,\middle|\,\begin{matrix}\Delta(r,1-h),\ a_p,\ \Delta(r,1-h)\\ \Delta(r,\frac{1}{2})\ b_q,\ \Delta(r,1)\end{matrix}\right]\cos(2h\,\varphi)\,,$$

where r is a positive integer, $c^* > 0$, $|\arg z| < c^*\pi, (0 < \varphi < \pi)$.

(Bajpai, 1969 [29]).

Taking $\varphi = \frac{\theta}{2}$, $h = 1$ in the above, show that it reduces to a known result recently given by Keserwani [formerly Narain, R. ((1966),[218]).

4.8. Show that

$$\sum_{r=0}^{\infty}\sin[(2r+1)\theta]\,G^{m+1,n+1}_{p+2,q+2}\left[z\,\middle|\,\begin{matrix}1-r,a_p,\ 2+r\\ \frac{3}{2},\ b_q,1\end{matrix}\right] = \frac{\sqrt{\pi}}{2}\sin\theta\,G^{m,n}_{p,q}\left[\frac{z}{\sin^2\theta}\,\middle|\,\begin{matrix}a_p\\ b_q\end{matrix}\right],$$

where $0 \leq \theta \leq \pi$, $|\arg z| < c^*\pi$, $c^* = m+n-\frac{p}{2}-\frac{q}{2} > 0$, (Keserwani, 1966, [218]).

4.9. Show that

$$G^{m,n+1}_{p+2,q}\left[za^{-2}\operatorname{cosec}^2\frac{\theta}{2}\,\middle|\,\begin{matrix}1-\frac{\rho}{2},\ a_p,\ 1-\frac{\rho}{2}+\nu\\ b_q\end{matrix}\right]$$

$$= \pi^{-\frac{1}{2}}\,2^{2\nu}\sin^{\rho}(\tfrac{\theta}{2})\,\Gamma(\nu)\sum_{r=0}^{\infty}(\nu+r)$$

$$\times\, C^r_n(\cos\theta)\,G^{m+1,n+1}_{p+3,q+1}\left[\frac{z}{a^2}\,\middle|\,\begin{matrix}1-r-\frac{\rho}{2},\ a_p,\ 1-\frac{\rho}{2}+r,1+2\nu+r-\frac{\rho}{2}\\ \frac{1-\rho}{2}+r,\ b_q\end{matrix}\right]$$

where $C^r_n(\cos\theta)$ is the Gegenbauer polynomial (Gupta and Rathie, 1968, [115]).

4.10 Establish the following summation formulas.

(i) $$\sum_{r=0}^{s} \frac{(-s)_r}{r!} G^{m+k,n+k}_{p+2k,q+2k}\left[\lambda z \,\middle|\, \begin{matrix} \Delta(k,\alpha+r),\ a_p,\ \Delta(k,\gamma-r) \\ \Delta(k,\beta+r),\ b_q,\ \Delta(k,\delta-r) \end{matrix}\right] = (\tfrac{2}{k})^s (\alpha-\beta)_s$$

$$\times G^{m+k+s,n+k}_{p+2k+s,q+2k+s}\left(\lambda z \,\middle|\, \begin{matrix} \Delta(k,\alpha+s),\ a_p,\ \Delta(k,\gamma),\ \Delta_s(2k,\alpha+\gamma-1) \\ \Delta(k,\beta),\ \Delta_s(2k,\alpha+\gamma+2k-1), b_q,\ \Delta(k,\delta-s) \end{matrix}\right),$$

$0 \le m \le q$, $0 \le n \le p$, $|\arg \lambda z| < c^* \pi$, $c^* = m+n-\frac{p}{2}-\frac{q}{2} > 0$, and $\alpha+\gamma+s-\beta-\delta-1 = 0$.

(ii) $$\sum_{r=0}^{s} \frac{(h+2r)\,\Gamma(h+r)\,(-s)_r}{\Gamma(h+r+s+1)\, r!}$$

$$\times G^{m,n}_{p,q}\left(\lambda z \,\middle|\, \begin{matrix} \Delta(k,\alpha+r),\ a_{k+1},\dots,a_{p-k},\ \Delta(k,\alpha-h-r) \\ \Delta(k,\beta+r),\ b_{k+1},\dots,b_{q-k},\ \Delta(k,\beta-h-r) \end{matrix}\right) = (-1)^s k^{-2s} (\alpha-\beta)_s$$

$$\times G^{m,n}_{p,q}\left[\lambda z \,\middle|\, \begin{matrix} \Delta(k,\alpha+s),\ a_{k+1},\dots,a_{p-k},\ \Delta(k,\alpha-h) \\ \Delta(k,\beta),\ b_{k+1},\dots,b_{q-k},\ \Delta(k,\beta-h-s) \end{matrix}\right],$$

$k \le m \le q-k$, $k \le n \le p-k$, $c^* = m+n-\frac{p}{2}-\frac{q}{2} > 0$, $|\arg z| < c^*\pi$, $R(\alpha-\beta+s) > 0$.

(iii) $$\sum_{r=0}^{s} \frac{(-s)_r}{r!} G^{m+k,n+k}_{p+2k,q+2k}\left(\lambda z \,\middle|\, \begin{matrix} \Delta(k,\alpha+r),\ a_p,\ \Delta(k,\gamma-r) \\ \Delta(k,\beta-r),\ b_q,\ \Delta(k,\delta+r) \end{matrix}\right) = (-1)^s (\tfrac{2}{k})^s (\alpha-\delta)_s$$

$$\times G^{m+k+s,n+k}_{p+2k+s,q+2k+s}\left(\lambda z \,\middle|\, \begin{matrix} \Delta(k,\alpha+s), a_p,\ \Delta(k,\gamma),\ \Delta_s(2k,\alpha+\gamma-1) \\ \Delta_s(2k,\alpha+\gamma+2k-1),\ \Delta(k,\beta-s), b_q,\ \Delta(k,\delta) \end{matrix}\right),$$

$0 \le m \le q$, $0 \le n \le p$, $c^* = m+n-\frac{p}{2}-\frac{q}{2} > 0$, $|\arg \lambda z| < c^*\pi$ and $\alpha+\gamma-\beta-\delta+s-1 = 0$.

(iv) $$\sum_{r=0}^{s} \frac{(h+2r)\,\Gamma(h+r)\,(-s)_r}{\Gamma(h+r+s+1)\, r!}$$

$$\times G^{m,n}_{p,q}\left[\lambda z \middle| \begin{matrix} \Delta(k,\alpha+r), a_{k+1},\dots,a_{p-k}, \Delta(k,\alpha-h-r) \\ \Delta(k,\beta-h-r), b_{k+1},\dots,b_{q-k}, \Delta(k,\beta+r) \end{matrix}\right] = k^{-2s}(\alpha-\beta)_s$$

$$\times G^{m,n}_{p,q}\left[\lambda z \middle| \begin{matrix} \Delta(k,\alpha+s), a_{k+1},\dots,a_{p-k}, \Delta(k,\alpha-h) \\ \Delta(k,\beta-h-s), b_{k+1},\dots,b_{q-k}, \Delta(k,\beta) \end{matrix}\right],$$

$k \leq m \leq q-k$, $k \leq n \leq p-k$, $c^* = m+n- \frac{p}{2} - \frac{q}{2} > 0$, $|\arg \lambda z| < c^*\pi$, $R(\alpha-\beta+s) > 0$.

(v) $$\sum_{r=o}^{s} \frac{(-s)_r}{r!} G^{m+2k,n}_{p+2k,q+2k}\left(\lambda z \middle| \begin{matrix} a_p, \Delta(k, \alpha+r), \Delta(k,\gamma-r) \\ \Delta(k,\beta+r), \Delta(k,\delta-r), b_q \end{matrix}\right) = (\tfrac{2}{k})^s (\alpha-\beta)_s$$

$$\times G^{m+2k+s,n}_{p+2k+s,q+2k+s}\left(\lambda z \middle| \begin{matrix} a_p, \Delta(k,\alpha+s), \Delta(k,\gamma), \Delta_s(2k,\alpha+\gamma-1) \\ \Delta(k,\beta), \Delta(k,\delta-s), \Delta_s(2k,\alpha+\gamma+2k-1), b_q \end{matrix}\right),$$

$0 \leq m \leq q$, $0 \leq n \leq p$, $c^* = m+n- \frac{p}{2} - \frac{q}{2} > 0$ $|\arg \lambda z| < c^*\pi$ and $\alpha+\beta+\gamma-\delta+s-1 = 0$.

(vi) $$\sum_{r=o}^{s} \frac{\Gamma(h+2r)\,\Gamma(h+s)\,(-s)_r}{\Gamma(h+r+s+1)\, r!}$$

$$\times G^{m+2k,n+2k}_{p+4k,q+4k}\left[\lambda z \middle| \begin{matrix} \Delta(k,\alpha+r), \Delta(m,\alpha-h-r), a_p, \Delta(k,\gamma+r), \Delta(k,\gamma-h-r) \\ \Delta(k,\beta+r), \Delta(k,\beta-h-r), b_q, \Delta(k,\delta+r), \Delta(k,\delta-h-r) \end{matrix}\right]$$

$$= 2^s (-k)^{-3s} (\alpha-\beta)_s (\gamma-\beta)_s$$

$$\times G^{m+2k+s,n+2k}_{p+4k+s,q+4k+s}\left[\lambda z \middle| \begin{matrix} \Delta(k,\alpha+s), \Delta(k,\alpha-h), a_p, \\ \Delta(k,\beta), \Delta(k,\beta-h-s), \end{matrix}\right.$$

$$\left. \begin{matrix} \Delta(k,\gamma+s), \Delta(k,\gamma-h), \Delta_s(2k,\alpha+\gamma-h-1) \\ \Delta_s(2k,\alpha+\gamma-h+2k-1), b_q, \Delta(k,\delta), \Delta(k,\delta-h-s) \end{matrix}\right],$$

$0 \le m \le q$, $0 \le n \le p$, $c^* = m+n - \frac{p}{2} - \frac{q}{2} > 0$, $|\arg \lambda z| < c^*\pi$, $\alpha-\beta+\gamma-\delta+s-1 = 0$.

(vii) $$\sum_{r=o}^{s} \frac{(-1)^r(-s)_r}{r\,!} G^{m,n}_{p,q}\left[\lambda z \middle| \begin{matrix} a_1,\dots,a_{p-k}, \Delta(k,\alpha+r) \\ b_1,\dots,b_{q-k}, \Delta(k,\beta+r) \end{matrix}\right]$$

$$= k^{-s}(\alpha-\beta)_s\, G^{m,n}_{p,q}\left[\lambda z \middle| \begin{matrix} a_1,\dots,a_{p-k}, \Delta(k,\alpha+s) \\ b_1,\dots,b_{q-k}, \Delta(k,\beta) \end{matrix}\right],$$

$0 \le m \le q-k$, $0 \le n \le p-k$, $c^* = m+n- \frac{p}{2} - \frac{q}{2} > 0$, $|\arg \lambda z| < c^*\pi$ and $R(\alpha-\beta+s) > 0$.

(Swaroop, 1965, [327]).

4.11. Prove that if $k \le m \le q-k$, $k \le n \le p-k$, $c^* = m+n- \frac{p}{2} - \frac{q}{2} > 0$, $|\arg \lambda z| < c^*\pi$ and $R(\alpha-\beta+s) > 0$, then

$$\sum_{r=o}^{\infty} \frac{(h+2r)\,\Gamma(h+r)(s)_r}{\Gamma(h+r-s+1)\; r!}$$

$$\times G^{m,n}_{p,q}\left[\lambda z \middle| \begin{matrix} \Delta(k,\alpha+r),\, a_{k+1},\dots,a_{p-k}, \Delta(k, \alpha-h-r) \\ \Delta(k,\beta+r),\, b_{k+1},\dots,b_{q-k}, \Delta(k,\beta-h-r) \end{matrix}\right]$$

$$= \frac{k^{2s}}{(1-\alpha+\beta)_s} G^{m,n}_{p,q}\left[\lambda z \middle| \begin{matrix} \Delta(k,\alpha-s),\, a_{k+1},\dots,a_{p-k}, \Delta(k,\alpha-h) \\ \Delta(k,\beta),\, b_{k+1},\dots,b_{q-k}, \Delta(k,\beta-h+s) \end{matrix}\right].$$

and $$\sum_{r=o}^{\infty} \frac{\Gamma(h+2r)\,\Gamma(h+r)\;(s)_r}{\Gamma(h+r-s+1)\; r!}$$

$$\times G^{m,n}_{p,q}\left[\lambda z \middle| \begin{matrix} \Delta(k,\alpha+r),\, a_{k+1},\dots,a_{p-k}, \Delta(k,\alpha-h-r) \\ \Delta(k,\beta-h-r),\, b_{k+1},\dots,b_{q-k}, \Delta(k,\beta+r) \end{matrix}\right] = \frac{(-1)^s\, k^{2s}}{(1-\alpha+\beta)_s}$$

$$\times\, G^{m,n}_{p,q}\left[\lambda x \,\middle|\, \begin{matrix} \Delta(k,\alpha-s),\ a_{k+1},\dots,a_{p-k},\ \Delta(k,\alpha-h) \\ \Delta(k,\beta-h+s),\ b_{k+1},\dots,b_{q-k},\ \Delta(k,\beta) \end{matrix}\right] .$$

4.12. Show that

$$\sum_{r=o}^{k} \frac{(-1)^r \binom{k}{r} (\alpha+2r)\,\Gamma(\alpha+r)}{\Gamma(\alpha+k+r+1)}$$

$$\times\, G^{m+4\lambda,n}_{p+4\lambda,q+4\lambda}\left[x \,\middle|\, \begin{matrix} a_p,\ \Delta(\lambda,-r+\gamma),\ \Delta(\lambda,\epsilon-r),\ \Delta(\lambda,\alpha+r+\gamma),\ \Delta(\lambda,\alpha+r+\epsilon) \\ \Delta(\lambda,r+\beta),\ \Delta(\lambda,r+\delta),\ \Delta(\lambda,\beta-\alpha-r),\ \Delta(\lambda,\delta-\alpha-r),\ b_q \end{matrix}\right]$$

$$= \left(\frac{\lambda}{\lambda^3}\right)^k (\alpha-\beta+\gamma)_k (\alpha+\gamma-\delta)_k$$

$$\times\, G^{m+6\lambda,n}_{p+6\lambda,q+6\lambda}\left[x \,\middle|\, \begin{matrix} a_p,\ \Delta(\lambda,\gamma), \Delta(\lambda,\alpha+k+ \begin{smallmatrix}\gamma\\ \epsilon\end{smallmatrix}),\ \Delta(\lambda,\epsilon),\ \Delta(2\lambda,\beta+\delta-\alpha-k) \\ \Delta(\lambda,\beta),\ \Delta(\lambda,-\alpha-k+1 \,|\, \begin{smallmatrix}\beta\\ \delta\end{smallmatrix}),\ \Delta(\lambda,\delta),\ \Delta(2\lambda,\beta+\delta-\alpha),\ b_q \end{matrix}\right] .$$

where λ, k and r are positive integers, $2\alpha+\gamma+\epsilon+k = \beta+\delta+1$ and $m+n > \frac{p}{2}+\frac{q}{2}$.

Hint: Use Dougall's first theorem (4.1.19) [Chhabra (1966), p.580 (3.6), [63]].

4.13. If k,λ,μ,t and r are positive integers, $t = \lambda-\mu$, $R(a-b) > k$ and $2m+2n-p-q+t+4\mu > 0$, then prove that

$$\sum_{r=o}^{k} \binom{k}{r} \frac{(-4\lambda\mu)^r}{(-2k)_r\, t^r}\, G^{m+2\lambda,n+2\mu}_{p+\lambda+\mu,q+2\lambda}\left(x \,\middle|\, \begin{matrix} \Delta(2\mu,2a-r),\ a_p,\ \Delta(t,1-a+b+r) \\ \Delta(2\lambda,2b+r),\ b_q \end{matrix}\right)$$

$$= \Gamma(\tfrac{1}{2})\,\Gamma(\tfrac{1}{2}-k)\,(2\pi)^{2\mu-1} \left(\frac{\lambda\mu}{t}\right)^k$$

$$\times\, G^{m+2\lambda,n+\lambda+\mu}_{p+2\lambda+2\mu,q+3\lambda+\mu}\left[x\,\middle|\,\begin{matrix}\Delta(2\mu,2a),\ \Delta(t,1-a+b+k),\ a_p,\ \Delta(\lambda,\tfrac{1}{2}+b),\Delta(\mu,a-k)\\ \Delta(2\lambda,2b),\ b_q,\ \Delta(\lambda,\tfrac{1}{2}+b+k),\ \Delta(\mu,a)\end{matrix}\right]$$

Hint: Use Whipple's theorem (4.1.16) (Bhise, 1963, p.16(4.7) [38]).

4.14. Establish the following results:

(i)
$$\sum_{r=o}^{s}(-1)^s\binom{s}{r}G^{m,n}_{p,q}\left(x\,\middle|\,\begin{matrix}a_1-\frac{s}{2}+r,\ a_2-\frac{s}{2},\ a_3,\dots,a_p\\ b_1,\dots,b_{q-1},\ b_q-\frac{s}{2}+r\end{matrix}\right)$$

$$=(a_1-b_q)_s\,G^{m,n}_{p,q}\left(x\,\middle|\,\begin{matrix}a_1+\frac{s}{2},\ a_2-\frac{s}{2},\ a_3,\dots,a_p\\ b_1,\dots,b_{q-1},\ b_q-\frac{s}{2}\end{matrix}\right)$$

where $n \geq 2$ and $q \geq 2$;

(ii)
$$\sum_{r=o}^{s}(-1)^{s-r}\binom{s}{r}\frac{\Gamma(a_1-b_q-r)}{\Gamma(a_1-b_q-s)}G^{m,n}_{p,q}\left(x\,\middle|\,\begin{matrix}a_1,\dots,a_p\\ b_1,\dots,b_{q-1},b_q+r\end{matrix}\right)$$

$$=G^{m,n}_{p,q}\left(x\,\middle|\,\begin{matrix}a_1-s,\ a_2,\dots,a_p\\ b_1,\dots,b_q\end{matrix}\right)$$

where $n \geq 2$ and $q \geq 2$;

(iii)
$$\sum_{r=o}^{s}(-1)^{s-r}\binom{s}{r}\frac{\Gamma(a_p-b_2-r)}{\Gamma(a_p-b_2-s)}G^{m,n}_{p,q}\left(x\,\middle|\,\begin{matrix}a_1,\dots,a_{p-1},a_p-r\\ b_1,\dots,b_q\end{matrix}\right)$$

$$=G^{m,n}_{p,q}\left(x\,\middle|\,\begin{matrix}a_1,\dots,a_{p-1},\ a_p\\ b_1,b_2+s,\ b_3,\dots,b_q\end{matrix}\right),$$

where $n \geq 1$, $p \geq 2$ and $q \geq 3$;

(iv)
$$\sum_{r=o}^{s}(-1)^{s-r}\binom{s}{r}G^{m,n}_{p,q}\left(x\,\middle|\,\begin{matrix}a_1,\dots,a_{p-1},\ a_p-r-1\\ b_1,b_2-r,\ b_3,\dots,b_q\end{matrix}\right)$$

$$= (a_p-b_2-1)_s \; G^{m,n}_{p,q}\left(x \,\middle|\, \begin{matrix} a_1,\ldots,a_{p-1},a_p-1 \\ b_1,b_2-s,b_3,\ldots,b_q \end{matrix}\right),$$

where $n \geq 1$, $p \geq 2$ and $q \geq 3$. (Olkha, 1970, p.428, [225]).

Also deduce the following expansions for Gauss's hypergeometric functions from the above results.

(v) $$\sum_{r=o}^{s} (-1)^{s-r} \binom{s}{r} \frac{\Gamma(1-a+\frac{s}{2}-r)}{\Gamma(1-c+\frac{s}{2}-r)} \; {}_2F_1\left(\begin{matrix} 1-a+\frac{s}{2}-r,\ 1-b+\frac{s}{2} \\ 1-c+\frac{s}{2}-r;\ -x \end{matrix}\right)$$

$$= (a-c)_s \frac{\Gamma(1-a-\frac{s}{2})}{\Gamma(1-c-\frac{s}{2})} \; {}_2F_1\left(\begin{matrix} 1-a-\frac{s}{2},\ 1-b+\frac{s}{2} \\ 1-c+\frac{s}{2};\ -x \end{matrix}\right);$$

(vi) $$\sum_{r=o}^{s} (-1)^{s-r} \binom{s}{r} \frac{\Gamma(1-a)\,\Gamma(a-c-r)}{\Gamma(1-c-r)\,\Gamma(a-c-s)}$$

$$\times\, {}_2F_1\left(\begin{matrix} 1-a,\ 1-b \\ 1-c-r;\ -x \end{matrix}\right) = \frac{\Gamma(1-a+s)}{\Gamma(1-c)} \; {}_2F_1\left(\begin{matrix} 1-a+s,\ 1-b \\ 1-c\ ;\ -x \end{matrix}\right);$$

(vii) $$\sum_{r=o}^{s} (-1)^{s-r} \binom{s}{r} \frac{\Gamma(b)\,\Gamma(c-b-r)}{\Gamma(c-r)\,\Gamma(c-b-s)} \; {}_2F_1\left(\begin{matrix} a,b \\ c-r;\ -x \end{matrix}\right) = \frac{\Gamma(b+s)}{\Gamma(c)} \; {}_2F_1\left(\begin{matrix} a,b+s \\ c;\ -x \end{matrix}\right);$$

(viii) $$\sum_{r=o}^{s} (-1)^{s-r} \binom{s}{r} \frac{\Gamma(b-r)}{\Gamma(c-r-1)} \; {}_2F_1\left(\begin{matrix} a,b-r \\ c-r-1 \end{matrix};\ -x\right)$$

$$= (c-b-1)_s \frac{\Gamma(b-s)}{\Gamma(c-1)} \; {}_2F_1\left(\begin{matrix} a,b-s \\ c-1 \end{matrix};\ -x\right)$$

(Olkha, 1970, p.429, [225]).

4.15 Prove the following four multiplication theorems for the G-function due to Meijer (1941, [207]).

$$G^{m,n}_{p,q}\left(\lambda x \,\middle|\, \begin{matrix} a_p \\ b_q \end{matrix}\right)$$

$$\text{(i)} \quad = \lambda^{b_1} \sum_{r=o}^{\infty} \frac{1}{r!} (1-\lambda)^r \, G^{m,n}_{p,q}\left(x \,\middle|\, \begin{matrix} a_p \\ b_1+r,\ b_2,\ldots,b_q \end{matrix}\right), \quad |\lambda-1| < 1,\ m \geq 1;$$

$$\text{(ii)} \quad = \lambda^{b_q} \sum_{r=o}^{\infty} \frac{1}{r!} (\lambda-1)^r \, G^{m,n}_{p,q}\left(x \,\middle|\, \begin{matrix} a_p \\ b_1,\ldots,b_{q-1},\ b_q+r \end{matrix}\right), \quad m < q,\ |\lambda-1| < 1;$$

$$\text{(iii)} \quad = \lambda^{a_1-1} \sum_{r=o}^{\infty} \frac{1}{r!} \left(1-\frac{1}{\lambda}\right)^r G^{m,n}_{p,q}\left(x \,\middle|\, \begin{matrix} a_1-r,\ a_2,\ldots,a_p \\ b_q \end{matrix}\right), \quad n \geq 1,\ R(\lambda) > \frac{1}{2};$$

$$\text{(iv)} \quad = \lambda^{a_p-1} \sum_{r=o}^{\infty} \frac{1}{r!} \left(\frac{1}{\lambda}-1\right)^r G^{m,n}_{p,q}\left(x \,\middle|\, \begin{matrix} a_1,\ldots,a_{p-1},\ a_p-r \\ b_q \end{matrix}\right).$$

Remark. If $p < q$ and $m = 1$, the condition $|\lambda-1| < 1$ may be omitted in (i) and similar remark applies to (iii) when $n = 1$ and $p > q$.

For a generalization of these theorems the reader is referred to the work of Skibinski (1970, [300].

4.16 Establish the following expansion:

$$\text{(i)} \quad G^{m,n}_{p,q}\left(z \,\middle|\, \begin{matrix} a_p \\ b_q \end{matrix}\right) = \sum_{r=o}^{\infty} (h+2r)\,(-z)^{-r}$$

$$\times\, G^{m+1,n}_{p+1,q+1}\left(z \,\middle|\, \begin{matrix} a_1, a_2+r,\ldots,a_p+r,\ h+a_1+2r \\ h+a_1+r-1,\ b_1+r,\ldots,b_q+r \end{matrix}\right)$$

where k is a positive integer, $c^* = m+n-\frac{p}{2}-\frac{q}{2} >$, $|\arg z| < c^*\pi$, $R(h) > 0$.

Hence deduce that

$$\text{(ii)} \quad G^{m+1,n}_{p+1,q+1}\left(z \,\middle|\, \begin{matrix} a_1, a_2+k,\ldots,a_p+k,\ h+a_1+2k-1 \\ h+a_1+k-1,\ b_1+k,\ldots,b_q+k \end{matrix}\right) = \sum_{r=o}^{\infty} (h+2r+2k)(-z)^{-r}$$

$$\times\, G^{m+1,n}_{p+1,q+1}\left(z \,\middle|\, \begin{matrix} a_1, a_2+r+k, \dots, a_p+r+k,\ h+a_1+2r+2k \\ h+a_1+r+k-1,\ b_1+r+k, \dots, b_q+r+k \end{matrix}\right)$$

where $|\arg z| < c^*\pi$ and $R(h+2k) > 0$.

Finally show that

$$\text{(iii)}\quad \sum_{r=o}^{k-1} (h+2r)(-z)^{-r}\, G^{m+1,n}_{p+1,q+1}\left(z \,\middle|\, \begin{matrix} a_1, a_2+r, \dots, a_p+r,\ h+a_1+2r \\ h_1+a_1+r-1, b_1+r, \dots, b_q+r \end{matrix}\right) = G^{m,n}_{p,q}\left(z \,\middle|\, \begin{matrix} a_p \\ b_q \end{matrix}\right)$$

$$-(-z)^{-k}\, G^{m+1,n}_{p+1,q+1}\left(z \,\middle|\, \begin{matrix} a_1, a_2+k, \dots, a_p+k,\ h+a_1+2k-1 \\ h+a_1+k-1,\ b_1+k, \dots, b_q+k \end{matrix}\right).$$

Hint: To prove (i), use the modified form of Whipple's theorem, namely (4.1.18). (iii) can be established on multiplying both sides of (ii) above by $(-z)^{-k}$, replacing r+k by r and finally subtracting the result thus obtained from (ii). (Jain, R.N. 1966, pp. 129, 130, [121]).

4.17. Establish the expansion

$$\text{(i)}\quad x^{\mu} = \frac{\prod_{j=p+1}^{s} \Gamma(1-\beta_j-\mu) \prod_{j=q+1}^{r} \Gamma(\alpha_j+\mu)}{\prod_{j=1}^{p} \Gamma(\beta_j+\mu) \prod_{j=1}^{q} \Gamma(1-\mu-\alpha_j)}$$

$$\times \sum_{n=o}^{\infty} \frac{(2\lambda+2n)\ \Gamma(\lambda+\mu+n)}{\Gamma(\lambda-\mu+n+1)}\, G^{p,q+1}_{r+2,s}\left(\frac{1}{x} \,\middle|\, \begin{matrix} 1-\lambda-n, \alpha_r,\ 1+\lambda+n \\ \beta_s \end{matrix}\right)$$

and hence deduce that

$$\text{(ii)}\quad x^{\rho} = 2^{\rho} \sum_{n=o}^{\infty} \frac{(2n+\lambda)\ \Gamma(n+\frac{\rho}{2}+\frac{\lambda}{2})}{\Gamma(n+1+\frac{\lambda}{2}-\frac{\rho}{2})}\, J_{\lambda+2n}(x)$$

and

$$\text{(iii)}\quad x^{2\mu}\, \frac{\Gamma(1+\mu)\ \Gamma(\frac{1}{2}+\mu)}{\Gamma(\frac{1}{2})\ \Gamma(1+\mu+\frac{\nu}{2}-\frac{\rho}{2})\ \Gamma(1+\mu+\frac{\rho}{2}-\frac{\nu}{2})}$$

$$= \sum_{n=o}^{\infty} \frac{(\rho+\nu+2n)}{(1+n-\mu+\frac{\rho}{2}+\frac{\nu}{2})}\, \Gamma(\mu+n+\frac{\rho}{2}+\frac{\nu}{2})\ J_{\rho+n}(x)\ J_{\nu+n}(x).$$

Also show that

$$\text{(iv)}\quad x^{\mu}\, G^{p_1+q,\,q_1+p}_{r_1+s,\,s_1+r}\left(ax \,\middle|\, \begin{matrix} 1-\beta_1-\mu,\ldots,1-\beta_p-\mu,\gamma_{r_1},1-\beta_{p+1}-\mu,\ldots,1-\beta_s-\mu \\ 1-\alpha_1-\mu,\ldots,1-\alpha_q-\mu,\,\delta_{s_1},1-\alpha_{q+1}-\mu,\ldots,1-\alpha_r-\mu \end{matrix}\right)$$

$$= \sum_{n=o}^{\infty} (2\lambda+2n)\, G^{p_1,q_1+1}_{r_1+2,s_1}\left(a \,\middle|\, \begin{matrix} 1-\lambda-\mu-n,\ \gamma_{r_1},1+\lambda-\mu+n \\ \delta_{s_1} \end{matrix}\right)$$

$$\times\, G^{p,q+1}_{r+2,s}\left(x^{-1} \,\middle|\, \begin{matrix} 1-\lambda-n,\ \alpha_r,\ 1+\lambda+n \\ \beta_s \end{matrix}\right),$$

provided that $p+q > \frac{r}{2}+\frac{s}{2}$, $p_1+q_1 > \frac{r_1}{2} + \frac{s_1}{2}$, $|\arg a| < (p_1 + q_1 - \frac{r_1}{2} - \frac{s_1}{2})\pi$ and $|\arg x| < (p+q- \frac{r}{2} - \frac{s}{2})\pi$.

Hint: Use the definition of the G-function and (i) above to prove (iv).

By specializing the parameters in (iv) above obtain the following expansions associated with Bessel functions.

$$\text{(v)}\quad \pi^{-\frac{1}{2}}\, G^{2,0}_{2,2}\left(a^2x^2 \,\middle|\, \begin{matrix} 1-\frac{\nu}{2}+\frac{\mu}{2},\ 1-\frac{\mu}{2}+\frac{\nu}{2} \\ 1,\ \frac{1}{2} \end{matrix}\right)$$

$$= \sum_{n=o}^{\infty} (\mu+\nu+2n)\, J_{\mu+\nu+2n}(2x)\, J_{\mu+n}(a)\, J_{\nu+n}(a)\,.$$

$$\text{(vi)}\quad \sum_{n=o}^{\infty} (\mu+\nu+2n)\, J_{\mu+\nu+2n}(2x)\, I_{\nu+n}(a) K_{\mu+n}(a)$$

$$= \frac{(ax)^{\nu-\mu}\,\Gamma(1+\frac{\mu}{2}-\frac{\nu}{2})\,\Gamma(1+\frac{\mu}{2}-\frac{\nu}{2})}{2\,\pi^{\frac{1}{2}}\,\Gamma(1-\nu)}$$

$$\times\, {}_2F_1\left(\begin{matrix} 1+\frac{\mu}{2}-\frac{\nu}{2},\ \frac{1}{2}-\frac{\nu}{2}+\frac{\mu}{2}\ ;\ -(ax)^{-2} \\ 1-\nu\ ; \end{matrix}\right)$$

(Abiodun and Sharma, B.L. 1971; pp.260 (27,28,29);262(33,34,35); 263(36) [1]).

4.18. Establish the expansion formula.

$$x^{\lambda}\cdot {}_{p_1}F_{q_1}(a_{p_1};b_{q_1};ax) = \frac{h\Gamma(1+\lambda)\prod_{j=1}^{p}\Gamma(\alpha_j)\prod_{j=1}^{q}\Gamma(\lambda+\beta_j)}{\prod_{j=1}^{q}\Gamma(\beta_j)\prod_{j=1}^{p}\Gamma(\lambda+\alpha_j)}$$

$$\times\sum_{n=o}^{\infty}\frac{(-1)^n\,\Gamma(h+\lambda-n\,\alpha)}{n!\,\Gamma(1-n\,\alpha+h)\,\Gamma(1+\lambda-n)}$$

$$\times\,{}_{p+2}F_{q+2}\left(\begin{matrix}-n,1+h(1-\alpha)^{-1},\,\alpha_p;x\\ 1+h-n\,\alpha,h(1-\alpha)^{-1},\beta_q\end{matrix}\right){}_{p_1+q+2}F_{q_1+p+1}\left(\begin{matrix}1+\lambda,\lambda+\beta_q,h+\lambda-n\,\alpha,\,a_{p_1};a\\ 1+\lambda-n,\lambda+\alpha_p,\,b_{q_1}\end{matrix}\right)$$

(Abiodun and Sharma, B.L. 1971, p.258 (18)[1]).

Note: When $\lambda = 0$ this example reduces to the formula (4.4.1).

4.19. Prove the following results.

$$G_{1,3}^{3,0}\left(x^2\left|\begin{matrix}\frac{1}{2}\\ 0,\nu,\ 2n-\nu\end{matrix}\right.\right) = 2^{2-2n}\,\pi^{-\frac{1}{2}}\,x^{2n}\sum_{r=o}^{n}\binom{2n}{n-r}K_{\nu-n-r}(x),$$

valid for $|\arg x| < \pi$, all values of ν, and integral values of n including zeros

and

$$G_{1,3}^{2,1}\left(x^2\left|\begin{matrix}\frac{1}{2}\\ \nu,2n-\nu,\ 0\end{matrix}\right.\right) = 2\pi^{\frac{3}{2}}\,\{\sin 2(\nu-n)\pi\}^{-1}\left(\frac{x}{2}\right)^{2n}$$

$$\times\sum_{r=o}^{n}{}'\binom{2n}{n-r}\left[I_{n-\nu-r}(x)\,I_{n-\nu+r}(x) - I_{\nu-n+r}(x)\,I_{\nu-n-r}(x)\right]$$

valid for $n-\nu < \frac{1}{2}$, ν not an integer, and 2ν not an odd integer or zero. Here the prime indicates that the first term is to be divided by 2, $\binom{n}{r}$ stands for binomial coefficient, and $I_\nu(x)$, $K_\nu(x)$ are modified Bessel functions of the first and second kind respectively. (Varma, V.K. (1966) [340]).

CHAPTER V

COMPUTABLE REPRESENTATIONS OF A G-FUNCTION IN THE LOGARITHMIC CASE

In the previous chapters we concentrated mainly on G-functions when the poles of the integrand in (1.1.1) are simple. Several properties and different types of expansion theorems are already discussed. But in practical problems, especially in problems in Statistics, it is seen that the properties discussed in the previous chpaters have only limited applications. The most frequently occurring cases are the ones where the poles of the integrand in (1.1.1) are of higher orders. For the sake of illustration we will consider two simple examples, namely, the product of independent Gamma and Beta Variates in Statistics.

5.1. Independent Gamma Variates.

A real stochastic random variables X is said to be a Gamma Variate with parameters (α,β) if X has the density function ,

$$g(x) = \frac{1}{\beta^{\alpha}\Gamma(\alpha)} x^{\alpha-1} e^{-x/\beta} , \alpha > 0, x > 0. \tag{5.1.1}$$

If $X_1,\ldots,X_k$ are k statistically independent Gamma Variates with parameters $(\alpha_1+1,1),\ldots,(\alpha_k+1,1)$ then the (s-1)th moment of the product $U = X_1 \ldots X_k$ is

$$E[(X_1 \ldots X_k)]^{s-1} = \prod_{j=1}^{k} E(X_j^{s-1}) = \prod_{j=1}^{k} \frac{\Gamma(\alpha_j+s)}{\Gamma(\alpha_j+1)} , \tag{5.1.2}$$

where E denotes the operator "mathematical expectation " and it is defind as follows. Let $\psi(X)$ be a measurable function of the stochastic variable X. Then the expected value of $\psi(X)$, denoted by $E[\psi(X)]$, is defined as,

$$E[\psi(X)] = \int_x \psi(x)f(x)dx , \tag{5.1.3}$$

where f(x) denotes the density function of X. Thus $E[\psi(x)]$ exists if the integral on the R.H.S. of (5.1.3) exists.

From (5.1.2) one can obtain the density function of $U = X_1 \ldots X_k$, denoted by h(u), by taking the inverse Mellin transform of (5.1.2). That is,

$$h(u) = \frac{1}{\prod_{j=1}^{k} \Gamma(\alpha_j+1)} \frac{1}{2\pi i} \int_{c-i\infty}^{c+i\infty} \{\prod_{j=1}^{k} \Gamma(\alpha_j+s)\} u^{-s} ds$$

$$= \frac{1}{\prod_{j=1}^{k} \Gamma(\alpha_j+1)} G_{0,k}^{k,0}[u|\alpha_1,\ldots,\alpha_k], \quad 0 < u < \infty. \tag{5.1.4}$$

The density h(u) exists if the G-function on the R.H.S. of (5.1.4) exists. In statistical theory a wide variety of problems are associated with Wilks' concept of generalized variance which is nothing but the determinant of variance covariance matrix. In the multivariate normal case it can be seen that the determinant of a central Wishart matrix is a product of independent Gamma variates having the structure of the variable U considered above. In these problems it is seen that usually the parameters $\alpha_1,\ldots,\alpha_k$ differ by integers thereby the poles of the integrand in (5.1.4) are not simple. In order to apply the results to practical problems one needs a representation of (5.1.4) in a form suitable for computation when the parameters $\alpha_1,\ldots,\alpha_k$ are known.

5.2 INDEPENDENT BETA VARIATES

A number of statistical problems are associated with product of independent Beta variates. A real stochastic Variable Y is said to have a Beta distribution with parameters (α,β) if its density function is of the form,

$$f_1(y) = \frac{\Gamma(\alpha+\beta)}{\Gamma(\alpha)\,\Gamma(\beta)} y^{\alpha-1}(1-y)^{\beta-1}, \quad 0 < y < 1,\ \alpha > 0,\ \beta > 0. \tag{5.2.1}$$

In problems of testing statistical hypotheses on multivariate normal distributions the likelihood ratio test statistics or one to one functions of these test statistics in the null case, are structurally ratios of independent Wishart determinants or products of independent Beta variates of the form,

$$V = Y_1 \cdots Y_k \tag{5.2.2}$$

where $Y_1,\ldots,Y_k$ are Beta variates with parameters (α_j,β_j), $j = 1,2,\ldots,k$. Then the h-th moment of V is given by,

$$E(V^h) = \prod_{j=1}^{k} E(X_j^h) = \prod_{j=1}^{k} \frac{\Gamma(\alpha_j+\beta_j)}{\Gamma(\alpha_j)} \prod_{j=1}^{k} \frac{\Gamma(\alpha_j+h)}{\Gamma(\alpha_j+\beta_j+h)} . \tag{5.2.3}$$

Therefore the density function of V, denoted by $f_2(V)$, is obtained by taking the inverse Mellin transform of (5.2.3), as,

$$f_2(V) = \prod_{j=1}^{k} \frac{\Gamma(\alpha_j+\beta_j)}{\Gamma(\alpha_j)} V^{-1} \int_{c-i\infty}^{c+i\infty} \frac{\Gamma(\alpha_j+h)}{\Gamma(\alpha_j+\beta_j+h)} v^{-h}\, dh$$

$$= \prod_{j=1}^{k} \frac{\Gamma(\alpha_j+\beta_j)}{\Gamma(\alpha_j)} v^{-1} G_{k,k}^{k,0}\left[v \,\Big|\, {\alpha_1+\beta_1,\ldots,\alpha_k+\beta_k \atop \alpha_1,\ldots,\alpha_k}\right], \quad 0 < v < 1 \cdot \tag{5.2.4}$$

Again the density function $f_2(v)$ exists provided the G-function on the R.H.S of (5.2.4) exists. If these test statistics are to be of any use one should be able to represent $\int_0^y f_2(v)dv$ in a computable form once the parameters (α_j,β_j), $j = 1,2,\ldots k$ are known. Again, in statistical problems usually the α's and β's differ by integers, thereby the poles of (5.2.4) are of higher orders.

A number of specific statistical problems are discussed in Chapter VI. Keeping in mind the two main streams of statistical problems, namely, the ones associated with a G-function of the type $G^{p,0}_{0,p}(\cdot)$ and the other type associated with a G-function of the type $G^{p,0}_{p,p}(\cdot)$, we will consider computable representations of $G^{p,0}_{0,p}(\cdot)$, $G^{p,0}_{p,p}(\cdot)$ and $G^{m,n}_{p,q}(\cdot)$ separately. There are some results available on the Gauss' hypergeometric functions and MacRobert E-functions when the parameters differ by integers. A few results on ${}_2F_1(\cdot)$ will be listed in Section 5.4. The methods used by MacRobert ([163],[165]) and Luke [152], have limitations that they become unmanagable for higher order poles and thus the practical utility of these methods is limited. Computable representations in the general cases are worked out by Mathai ([175],[178],[179]) and the discussion in this chapter is mainly based on these results. Computability is illustrated by computing numerical tables connected with some statistical problems which will be mentioned in Chapter VI.

5.3. SOME PRELIMINARY RESULTS.

Here we list a few definitions and preliminary results which are needed in the developments in later sections. Throughout this book log z stands for the natural logarithm or z.

Definition 5.3.1. The Psi-function $\psi(z)$. The logarithmic derivative of a Gamma function is defined as a Psi-function. That is,

$$\psi(z) = \frac{d}{dz}\log\Gamma(z) = \frac{\Gamma'(z)}{\Gamma(z)} \quad \text{or} \quad \log\Gamma(z) = \int_1^z \psi(z)dz\,. \qquad (5.3.1)$$

From the definition 5.3.1 the following properties are available.

$$\psi(z) = -\gamma + (z-1)\sum_{n=o}^{\infty}\frac{1}{[(n+1)\ (z+1)]}\,, \qquad (5.3.2)$$

where γ is the Euler's constant; $\gamma = 0.5772156649\ldots$ Evidently $\psi(z)$ is a meromorphic function with simple poles at $z = 0, -1, -2, \ldots$

$$\psi(1) = -\gamma . \tag{5.3.3}$$

$$\psi(z+1) = \frac{1}{z} + \psi(z) . \tag{5.3.4}$$

$$\psi(n+1) = 1 + \frac{1}{2} + \frac{1}{3} + \ldots + \frac{1}{n} - \gamma \tag{5.3.5}$$

$$\psi(z+n) = \frac{1}{z} + \frac{1}{z+1} + \ldots + \frac{1}{z+n-1} + \psi(z), \quad n = 1,2,\ldots \tag{5.3.6}$$

$$\psi(z) - \psi(1-z) = -\pi \cot(\pi z) \tag{5.3.7}$$

$$\psi(z) - \psi(-z) = -\pi \cot(\pi z) - \frac{1}{z} \tag{5.3.8}$$

$$\psi(1+z) - \psi(1-z) = \frac{1}{z} - \pi \cot(\pi z) \tag{5.3.9}$$

$$\psi(\frac{1}{2} + z) - \psi(\frac{1}{2} - z) = \pi \tan(\pi z) \tag{5.3.10}$$

$$\psi(mz) = \log m + \frac{1}{m} \sum_{r=o}^{m-1} \psi(z + \frac{r}{m}), \quad m = 1,2,\ldots \tag{5.3.11}$$

Definition 5.3.2. The Generalized Riemann Zeta Function.

$$\zeta(s,v) = \sum_{n=o}^{\infty} \frac{1}{(v+n)^s}, \quad v \neq 0, -1, -2, \ldots, R(s) > 1 . \tag{5.3.12}$$

From the definitions 5.3.1 and 5.3.2 the following result follows:

$$\frac{\partial^k}{\partial z^k} \log \{ \prod_{j=1}^{m} \Gamma(a_j + z) \} = \sum_{j=1}^{m} \psi(a_j + z) \text{ for } k = 1$$

$$= (-1)^k (k-1)! \sum_{j=1}^{m} \zeta(k, a_j + z) \text{ for } k \geq 2. \tag{5.3.13}$$

In order to obtain the simplifications of the representations in later sections the results (5.3.1) to (5.3.13) will be helpful. A procedure of computing the residues will also be illustrated here for convenience. Consider the Gamma product $\Gamma(z)\,\Gamma(z+m)$, $m = 0,1,\ldots$. The poles of $\Gamma(z)$ and $\Gamma(z+m)$ are given by the equations.

$$z = -v, \quad v = 0,1,2,\ldots .$$

and

$$z = -v, \quad v = m, m+1, \ldots \tag{5.3.14}$$

respectively. That is, for $v = 0,1,2,\ldots, m-1$ the poles of $\Gamma(z)\,\Gamma(z+m)$ are of order one and for $v = m, m+1,\ldots$ poles are of order two. The residues are given by,

$$R_v = \lim_{z \to -v} (z+v)\,\Gamma(z)\,\Gamma(z+m),\quad v = 0,1,\ldots,m-1$$

$$= \frac{(-1)^v\,\Gamma(-v+m)}{v\,!},\quad v = 0,1,\ldots,m-1\ ; \tag{5.3.15}$$

and

$$R'_v = \lim_{z \to -v} \frac{d}{dz}\,[(z+v)^2\,\Gamma(z)\,\Gamma(z+m)],\quad v = m, m+1,\ldots$$

$$= \lim_{z \to -v} \frac{\partial}{\partial z}\,\frac{\Gamma^2(z+v+1)}{(z+v-1)^2(z+v-2)^2\ldots(z+m)^2(z+m-1)\ldots(z)},$$

$$= A_o Z_o = \frac{(-1)^m}{v!\,(v-m)!}\,[2\,\psi(1) + 2\,(\frac{1}{1} + \frac{1}{2} + \ldots + \frac{1}{m-v}\,)$$

$$+ (\frac{1}{v-m-1} + \ldots + \frac{1}{v}\,)]$$

$$= \frac{(-1)^m}{v!\,(v-m)!}\,[\,\psi(v+1) + \psi(v-m+1)],\quad v = m, m+1,\ldots, \tag{5.3.16}$$

where,

$$Z_o = (z+v)^2\,\Gamma(z)\,\Gamma(z+m),\ \text{at}\ \ z \to -v, \tag{5.3.17}$$

and

$$A_o = \frac{\partial}{\partial z}\,\log\,[(z+v)^2\,\Gamma(z)\,\Gamma(z+m)]\ ,\ \text{at}\ z \to v\ . \tag{5.3.18}$$

The simplification in the ψ-function is done by using (5.3.6)

In order to evaluate higher order poles and to write the resulting expressions in a form suitable for computer programming we use the following theorem.

Theorem 5.3.1. If $\Delta(z)$ is a Gamma product with a pole of order k at $z = a$ then the residue of $\Delta(z)\,x^{-z}$ at $z = a$ is given by

$$R = \frac{x^{-a}}{(k-1)!} \sum_{r=o}^{k-1} \binom{k-1}{r} (-\log x)^{k-1-r} \{ \sum_{r_1=o}^{r-1} \binom{r-1}{r_1} A_o^{(r-1-r_1)}$$

$$\times \sum_{r_2=o}^{r_1-1} \binom{r_1-1}{r_2} A_o^{(r_1-1-r_2)} \cdots \} \quad B_o , \qquad (5.3.19)$$

where,

$$B_o = (z-a)^k \; \Delta(z) \quad \text{at} \quad z \to a \qquad (5.3.20)$$

and

$$A_o^{(t)} = \frac{\partial^t}{\partial z^t} [\log (z-a)^k \; \Delta(z)], \text{ at } z \to a . \qquad (5.3.21)$$

<u>Proof:</u> From Calculus of residues,

$$R = \frac{1}{(k-1)!} \frac{\partial^{k-1}}{\partial z^{k-1}} [(z-a)^k \; \Delta(z) \; x^{-z}] .. \qquad (5.3.22)$$

It is easy to verify that,

$$\frac{\partial^{k-1}}{\partial z^{k-1}} [(z-a)^k \; \Delta(z) \; x^{-z}] = x^{-z}[\frac{\partial}{\partial z} + (-\log x)]^{k-1} \; (z-a)^k \; \Delta(z)$$

$$= x^{-z} \sum_{r=o}^{k-1} \binom{k-1}{r} (-\log x)^{k-1-r} \; \frac{\partial^r}{\partial z^r} [(z-a)^k \; \Delta(z)], \qquad (5.3.23)$$

Also,

$$\frac{\partial^r}{\partial z^r} [(z-a)^k \; \Delta(z)] = \frac{\partial^{r-1}}{\partial z^{r-1}} \{[(z-a)^k \; \Delta(z)] [\frac{\partial}{\partial z} \log(z-a)^k \; \Delta(z)]\}$$

$$= \{ \sum_{r_1=o}^{r-1} \binom{r-1}{r_1} A^{(r-1-r_1)} \sum_{r_2=o}^{r_1-1} \binom{r_1-1}{r_2} A^{(r_1-1-r_2)} \cdots \} \; B ,$$

where,

$$A = \frac{\partial}{\partial z} \log B \quad , \quad B = (z-a)^k \; \Delta(z), \qquad (5.3.24)$$

and $A^{(t)}$ denotes the t-th derivative of A with respect to z. The proof follows from (5.3.23) and (5.3.24). Theorem 5.3.1 and (5.3.13) will be used to work out the computable representations of G-functions.

Remark: In Theoerem 5.3.1 we have assumed that $\Delta(z)$ is a Gamma product. This assumption is not needed but in all practical problems discussed in Chapter VI it will be seen that the expressions corresponding to $\Delta(z)$ are all Gamma products. Thus, with the help of (5.3.13), $A^{(t)}$ can be written in terms of computable functions, namely ψ and generalized Zeta functions.

The technique used in Theorem 5.3.1. is the simple manipulation of the operator $G_v \equiv \{\frac{\partial}{\partial z} + (-\log x)\}^v$. In order to enhance the practical utility of Theorem 5.3.1 we will list some particular cases of the operator G_v.

$$G_o = 1 ;$$

$$G_1 = (-\log x) + A ;$$

$$G_2 = (-\log x)^2 + 2(-\log x)A + A^{(1)} + A^2 ;$$

$$G_3 = (-\log x)^3 + (-\log x)^2 A + 3(-\log x)(A^{(1)} + A^2) + (A^{(2)} + 3A^{(1)}A + A^3) ;$$

$$G_4 = (-\log x)^4 + 4(-\log x)^3 A + 6(-\log x)^2(A^{(1)} + A^2) + 4(-\log x)(A^{(2)} + 3A^{(1)}A + A^3) + [A^{(3)} + 4A^{(2)}A + 3(A^{(1)})^2 + 6A^2A^{(1)} + A^4] . \tag{5.3.25}$$

5.4. COMPUTABLE REPRESENTATION OF A HYPERGEOMETRIC FUNCTION IN THE LOGARITHMIC CASE

The result given in this section are available by considering the Barnes representation for a hypergeometric function, namely,

$$\frac{\Gamma(a)\Gamma(b)}{\Gamma(c)} \; {}_2F_1(a,b;c;z) = \frac{1}{2\pi i} \int_{-i\infty}^{i\infty} \frac{\Gamma(a+s)\Gamma(b+s)\Gamma(-s)}{\Gamma(c+s)} (-z)^s \, ds, \tag{5.4.1}$$

where $|\arg(-z)| < \pi$ and the path of integration is indented if necessary to separate the poles at $s = 0,1,2,\dots$ from the poles at $s = -a-n$, $s = -b-n$, $(n = 0,1,2,\dots)$ of the integrand. The following results are obtained by evaluating (5.4.1) as the sum of residues of the integrand at the poles $s = -a-r$, $s = -b-k$, where $r,k = 0,1,2,\dots$. A simple application of (5.3.13), (5.3.15) and (5.3.16) will yield the following results. The convention, $\sum_{j=a}^{b} (\cdot) = 0$ if $b < a$, will be used throughout.

$$\frac{\Gamma(\alpha+m)}{\Gamma(\gamma)}\ {}_2F_1(\alpha,\ \alpha+m;\gamma;z) = \frac{(-z)^{-\alpha-m}}{\Gamma(\gamma-\alpha)} \sum_{n=o}^{\infty} \frac{(\alpha)_{n+m}(1-\gamma+\alpha)_{n+m}}{n!\ (n+m)}$$

$$\times z^{-n}\,[\log(-z) + h_n] + (-z)^{-\alpha} \sum_{n=o}^{m-1} \frac{(\alpha)_n\,\Gamma(m-n)z^{-n}}{\Gamma(\gamma-\alpha-n)\ n!},$$

$\alpha \neq 0, -1,-2,\dots, m = 0,1,2,\dots,$ $|\arg(-z)| < \pi$ and

$$h_n = \psi(1+m+n) + \psi(1+n) - \psi(\alpha+m+n) - \psi(\gamma-\alpha-m-n). \tag{5.4.2}$$

$$\frac{\Gamma(\alpha+m)}{\Gamma(\alpha+m+n)}\ {}_2F_1(\alpha,\alpha+m;\ \alpha+m+n;\ z) = (-1)^{m+n}\,(-z)^{-\alpha-m} \sum_{j=n}^{\infty} \frac{(\alpha)_{j+m}(j-n)!}{(j+m)!\ \ j!}\, z^{-j}$$

$$+ \frac{(-z)^{-\alpha-m}}{(n+m-1)!} \sum_{j=o}^{n-1} \frac{(\alpha)_{j+m}(1-m-n)_{j+m}}{j!\ (j+m)!}\, z^{-j}[\log(-z) + h'_j]$$

$$+ (-z)^{-\alpha} \sum_{j=o}^{m-1} \frac{(m-j-1)!\ (\alpha)_j\, z^{-j}}{(m+n-j-1)!\ j!},$$

$\alpha + m \neq 0, -1,-2,\dots,$ $|\arg(-z)| < \pi$ and

$$h'_j = \psi(1+m+j) + \psi(1+j) - \psi(\alpha+m+j) - \psi(n-j). \tag{5.4.3}$$

$${}_2F_1(\alpha,\beta;\ \alpha+\beta+m;\ z) = \frac{\Gamma(m)\,\Gamma(\alpha+\beta+m)}{\Gamma(\alpha+m)\Gamma(\beta+m)} \sum_{n=o}^{m-1} \frac{(\alpha)_n\,(\beta)_n}{(1-m)_n\ n!}\,(1-z)^n$$

$$+\ (1-z)^m(-1)^m \frac{\Gamma(\alpha+\beta+m)}{\Gamma(\alpha)\,\Gamma(\beta)} \sum_{n=o}^{\infty} \frac{(\alpha+m)_n(\beta+m)_n}{n\,!\ (n+m)!} [\,k_n - \log(1-z)]\ (1-z)^n,$$

where, $m = 0,1,\ldots$ and

$$k_n = \psi(n+1) + \psi(n+1+m) - \psi(\alpha+n+m) - \psi(\beta+n+m). \qquad (5.4.4)$$

$$z^{1-\alpha-\beta-m}\ {}_2F_1(1-\beta-m,\ 1-\alpha-m;\ 2-\alpha-\beta-m;\ z)$$

$$= \frac{\Gamma(m)\ \Gamma(2-\alpha-\beta-m)}{\Gamma(1-\alpha)\ \Gamma(1-\beta)}\ z^{1-\alpha-\beta-m} \sum_{n=o}^{m-1} \frac{(1-\beta-m)_n(1-\alpha-m)_n}{(1-m)_n\ \ n!}$$

$$\times (1-z)^n + z^{1-\alpha-\beta-m}\ (1-z)^m(-1)^m \frac{\Gamma(2-\alpha-\beta-m)}{\Gamma(1-\beta-m)\ \Gamma(1-\alpha-m)}$$

$$\cdot \times \sum_{n=o}^{\infty} \frac{(1-\alpha)_n(1-\beta)_n}{n!\ (n+m)!}\ [K'_n - \log(1-z)]\ (1-z)^n\ ,$$

where $m = 0,1,\ldots$ and

$$K'_n = \psi(n+1) + \psi(n+1+m) - \psi(1-\beta+n) - \psi(1-\alpha+n). \qquad (5.4.5)$$

$${}_2F_1\ (\alpha,\beta;\ \alpha+\beta-m;z) = (1-z)^{-m} \frac{\Gamma(m)\ \Gamma(\alpha+\beta-m)}{\Gamma(\alpha)\ \Gamma(\beta)}$$

$$\times \sum_{n=o}^{m-1} \frac{(\beta-m)_n(\alpha-m)_n}{n!\ (1-m)_n}\ (1-z)^n + (-1)^m \frac{\Gamma(\alpha+\beta-m)}{\Gamma(\alpha-m)\ \Gamma(\beta-m)} \sum_{n=o}^{\infty} \frac{(\alpha)_n(\beta)_n}{n!\ (n+m)!}$$

$$\times [K''_n - \log(1-z)]\ (1-z)^n,$$

where $m = 0,1,2,\ldots$ and

$$K''_n = \psi(1+m+n) + \psi(1+n) - \psi(\alpha+n) - \psi(\beta+n). \qquad (5.4.6)$$

Remark: It goes without saying that when the poles of the integrand in (5.4.1) are simple then one gets,

$$ {}_2F_1\,(a,b;c;z) = \sum_{r=0}^{\infty} \frac{(a)_r\,(b)_r\,z^r}{(c)_r\quad r!} \,. \qquad (5.4.7)$$

Due to lack of distinction made in the literature about the case (5.4.7) and the various logarithmic cases discussed in (5.4.2) to (5.4.6) some confusion is created in applying these various cases to statistical problems, for example, see Consul ([68], [69]).

5.5. COMPUTABLE REPRESENTATION OF A $G^{p,0}_{0,p}(\cdot)$ IN THE LOGARITHMIC CASE .

By using the techniques developed in section 5.3 we can easily write down the computable series representation of a G-function. Start with the Gamma product $\Gamma(\alpha_1+s)\ \ldots\ \Gamma(\alpha_p+s)$. Now consider the most general way in which the α's can differ by integers. Let,

$$\Gamma(\alpha_1+s)\ldots\ \Gamma(\alpha_p+s) = \prod_{r=1}^{k}\ \prod_{j=1}^{q_r}\ \{\Gamma(\alpha_{jr1}+s)\ldots\Gamma(\alpha_{jrr}+s)\}, \qquad (5.5.1)$$

where , for $r \geq 2$ let,

$$\alpha_{jr1}-\alpha_{jr2} = n_{jr1},\ \alpha_{jr2}-\alpha_{jr3} = n_{jr2},\ldots,\alpha_{jrr-1}-\alpha_{jrr} = n_{jrr-1}\ , \qquad (5.5.2)$$

where $n_{jr1}, n_{jr2},\ldots, n_{jrr-1}$ are non-negative integers and for fixed r,

$$\alpha_{jri}-\alpha_{mrh} \neq \pm\, v,\ v = 0,1,\ldots,;\ j \neq m,\ i,\ h= 1,2,\ldots,r. \qquad (5.5.3)$$

An empty product is always interpreted as unity. Now consider a G-function,

$$G^{p,0}_{0,p}\,(x|\,\alpha_1,\ldots,\alpha_p) = \frac{1}{2\pi i}\int_{c-i\infty}^{c+i\infty} \{\prod_{j=1}^{p}\Gamma(\alpha_j+s)\}\ x^{-s}ds\ , \qquad (5.5.4)$$

where $\prod_{j=1}^{p}\Gamma(\alpha_j+s)$ has the structure in (5.5.1). Then the G-function in (5.5.4) can be evaluated by using Braaksma ([51],[6.1]). For the purpose of illustration, for fixed j and r, let $\Gamma(\alpha_{jrk}+s)$ be denoted as $\Gamma(\beta_k+s)$. Then

$$\Gamma(\alpha_{jr1}+s) \ldots \Gamma(\alpha_{jrr}+s) = \Gamma(\beta_1+s)\ldots\Gamma(\beta_r+s) \qquad (5.5.5)$$

Let

$$\beta_1-\beta_2 = m_1,\ \beta_2-\beta_3 = m_2, \ldots, \beta_{r-1}-\beta_r = m_{r-1}, \qquad (5.5.6)$$

and $m_1,\ldots,m_{r-1}$ being non-negative integers. If $\beta_1,\ldots,\beta_r$ exhaust all the α's then the poles of order one are coming from the equation,

$$\beta_r + s+i = 0 \quad \text{for } i = 0,1,\ldots, m_{r-1}\ ; \qquad (5.5.7)$$

poles of order two are coming from the equation ,

$$\beta_{r-1} + s+i = 0 \quad \text{for } i = 0,1,\ldots,m_{r-2}-1; \qquad (5.5.8)$$

and so on and the poles of order r are coming from,

$$\beta_1 + s + i = 0, \quad \text{for } i = 0,1,\ldots \qquad (5.5.9)$$

The residues R_{1i} corresponding to the poles of order one are available as,

$$R_{1i} = \lim_{s\to-(\beta_r+i)} \{ (s+\beta_r+i) \prod_{j=1}^{r} \Gamma(\beta_j+s)\, x^{-s} \}$$

$$= \frac{(m_1+\ldots+m_{r-1}-i-1)!\,(m_2+\ldots+m_{r-1}-i-1)!\ldots(m_{r-1}-i-1)!\ \ x^{\beta_r+i}}{(-1)^i\ i\,!}$$

for $i = 0,1,\ldots, m_{r-1}-1$, $r \geq 2$ and

$$R_{1i} = \frac{(-1)^i}{i!} x^{\beta_1+i} \quad \text{for } i = 0,1,\ldots \text{ and } r = 1. \qquad (5.5.10)$$

In general the residue corresponding to the pole of order ν is given by

$$R_{\nu i} = \lim_{s \to -(\beta_{r-\nu+1}+i)} \{ \frac{1}{(\nu-1)!} \frac{\partial^{\nu-1}}{\partial s^{\nu-1}} (s+\beta_{r-\nu+1}+i)^{\nu}$$

$$\times \Gamma(\beta_1+s) \ldots \Gamma(\beta_r+s)x^{-s} \} . \qquad (5.5.11)$$

This can be evaluated by using Theorem 5.3.1 and (5.3.13). Thus after identifying the poles in (5.5.1) one can write down a computable expansion of (5.5.4) easily. Hence the proof for the following theorem is omitted.

Theorem 5.5.1: When $\prod_{j=1}^{p} \Gamma(\alpha_j+s)$ has the structural set up in (5.5.4)

$$G^{p,0}_{0,p} [x| \alpha_1,\ldots,\alpha_p] = \sum_{r=1}^{k} \sum_{j=1}^{q_r} \sum_{\nu=1}^{r} \sum_{i=o}^{n_{jrr-\nu}-1} \frac{[x^{\alpha_{jrr-\nu+1}+i}}{(\nu-1)!}$$

$$\times \sum_{n=o}^{\nu-1} \binom{\nu-1}{n}(-\log x)^{\nu-1-n} \sum_{\nu_1=o}^{n-1} \binom{n-1}{\nu_1} A^{(n-1-\nu_1)}_{jro} \sum_{\nu_2=o}^{\nu_1-1} \binom{\nu_1-1}{\nu_2}$$

$$\times A^{(\nu_1-1-\nu_2)}_{jro} \ldots] \; Z_{rjo} , \qquad (5.5.12)$$

where Z_{jro}, A_{jro} and $A^{(t)}_{jro}$, $t \geq 1$ are given below with the following convention that $n_{jro} = \infty$ for all $j = 1,\ldots,q_r$; $r = 1,\ldots,k$; and $(\nu-1)!$, $\psi(\nu)$ and $\zeta(t+1,\nu)$ are tacitly omitted whenever $\nu = o,\infty$ or a negative integer. The ψ and Zeta functions are defined in Section 5.3.

$$Z_{jro} = \prod_{\substack{u=1 \\ (m,u) \neq (j,r)}}^{k} \prod_{m=1}^{q_u} \{[\Gamma(\alpha_{mu1} - \alpha_{jrr-\nu+1}-i) \ldots \Gamma(\alpha_{muu} - \alpha_{jrr-\nu+1}-i)$$

$$\times (n_{jr1} + n_{jr2} +\ldots+ n_{jrr-\nu}-i-1)! \, (n_{jr2}+ n_{jr3} +\ldots+ n_{jrr-\nu}-i-1) \, !$$

$$\ldots (n_{jrr-\nu}-i-1)! \, (-1)^{\nu i+(\nu-1)n_{jrr-\nu+1}+\ldots+ n_{jrr-1}}]/[(n_{jrr-\nu+1}$$

$$+ n_{jrr-\nu+2} +\ldots+ n_{jrr-1} +i)!\,(n_{jrr-\nu+1} +\ldots+ n_{jrr-2} +i)\,!\ldots(n_{jrr-\nu+1} + i)!\; i!\;]\};$$

(5.5.13)

$$A_{jro} = \sum_{\substack{n=1\\(m,n)\neq(j,r)}}^{k} \sum_{m=1}^{q_n} \sum_{u=1}^{n} \psi(\alpha_{mru} - \alpha_{jrr-\nu+1} - i)$$

$$+ \nu\,\psi(1) + \psi(n_{jr1} + n_{jr2} +\ldots+ n_{jrr-\nu} - i) + \psi(n_{jr2} +\ldots+ n_{jrr-\nu} - i)$$

$$+\ldots+ \psi(n_{jrr-\nu} - i) + \nu(\frac{1}{1} + \frac{1}{2} +\ldots+ \frac{1}{i}) + (\nu-1)[\frac{1}{i+1} +\ldots+ \frac{1}{i+n_{jrr-\nu+1}}]$$

$$+\ldots+ \frac{1}{(i+n_{jrr-\nu+1} +\ldots+ n_{jrr-1})}\;; \qquad (5.5.14)$$

$$A^{(t)}_{jro},\; t \geq 1 = (-1)^{t-1}\, t\,!\,\{ \sum_{\substack{n=1\\(m,n)\neq(j,r)}}^{k} \sum_{m=1}^{q_r} \sum_{u=1}^{r} \zeta(t+1\,,\, \alpha_{mnu} - \alpha_{jrr-\nu+1} - i)$$

$$+ \nu\,\zeta(t+1,1) + (t+1, n_{jr1} + n_{jr2} +\ldots+ n_{jrr-\nu} - i)$$

$$+ \zeta(t+1, n_{jr2} +\ldots+ n_{jrr-\nu} - i) +\ldots+ \zeta(t+1, n_{jrr-\nu} - i)$$

$$+ t!\,(\nu\,[\frac{1}{1^{t+1}} +\ldots+ \frac{1}{i^{t+1}}] + (\nu-1)[\frac{1}{(i+1)^{t+1}} +\ldots+ \frac{1}{(i+n_{jrr-\nu+1})^{t+1}}]$$

$$+\ldots+ \frac{1}{(i+ n_{jjr-\nu+1} +\ldots+ n_{jrr-1})^{t+1}})\;\}, \qquad (5.5.15)$$

$\nu = 1,2,\ldots,r;\;\; j = 1,\ldots,q_r;\;\; r = 1,\ldots,k\;;\;\; i = 0,1,\ldots, n_{jrr-\nu} - 1$.

Som particular cases are given in the exercises at the end of this chapter.

5.6. COMPUTABLE REPRESENTATION OF A $G^{p,0}_{p,p}(\cdot)$ IN THE LOGARITHMIC CASE.

In this section we will consider the computable representation of a G-function of the type,

$$G^{m+n+p,0}_{m+n+p,m+n+p}\left[x\left|\begin{matrix}\alpha_1+\beta_1,\ldots,\alpha_{m+n+p}+\beta_{m+n+p}\\ \alpha_1,\ldots,\alpha_{m+n+p}\end{matrix}\right.\right],\ |x| < 1. \tag{5.6.1}$$

The case for $|x| > 1$ can be dealt with by using the result (1.2.3) and the technique used in the following Theorem 5.6.1. Evidently (5.6.1) is the inverse Mellin transform of the Gamma product,

$$\prod_{j=1}^{m+n+p} \frac{\Gamma(\alpha_j+s)}{\Gamma(\alpha_j+\beta_j+s)} = \Delta_1\ \Delta_2\ \Delta_3\ ,\ \text{say}, \tag{5.6.2}$$

where,

$$\Delta_1 = \prod_{j=1}^{m} \frac{\Gamma(\alpha_j+s)}{\Gamma(\alpha'_{1j}+\beta'_{1j}+s)}$$

$$= \frac{\prod_{r=1}^{k}\prod_{j=1}^{p_r} \Gamma(\alpha_{jr1}+s)\,\Gamma(\alpha_{jr2}+s)\ \ldots\ \Gamma(\alpha_{jrr}+s)}{\prod_{j=1}^{m}\Gamma\,(\alpha'_{1j}+\beta'_{1j}+s)}$$

where for $r \geq 2$, $\alpha_{jr1} - \alpha_{jr2} = n_{jr1}$, $\alpha_{jr2} - \alpha_{jr3} = n_{jr2}$, ...

$\alpha_{jrr-1} - \alpha_{jrr} = n_{jrr-1}$; $n_{jr1},\ldots,n_{jrr-1}$ are non-negative integers and for every fixed r, $\alpha_{jri} - \alpha_{mrh} \neq \pm\, \nu$, $\nu = 0,1,\ldots$, $j \neq m$; $i,h = 1,2,\ldots,r$;

$$\Delta_2 = \prod_{j=1}^{n} \frac{\Gamma(\alpha_{m+j}+s)}{\Gamma(\alpha'_{2j}+\beta'_{2j}+s)} = \prod_{k=1}^{n'}\prod_{j=1}^{q_k} (\alpha'_{2k}+\beta'_{2k}+s-j)^{-m_{kj}}\ ;$$

$$\Delta_3 = \prod_{j=1}^{p} \frac{\Gamma(\alpha_{m+n+j}+s)}{\Gamma(\alpha'_{3j}+\beta'_{3j}+s)} \tag{5.6.3}$$

The aim in (5.6.3) is to obtain the most general form of the Gamma product in (5.6.2). It is assumed that in Δ_1 none of the Gammas cancels out, the cancellation in Δ_2 is shown in (5.6.3) and in Δ_3 the Gammas cancel out having some factors in the numerator. Further,

$$0 \le m_{kj} \le n, \quad 0 \le n' \le n; \; \alpha'_{2i} + \beta'_{2i} - k \ne \alpha'_{2j} + \beta'_{2j} - \nu ,$$
$$i,j = 1,2,\ldots,n' , \; i \ne j ; \quad k, \nu = 1,2,\ldots, \quad \max_j (q_j); \; m_{kj} ,$$
$$j = 1,2,\ldots,q_k, \; k = 1,2,\ldots,n' \text{ are all non-negative integers.} \tag{5.6.4}$$

It is further assumed that none of the poles of Δ_1 coincides with any of the poles of Δ_2 or any of the zeros of Δ_3 and none of the poles of Δ_2 coincides with any of the zeros of Δ_3. By rearranging the Gammas in (5.6.2) this assumption can always be met. Now we will state the expansion in the form of a theorem.

<u>Theorem 5.6.1.</u> When the Gammas in (5.6.1) have the structure described in (5.6.2) to (5.6.4),

$$G^{m+n+p,0}_{m+n+p,m+n+p}\left[x\Big|{\alpha_1+\beta_1,\ldots,\alpha_{m+n+p}+\beta_{m+n+p} \atop \alpha_1,\ldots,\alpha_{m+n+p}}\right], \quad |x| < 1,$$

$$= \sum_{k=1}^{n'} \sum_{j=1}^{q_k} \frac{z^{\alpha'_{2k}+\beta'_{2k}-j}}{(m_{kj}-1)!} \Big[\sum_{u=o}^{m_{kj}-1} \binom{m_{kj}-1}{u}$$

$$\times (-\log z)^{m_{kj}-1-u} \sum_{v=o}^{u-1} \binom{u-1}{v} B_{kjo}^{(u-1-v)} \sum_{v_1=o}^{v-1} \binom{v-1}{v_1} B_{kjo}^{(v-1-v_1)} \cdots \Big] W_{kjo}$$

$$+ \sum_{r=1}^{k} \sum_{j=1}^{p_r} \sum_{v=1}^{r} \sum_{i=o}^{n_{jrr-v}-1} \frac{z^{\alpha_{jrr-v+1}+i}}{(v-1)!} \Big[\sum_{h=o}^{v-1} \binom{v-1}{h}$$

$$\times (-\log z)^{v-1-h} \sum_{v_1=o}^{h-1} \binom{h-1}{v_1} A_{jro}^{(h-1-v_1)} \sum_{v_2=o}^{v_1-1} \binom{v_1-1}{v_2} A_{jro}^{(v_1-1-v_2)} \cdots \Big] Z_{jro}$$

where B_{kjo}, $B^{(t)}_{kjo}$, $t \geq 1$, W_{kjo}, A_{jro}, $A^{(t)}_{jro}$, $t \geq 1$ and Z_{jro} are given below.

$$B_{kjo} = \sum_{i=1}^{m} \psi(\alpha_i - \alpha'_{2k} - \beta'_{2k} + j) - \sum_{i=1}^{m} \psi(\alpha'_{1i} + \beta'_{1i} - \alpha'_{2k} - \beta'_{2k} + j)$$

$$- \sum_{\substack{r=1 \\ (r,i) \neq (k,j)}}^{n'} \sum_{i=1}^{q_k} \left[\frac{m_{ri}}{\alpha'_{2k} + \beta'_{2k} - i - \alpha'_{2k} - \beta'_{2k} + j} \right]$$

$$+ \sum_{i=1}^{p} \psi(\alpha_{m+n+i} - \alpha'_{2k} - \beta'_{2k} + j) - \sum_{i=1}^{p} \psi(\alpha'_{3i} + \beta'_{3k} - \alpha'_{2k} - \beta'_{2k} + j);$$

$$B^{(t)}_{kjo},\ t \_ 1, = (-1)^{t-1}\, t\,!\ \left\{ \sum_{i=1}^{m} \zeta(t+1, \alpha_i - \alpha'_{2k} - \beta'_{2k} + j) \right.$$

$$- \sum_{i=1}^{m} \zeta(t+1, \alpha'_{1i} + \beta'_{1i} - \alpha'_{2k} - \beta'_{2k} + j)$$

$$+ \sum_{\substack{r=1 \\ (r,i) \neq (k,j)}}^{n'} \sum_{i=1}^{q_r} \left[\frac{m_{ri}}{(\alpha'_{2r} + \beta'_{2r} - \alpha'_{2k} - \beta'_{2k} - i + j)^{t+1}} \right]$$

$$\left. + \sum_{i=1}^{p} \zeta(t+1, \alpha_{m+n+i} - \alpha'_{2k} - \beta'_{2k} + j) - \sum_{i=1}^{p} \zeta(\alpha'_{3i} + \beta'_{3i} - \alpha'_{2k} - \beta'_{2k} + j) \right\}.$$

$$W_{jko} = \prod_{i=1}^{m} \frac{\Gamma(\alpha_i - \alpha'_{2k} - \beta'_{2k} + j)}{\Gamma(\alpha'_{1i} + \beta'_{1i} - \alpha'_{2k} - \beta'_{2k} + j)}$$

$$\times \prod_{\substack{r=1 \\ (r,i) \neq (k,j)}}^{n'} \prod_{i=1}^{q_r} (\alpha'_{2r} + \beta'_{2r} - i - \alpha'_{2k} - \alpha'_{2k} + j)^{-m_{ri}}$$

$$\times \prod_{i=1}^{p} \frac{\Gamma(\alpha_{m+n+i} - \alpha'_{2k} - \beta'_{2k} + j)}{\Gamma(\alpha'_{3i} + \beta'_{3i} - \alpha'_{2k} - \beta'_{2k} + j)} \quad ;$$

$$Z_{jro} = \{ [(n_{jr1} + n_{jr2} + \ldots + n_{jrr-v} - i-1)!$$

$$\times (n_{jr2} + n_{jr3} + \ldots + n_{jrr-v} - i-1)!$$

$$\ldots (n_{jrr-v} - i-1)! \, (-1)^{vi+(v-1)n_{jrr-v+1} + \ldots + n_{jrr-1}}]/[$$

$$[(n_{jrr-v+1} + n_{jrr-v+2} + \ldots + n_{jrr-1} + i)! \, (n_{jrr-v+1} + \ldots + n_{jrr-2} + i)!$$

$$\ldots (n_{jrr-1} + i)! \, i!]\} \{ \prod_{\substack{n=1 \\ (k,n) \neq (j,r)}}^{k} \prod_{m=1}^{p_n} \prod_{u=1}^{n} \Gamma(\alpha_{mnu} - \alpha_{jrr-v+1} - i) \}$$

$$\times \{ \prod_{k=1}^{n'} \prod_{\sigma=1}^{q_k} (\alpha'_{2k} + \beta'_{2k} - \alpha_{jrr-v+1} - i-\sigma)^{-m_{k\sigma}} ;$$

$$A_{jro} = \sum_{\sigma=1}^{p} \psi(\alpha_{m+n+\sigma} - \alpha_{jrr-v+1} - i) - \sum_{\sigma=1}^{p} \psi(\alpha'_{3\sigma} + \beta'_{3\sigma} -$$

$$- \alpha_{jrr-v+1} - i) - \sum_{k=1}^{n'} \sum_{\sigma=1}^{q_k} [\frac{m_{k\sigma}}{(\alpha'_{2k} + \beta'_{2k} - \sigma - \alpha_{jrr-v+1} - i)}$$

$$+ \sum_{\substack{n=1 \\ (m,n) \neq (j,r)}}^{k} \sum_{m=1}^{p_n} \sum_{u=1}^{n} \psi(\alpha_{mnu} - \alpha_{jrr-v+1} - i) + v\, \psi(1)$$

$$+ \psi(n_{jr1} + n_{jr2} + \ldots + n_{jrr-v} - i) + \psi(n_{jr2} + \ldots + n_{jrr-v} - i)$$

$$+ \ldots + \psi(n_{jrr-v} - i) + v(\frac{1}{1} + \frac{1}{2} + \ldots + \frac{1}{i}) + (v-1)[\frac{1}{i+1} +$$

$$\ldots + \frac{1}{(i+n_{jrr-v+1})} + \ldots + \frac{1}{(i+n_{jrr-v+1} + \ldots + n_{rr-1})}] ;$$

$$A^{(t)}_{jro}\ ,\ t \geq 1,\ = (-1)^{t-1}\ t!\ \{\ \sum_{\sigma=1}^{p} \zeta(t+1,\ \alpha_{m+n+\sigma} - \alpha_{jrr-v+1} - i)$$

$$- \sum_{\sigma=1}^{p} \zeta(t+1,\ \alpha'_{3\sigma} + \beta'_{3\sigma} - \alpha_{jrr-v+1} - i)$$

$$- \sum_{k=1}^{n'} \sum_{\sigma=1}^{q_k} \left[\frac{m_{k\sigma}}{(\alpha'_{2k} + \beta'_{2k} - \sigma - \alpha_{jrr-v+1} - i)^{t+1}} \right]$$

$$+ \sum_{\substack{n=1 \\ (m,n) \neq (j,r)}}^{k} \sum_{m=1}^{p_n} \sum_{u=1}^{n} \zeta(t+1,\ \alpha_{mnu} - \alpha_{jrr-v+1} - i) + v\ \zeta(t+1,1)$$

$$+ \zeta(t+1,\ n_{jr1} + n_{jr2} + \ldots + n_{jrr-v} - i) + \zeta(t+1, n_{jr2} + n_{jr3} +$$

$$\ldots + n_{jrr-v} - i) + \ldots + \zeta(t+1, n_{jrr-v} - i)\ \}$$

$$+ t!\{v\ \left[\frac{1}{1^{t+1}} + \frac{1}{2^{t+1}} + \ldots + \frac{1}{i^{t+1}} \right] + (v-1)\ \left[\frac{1}{(i+1)^{t+1}} \right.$$

$$\left. + \ldots + \frac{1}{(i+n_{jrr-v+1})^{t+1}} \right] + \ldots + \frac{1}{(i+n_{jrr-v+1} + \ldots + n_{jrr-1})^{t+1}}\ \};$$

Some particular cases are given in the exercises and the applications of several particular cases can be found in Chapter VI.

5.7. COMPUTABLE REPRESENTATION OF A $G^{m,n}_{p,q}(\cdot)$ IN THE LOGARITHMIC CASE.

In this section we consider a general G-function of the type $G^{m,n}_{p,q}(\cdot)$ and consider the most general way in which the parameters can differ by integers. Then we obtain a series representation of such a G-function in this logarithmic case. Since the existence conditions are already discussed in Chapter I they won't be discussed here. The G-function in (5.7.1) will be represented in a computable form by evaluating it as a sum of the residues at the poles of the Gamma product $\prod_{j=1}^{m} \Gamma(b_j + s)$. More cases can be dealt with in a similar manner with the help of (1.2.3). Now consider a general G-function,

$$G(x) = \frac{1}{2\pi i} \int_L h(s)\, x^{-s}\, ds,$$

where, $i = \sqrt{-1}$ and

$$h(s) = \frac{\prod_{j=1}^{m} \Gamma(b_j+s) \prod_{j=1}^{n} \Gamma(1-a_j-s)}{\prod_{j=m+1}^{q} \Gamma(1-b_j-s) \prod_{j=n+1}^{p} \Gamma(a_j+s)} \tag{5.7.1}$$

The contour L separates the poles of $\prod_{j=1}^{m} \Gamma(b_j+s)$ and $\prod_{j=1}^{n} \Gamma(1-a_j-s)$. But some of the Gammas of the denominator of (5.7.1) may cancel with Gammas in the numerator. For simplicity we may assume that,

$$1-b_k + b_j \neq -\nu, \quad \nu = 0,1,\ldots;\ j = m+1,\ldots,q;\ k = 1,\ldots,m, \tag{5.7.2}$$

so that none of the poles of $\prod_{j=m+1}^{q} \Gamma(1-b_j-s)$ and $\prod_{j=1}^{m} \Gamma(b_j+s)$ coincide. The most general form of the classification of $h(s)$ can be given as follows.

$$h(s) = \left\{\frac{\prod_{j=1}^{n} \Gamma(1-a_j-s)}{\prod_{j=m+1}^{q} \Gamma(1-b_j-s)}\right\}\left\{\frac{\prod_{r=1}^{t} \prod_{j=1}^{u_r} \Gamma(b_{jr1}+s)\ldots\Gamma(b_{jrr}+s)}{\prod_{j=n+1}^{n+n_1} \Gamma(a_j+s)}\right\}$$

$$\times \prod_{k=1}^{n'} \prod_{j=1}^{q_k} (a'_k + s - j)^{-n_{kj}} \qquad \frac{\prod_{j=m_1+m_2+1}^{m_1+m_2+m_3} \Gamma(b_j+s)}{\prod_{j=n_1+n_2+1+n}^{n+n_1+n_2+n_3} \Gamma(b_j+s)} , \qquad (5.7.3)$$

where $m = m_1+m_2+m_3$, $n_1+n_2+n_3 = p-n$, the last factor of (5.7.3) has only zeros, n', r_k, n_{kj} are all non-negative integers, a'_k are functions of a_j, $j = n+n_1+1,\ldots,n+n_1+n_2$ and b_j, $j = m_1+1,\ldots,m_1+m_2$, $t \leq m_1$, $b_{jr1} - b_{jr2} = m_{jr1}$, $b_{jr2} - b_{jr3} = m_{jr2},\ldots,b_{jrr-1} - b_{jrr} = m_{jrr-1}$ where $m_{jr1}, m_{jr2},\ldots,m_{jrr-1}$ are non-negative integers and for fixed r, $b_{jri} - b_{mrh} \neq \pm v$, $v = 0,1,\ldots; j \neq m$; $i,h = 1,2,\ldots,r$. The G-function of (5.7.1) is available as the sum of the residues at the poles of $\prod_{j=1}^{m} \Gamma(b_j+s)$, Braaksma [51], Theorem 1). Hence we state the following theorem.

<u>Theorem 5.7</u>.1. When the Gamma products admit a classification as in (5.7.3) a general G-function, under the conditions of its existence, admits a series expansion,

$$G^{m,n}_{p,q}\left[z\middle|{a_1,\ldots,a_p \atop b_1,\ldots,b_q}\right] = \sum_{k=1}^{n'} \sum_{j=1}^{q_k} A_{kj} + \sum_{r=1}^{t} \sum_{j=1}^{u_r} \sum_{v=1}^{r} \sum_{i=o}^{m_{jrr-v}-1} R_{vi} , \qquad (5.7.4)$$

with the convention that $m_{jro} = \infty$ for all $j = 1,2,\ldots,u_r$; $r = 1,2,\ldots,t$; $(v-1)!\ \psi(v)$, $\zeta(t+1,v)$ are tacitly omitted whenever v is $0,\infty$ or a negative integer, where,

$$A_{kj} = \frac{z^{a'_k - j}}{(n_{kj}-1)!} \sum_{u=o}^{n_{kj}-1} \binom{n_{kj}-1}{u} (-\log z)^{n_{kj}-1-u}$$

$$\times \left\{ \sum_{v=o}^{u-1} \binom{u-1}{v} M_{kj}^{(u-1-v)} \sum_{v_1=o}^{v-1} \binom{v-1}{v_1} M_{kj}^{(v-1-v_1)} \right\} N_{kj} \ ; \qquad (5.7.5)$$

$$N_{kj} = \left\{ \frac{\prod_{i=1}^{n} \Gamma(1-a_i+a_k'-j)}{\prod_{i=m+1}^{q} \Gamma(1-b_i+a_k'-j)} \quad \frac{\prod_{r=1}^{t} \prod_{i=1}^{u_r} \prod_{v=1}^{r} \Gamma(b_{irv}-a_k'+j)}{\prod_{i=n+1}^{n+n_1} \Gamma(a_i-a_k'+j)} \right\}$$

$$* \left\{ \prod_{\substack{r=1 \\ (r,i) \neq (k,j)}}^{n'} \prod_{i=1}^{q_r} \Gamma(a_r' - i - a_k' + j) \right\} \left\{ \frac{\prod_{j=m_1+m_2+1}^{m_1+m_2+m_3} \Gamma(b_i-a_k'+j)}{\prod_{i=n+n_1+n_2+1}^{n+n_1+n_2+n_3} \Gamma(a_i-a_k'+j)} \right\}; \qquad (5.7.6)$$

$$M_{kj}^{(\alpha)} = \sum_{i=1}^{n} \psi(1-a_i+a_k'-j) - \sum_{i=m+1}^{q} \psi(1-b_i+a_k'-j)$$

$$+ \sum_{r=1}^{t} \sum_{i=1}^{u_r} \sum_{v=1}^{r} \psi(b_{irv}-a_k'+j) - \sum_{i=n+1}^{n+n_1} \psi(a_i-a_k'+j)$$

$$- \sum_{\substack{r=1 \\ (r,i) \neq (k,j)}}^{n'} \sum_{i=1}^{q_r} \left[\frac{n_{ri}}{(a_r'-i-a_k'+j)} \right] + \sum_{i=m_1+m_2+1}^{m_1+m_2+m_3} \psi(b_i-a_k'+j)$$

$$- \sum_{i=n+n_1+n_2+1}^{n+n_1+n_2+n_3} \psi(a_i-a_k'+j), \quad \text{for} \quad \alpha = 0; \qquad (5.7.7)$$

$M_{kj}^{(\alpha)} = (-1)^{\alpha-1}\, \alpha\,!$ {the same expression as in (5.7.7) with $\psi(\cdot)$ replaced by $\zeta(\alpha+1, \cdot)$ and $-(a_r'-i-a_k'+j)$ replaced by $(a_r'-i-a_k'+j)^{\alpha+1}$ } for $\alpha \geq 1$,

$$R_{vi} = \frac{z^{b_{jrr-v+1}+i}}{(v-1)!} \left\{ \sum_{h=0}^{v-1} \binom{v-1}{h} (-\log z)^{v-1-h} \left[\sum_{h_1=0}^{h-1} \binom{h-1}{h_1} \right.\right.$$

$$\times B_{vi}^{(h-1-h_1)} \sum_{h_2=o}^{h_1-1} \binom{h_1-1}{h_2} B_{vi}^{(h_1-1-h_2)}] C_{vi} ; \qquad (5.7.8)$$

$$C_{vi} = \{ [(m_{jr1}+m_{jr2}+\ldots+m_{jrr-v}-i-1)!\,(m_{jr2}+m_{jr3}+\ldots$$

$$+m_{jrr-v}-i-1)!\ldots(m_{jrr-v}-i-1)!\,(-1)^{vi+(v-1)m_{jrr-v+1}+\ldots+m_{jrr-1}}]$$

$$\div [(m_{jrr-v+1}+m_{jrr-v+2}+\ldots+m_{jrr-1}+i)!$$

$$\times (m_{jrr-v+1}+\ldots+m_{jrr-2}+i)!\ldots(m_{jrr-1}+i)!\,i!\,]\}$$

$$\times \left\{ \frac{\prod_{\substack{n=1\\(m,n)\neq(j,r)}}^{t} \prod_{m=1}^{u_r} \prod_{w=1}^{r} \Gamma(b_{mnw}-b_{jrr-v+1}-i)}{\prod_{k=n+1}^{n+n_1} \Gamma(a_k-b_{jrr-v+1}-i)} \right\} \left\{ \frac{\prod_{k=1}^{n} \Gamma(1-a_k+b_{jrr-v+1}+i)}{\prod_{k=m+1}^{q} \Gamma(1-b_k+b_{jrr-v+1}+i)} \right\}$$

$$\times \{ \prod_{k=1}^{n'} \prod_{m=1}^{q_k} (a'_k-m-b_{jrr-v+1}-i)^{-n_{kn}} \prod_{k=m_1+m_2+1}^{m_1+m_2+m_3} \Gamma(b_k-b_{jrr-v+1}-i) /$$

$$\sum_{k=n+n_1+n_2+1}^{n+n_1+n_2+n_3} \Gamma(a_k-b_{jrr-v+1}-i) \}; \qquad (5.7.9)$$

$$B_{vi}^{(t)} = \sum_{k=1}^{n} \psi(1-a_k+b_{jrr-v+1}+i) - \sum_{k=n+1}^{q} \psi(1-b_k+b_{jrr-v+1}+i)$$

$$- \sum_{k=1}^{n'} \sum_{m=1}^{q_k} \frac{n_{km}}{(a'_k - m - b_{jrr-v+1} - i)} + \sum_{k=m_1+m_2+1}^{m_1+m_2+m_3} \psi(b_k - b_{jrr-v+1} - i)$$

$$- \sum_{k=n+n_1+n_2+1}^{n+n_1+n_2+n_3} \psi(a_k - b_{jrr-v+1} - i)$$

$$+ \sum_{\substack{n=1 \\ (m,n) \neq (j,r)}}^{t} \sum_{m=1}^{u_n} \sum_{k=1}^{n} \psi(b_{mnk} - b_{jrr-v+1} - i) - \sum_{k=n+1}^{n+n_1} \psi(a_k - b_{jrr-v+1} - i) +$$

$$+ v\,\psi(1) + \psi(m_{jr1} + m_{jr2} + \ldots + m_{nrr-v} - i) + \psi(m_{jr2} + \ldots + m_{jrr-v} - i) +$$

$$\ldots + \psi(m_{jrr-v} - i) + v\,[\tfrac{1}{1} + \tfrac{1}{2} + \ldots + \tfrac{1}{i}]$$

$$+(v-1)\,[\frac{1}{(i+1)} + \ldots + \frac{1}{i+m_{jrr-v+1}}] + \ldots + \frac{1}{(i+m_{jrr-v+1} + \ldots + m_{nrr-1})}) \tag{5.7.10}$$

for $t = 0$;

$B_{vi}^{(t)} = (-1)^{t+1}\, t!\ \{(5.7.10)$ with $\psi(\cdot)$ replaced by $\zeta(t+1,.)$

and the denominators of the terms not containing ψ are raised to the power $t+1$ and multiplied by $(-1)^{t+1}$, for $t \geq 1$ }

5.8. COMPUTABLE REPRESENTATION OF AN H-FUNCTION IN THE LOGARITHMIC CASE.

The H-function is the most generalized Special Function and a G-function is a special case of it. In this section we will discuss this function briefly and indicate a computable representation.

This function was introduced by Pincherle in 1888. Its theory is developed by Barnes [35] and Mellin [211]. Fox [94] studied it as a symmetrical Fourier kernel and a detailed study of its asymptotic behaviour is done by Braaksma [51].

Several of its applications are studied by Mathai and Saxena. Some of them will be discussed in the next chapter.

Definition:

$$H(z) = H^{m,n}_{p,q}\left[z\left|\begin{matrix}(a_1,\alpha_1),\ldots,(a_p,\alpha_p)\\(b_1,\beta_1),\ldots,(b_q,\beta_q)\end{matrix}\right.\right] = \frac{1}{2\pi i}\int_L \frac{\prod_{j=1}^{m}\Gamma(b_j+\beta_j s)\prod_{j=1}^{n}\Gamma(1-a_j-\alpha_j s)}{\prod_{j=m+1}^{q}\Gamma(1-b_j-\beta_j s)\prod_{j=n+1}^{p}\Gamma(a_j+\alpha_j s)} z^{-s}\,ds \tag{5.8.1}$$

where $i = \sqrt{-1}$, p,q, m,n are integers such that $1 \leq m \leq q$, $0 \leq n \leq p$, α_j $(j = 1,\ldots,p)$, $\beta_j (j = 1,\ldots,q)$ are positive numbers and $a_j (j = 1,\ldots,p)$, b_j $(j = 1,\ldots,q)$ are complex numbers such that,

$$\alpha_j\ (b_h + \nu\ ) \neq \beta_h\ (a_j-1-\lambda),\ \text{for}\ \ \nu,\lambda = 0,1,\ldots;\ \ h = 1,\ldots,m;\ j = 1,\ldots,n. \tag{5.8.2}$$

L is a contour separating the points,

$$-s = (b_j + \nu)/\beta_j,\quad j = 1,\ldots,m;\ \nu = 0,1,\ldots$$

and

$$-s = (a_j-1-\nu\ )/\alpha_j,\ j = 1,\ldots,n;\ \nu = 0,1,\ldots \tag{5.8.3}$$

One condition of existence of the H-function is that it exists, for every $z \neq 0$ if if $\mu > 0$ where,

$$\mu = \sum_1^q \beta_j - \sum_1^p \alpha_j \tag{5.8.4}$$

and for $0 < |z| < \beta^{-1}$, if $\mu = 0$ where

$$\beta = \prod_1^p \alpha_j^{\alpha_j} \prod_1^q \beta_j^{-\beta_j}. \tag{5.8.5}$$

A detailed discussion of (5.8.1) is available in Braaksma [51] . A computable representation of (5.8.1) will be obtained by evaluating (5.8.1) as the sum of residues at the poles of $\prod_{j=1}^{m}\Gamma(b_j+\beta_j s)$ after identifying the poles. The residue theorem is applicable according to Braaksma ([51], Theorem 1).

Before discussing the computable representation it should be remarked that when $\alpha_1 = \ldots = \alpha_p = 1 = \beta_1 = \ldots = \beta_q$ (5.8.1) reduces to a G-function. Also when α's and β's are rational numbers it can be easily seen that (5.8.1) can be represented as a G-function with the help of the multiplication formula for a Gamma function given in (1.2.6) and a transformation of the variable z to z^{γ} when γ is an appropriate rational number. In all practical problems, that are seen so far, α's and β's are rational numbers and hence the H-functions in these cases are nothing but G-functions.

In order to show the importance of the H-function in Statistics we will consider a simple problem associated with a generalized Gamma density. A real stochastic variable X is said to have a generalized Gamma distribution if its density function is of the form,

$$f_2(x) = \frac{\beta\, a^{\alpha/\beta}}{\Gamma(\alpha/\beta)}\, x^{\alpha-1}\, e^{-ax^{\beta}}, \quad x > 0,\ a > 0,\ \alpha > 0,\ \beta > 0. \tag{5.8.6}$$

A large number of statistical problems in Life Testing Models, Reliability Analysis and related fields are associated with (5.8.6). These may be seen from the Vast literature on these topics cited in the bibliographies, Mendenhall [212] and Govindarajulu [107]. In these fields as well as in other fields the distribution of a product of independent generalized Gamma variates is important. Let $X_1,\ldots,X_k$ be k independent Generalized Gamma Variables distributed according to (5.8.6) with parameters (a_j,α_j,β_j), $j = 1,2,\ldots,k$. Let,

$$Y = (X_1 \ldots X_m)/(X_{m+1} \ldots X_k)\,. \tag{5.8.7}$$

If the density of Y, denoted by $f_3(y)$, exists then it is given by,

$$f_3(y) = y^{-1}\, \frac{1}{2\pi i} \int_{c-i\infty}^{c+i\infty} E(Y^h)\, y^{-h}\, dh\,, \tag{5.8.8}$$

where $E(Y^h)$ is the h-th moment of Y, namely,

$$E(Y^h) = C \prod_{j=1}^{m} \frac{\Gamma(\frac{\alpha_j}{\beta_j} + \frac{h}{\beta_j})}{a_j^{h/\beta_j}} \prod_{j=m+1}^{k} \frac{\Gamma(\frac{\alpha_j}{\beta_j} - \frac{h}{\beta_j})}{a_j^{-h/\beta_j}}, \tag{5.8.9}$$

where,

$$C^{-1} = \prod_{j=1}^{k} \Gamma(\frac{\alpha_j}{\beta_j}) . \tag{5.8.10}$$

Evidently,

$$f_3(y) = C \quad y^{-1} H^{m,n}_{n,m} \left[(\tfrac{a}{a'})y \,\middle|\, \begin{matrix} (1-\frac{\alpha_{m+1}}{\beta_{m+1}}, \frac{1}{\beta_{m+1}}), \ldots, (1-\frac{\alpha_k}{\beta_k}, \frac{1}{\beta_k}) \\ (\frac{\alpha_1}{\beta_1}, \frac{1}{\beta_1}), \ldots, (\frac{\alpha_m}{\beta_m}, \frac{1}{\beta_m}) \end{matrix} \right],$$

where

$$n = k-m, \quad a = \prod_{j=1}^{m} a_j^{\frac{1}{\beta_j}}, \quad a' = \prod_{j=m+1}^{k} a_j^{\frac{1}{\beta_j}} . \tag{5.8.11}$$

In order to have practical utility of $f_3(y)$ one needs its representation in a computable form.

In this section we indicate a series representation of a general H-function in the logarithmic case. When the poles of the integrand in (5.8.1) are simple a series expansion is given in Braaksma [51] . Now consider $\Gamma(b_j+\beta_j s)$. The poles are given by the equation,

$$-s = (b_j + \nu)/\beta_j, \quad \nu = 0,1,\ldots \tag{5.8.12}$$

Similarly the poles of $\Gamma(b_h + \beta_h s)$, $h \neq j$, are obtained from

$$-s = (b_h + \lambda)/\beta_h, \quad \lambda = 0,1,\ldots . \tag{5.8.13}$$

There may exist a pair of values (ν_1,λ_1) such that

$$(b_j + \nu_1)/\beta_j = (b_h + \lambda_1)/\beta_h \tag{5.8.14}$$

Then evidently, the point $-s = (b_j + \nu_1)/\beta_j$ is a pole of order two if this point does not coincide with a pole of any other Gamma of (5.8.1). Now (5.8.1) can be evaluated as the sum of the residues at the poles of $\prod_{j=1}^{m} \Gamma(b_j + \beta_j s)$, after identifying the poles . Since the technique is the same as the ones discussed in the previous sections and since the final expression is lengthy we will give only the method of identifying the poles. The details can be found in Mathai [176].

In order to distinguish poles of all orders we will consider the following equations. For a fixed j consider the equations,

$$\frac{b_1 + v^{(j_1)}_{j_1\ldots j_m}}{\beta_1} = \frac{b_2 + v^{(j_2)}_{j_1\ldots j_m}}{\beta_2} = \ldots = \frac{b_m + v^{(j_m)}_{j_1\ldots j_m}}{\beta_m} . \tag{5.8.15}$$

The following convention is used in (5.8.15). For a fixed j, $j_r = 0$ or 1 for $r = 1,2,\ldots, m$. If $j_r = 0$ then $(b_r + v^{(j_r)}_{j_1\ldots j_m})/\beta_r$ is to be excluded from the equations in (5.8.15). $v^{(jj)}_{j_1\ldots j_m}$ is a value of v in (5.8.12).

Evidently the possible values are $0,1,2,\ldots$. $v^{(jr)}_{j_1\ldots j_m}$ denotes the number corresponding to $v^{(jj)}_{j_1\ldots j_m}$ when the equation,

$$\frac{b_j + v^{(jj)}_{j_1\ldots j_m}}{\beta_j} = \frac{b_r + v^{(jr)}_{j_1\ldots j_m}}{\beta_r} \tag{5.8.16}$$

is satisfied by some values of $v^{(jj)}_{j_1\ldots j_m}$ and $v^{(jr)}_{j_1\ldots j_m}$. Thus $v^{(jr)}_{j_1\ldots j_m}$ may or may not exist. Under these notations,

$$0 \leq j_1 + j_2 + \ldots + j_m \leq m, \tag{5.8.17}$$

for every fixed j and $j_1 + \ldots + j_m$ denotes the order of the pole at

$-s = (b_j + v^{(jj)}_{j_1\ldots j_m})/\beta_j$. If $j_1 + \ldots + j_m = r$ then in (5.8.15) there will be the elements of $(b_k + v^{(jk)}_{j_1\ldots j_m})/\beta_k$, $k = 1,2,\ldots,m$ are called the elements in (5.8.15). In the above notation a pole of order r will be considered r times. In order to avoid duplication we will always assume that,

$$j_1 = j_2 = \ldots j_{j-1} = 0, \tag{5.8.18}$$

while considering the points corresponding the (5.8.12). If $j_r = 0$ for $r = 1,\ldots,m$, $r \neq j$ then $-s = (b_j + v^{(jj)}_{j_1\ldots j_m})/\beta_j$ gives simple poles. If $j_1 + \ldots + j_m = 0$ then the corresponding points is not a pole to be considered.

Let

$$S^{(jj)}_{j_1\ldots j_m} = \{ v^{(jj)}_{j_1\ldots j_m} \} \tag{5.8.19}$$

That is, $S^{(jj)}_{j_1\ldots j_m}$ is the set of all values $v^{(jj)}_{j_1\ldots j_m}$ takes for given $j_1,\ldots,j_m$. With this notation the H-function in (5.8.1) can be written as follows.

$$H(z) = \sum_{j=1}^{m} \Sigma R_j , \tag{5.8.20}$$

where the second sum is over $S^{(jj)}_{j_1\ldots j_m}$.

Some particular cases are given in the exercises. Some of these connect H-function, G-function and some elementary Special Functions.

Remark 1: It is possible to obtain multiple integral representations for G and H-functions. These as well as some statistical techniques of obtaining new results in the theory of Special Functions can be found in Mathai and Saxena ([200],[201]). Some series expansions in the logarithmic case for a G-function of two variables are studied in Mathai and Saxena [202].

Remark 2: There are several applications of hypergeometric functions with matrix arguments. These functions are mainly applied in statistical problems. A G-function with a matrix argument is defined and its properties are studied in Mathai and Saxena [199]. Some applications of these functions will be mentioned in the next chapter.

EXERCISES

Prove the following results. Most of these deal with logarithmic cases.

5.1. $$H^{1,0}_{0,1}\left[\tfrac{x}{4}\,\middle|\,(2\alpha-1,2)\right] = \pi\, x^{\alpha-\frac{1}{2}}\, e^{-2x^{\frac{1}{2}}}.$$

5.2 (1) $$2\int_{o}^{\infty} x^{2r+1}\exp(-x^{2}-ux^{-1})dx = C_r(u)$$

where

$$C_r(u) = \Gamma(r+1)\left[\sum_{i=o}^{r}\frac{(r+i)!\,u^{2i}}{r!\,(2i)!}\right.$$

$$+\sum_{i=r+1}^{\infty}\frac{(-1)^{i-r-1}\,u^{2i}}{(i-r-1)!\,(2i)!}\left[2\,\psi(2i+1)+\psi(i-r)-2\log u\right]$$

$$\left.-\,\Gamma(r+\tfrac{1}{2})\left[u+\sum_{i=1}^{\infty}\frac{(-1)^{i}\,u^{2i+1}}{(2i+1)!\,\prod_{t=1}^{i}(t-r-\frac{1}{2})}\right]\right];$$ (Bagai, 1965 [19]).

(2) $$H^{2,0}_{0,2}\left[\tfrac{x}{4}\,\middle|\,(2\alpha-1,2),\ (\alpha-1,1)\right] = 2^{2(\alpha-1)}\, x^{\alpha-1}\, C_o(2x^{\frac{1}{2}}).$$

5.3. $$2^{2a-1}\, H^{2,0}_{1,2}\left[\tfrac{y}{4}\,\middle|\,\begin{matrix}(2a,2)\\(a,1),(a,1)\end{matrix}\right] = y^{a}(1-y)^{-\frac{1}{2}} = \pi^{\frac{1}{2}}\, G^{1,0}_{1,1}\left(y\,\middle|\,\begin{matrix}a+\frac{1}{2}\\ a\end{matrix}\right),\quad y<1.$$

5.4. $$H^{3,0}_{1,3}\left[\tfrac{y}{9}\,\middle|\,\begin{matrix}(3a,3)\\(a,1),(a-\frac{1}{2},1),(a-1,1)\end{matrix}\right] = 2^{3}\,\pi^{\frac{1}{2}}\,3^{-3a-\frac{1}{2}}\,y^{a-1}$$

$$\times(1-y)^{\frac{3}{2}}\ {}_2F_1(\tfrac{5}{6},\tfrac{7}{6};\tfrac{5}{2};1-y) = (2\pi)3^{-3a+\frac{1}{2}}\, G^{2,0}_{2,2}\left[y\,\middle|\,\begin{matrix}a+\frac{1}{3},\ a+\frac{3}{3}\\ a-\frac{1}{2},\ a-1\end{matrix}\right],\quad y<1.$$

5.5 $$H^{3,1}_{1,3}\left[\frac{y}{9}\,\middle|\,\begin{matrix}(3a,3)\\(a,1),(a,1),(a,1)\end{matrix}\right] = (2\pi)3^{-3a+\frac{1}{2}}\,y^{a}\;{}_2F_1\left(\tfrac{2}{3},\tfrac{1}{3};1;1-y\right)$$

$$= (2\pi)3^{-3a+\frac{1}{2}}\,G^{2,0}_{2,2}\left[y\,\middle|\,\begin{matrix}a+\frac{1}{3},a+\frac{2}{3}\\a,a\end{matrix}\right],\quad |y|<1.$$

Problems 5.1 to 5.5 are directly connected to statistical problems, see Mathai [172]. Problems 5.6-5.10 are also directly related to statistical problems, see Mathai and Saxena [200].

5.6 $$G^{2,0}_{2,2}\left(y\,\middle|\,\begin{matrix}\alpha_1+\beta_1-1,\ \alpha_2+\beta_2-1\\\alpha_1-1,\ \alpha_2-1\end{matrix}\right) = \frac{1}{\Gamma(\beta_1+\beta_2)}\,y^{\alpha_2-1}\,(1-y)^{\beta_1+\beta_2-1}$$

$$\times\;{}_2F_1\,(\alpha_2+\beta_2-\alpha_1,\ \beta_1;\ \beta_1+\beta_2;\ 1-y)\ .$$

5.7 $$\frac{1}{\Gamma(\beta_1+\beta_2)}\,y^{\alpha_1+m-1}\,(1-y)^{\beta_1+\beta_2-i}\;{}_2F_1\,(m+\beta_2,\ \beta_1;\ \beta_1+\beta_2;\ 1-y)$$

$$= y^{-1}\left\{\sum_{\nu=o}^{m-1}\frac{y^{\alpha_1+\nu}\,(-1)^{\nu}\,\Gamma(m-\nu)}{\nu!\,\Gamma(1-\nu+\beta_1)\,\Gamma(-\nu+m+\beta_2)}\right.$$

$$+\sum_{\nu=m}^{\infty}\frac{y^{\alpha_1+\nu}\,(-1)^{m}}{\nu!\,(\nu-m)!\,\Gamma(-\nu+\beta_1)\,\Gamma(-\nu+m+\beta_2)}\,[-\log y+\psi(\nu-m+1)$$

$$\left.+\ \psi(\nu+1)-\psi(\beta_1-\nu)-\psi(\beta_2+m-\nu)]\right\}\ ,\quad m=0,1,2,..$$

5.8. $$y^{-1}\,G^{2,0}_{2,2}\left[y\,\middle|\,\begin{matrix}a,a\\a-\frac{1}{2},\ a-1\end{matrix}\right] = y^{-\frac{1}{2}}\left[\,\Gamma(\tfrac{1}{2})\,y^{a-1}-\left(\frac{2}{\pi^{1/2}}\,y^{a-\frac{1}{2}}\right)\right.$$

$$\left.\times\,{}_2F_1\left(\tfrac{1}{2},\tfrac{1}{2};\ \tfrac{3}{2};\ y\right)\right] = \frac{1}{\Gamma(\frac{3}{2})}\,y^{a-2}\,(1-y)^{\frac{1}{2}}\;{}_2F_1\left(\tfrac{1}{2},\tfrac{1}{2};\tfrac{3}{2};\ 1-y\right),\quad |y|<1.$$

5.9 $\frac{1}{2}\, y^{a-2}\,(1-y)^4 \int_0^1 t^2(1-t)\,[1-t(1-y)]^{3/2}\;{}_2F_1(\tfrac{3}{2},1;3;t(1-y))\,dt$

$$= y^{-1}\, G^{3,0}_{3,3}\left[y \,\middle|\, \begin{matrix} a+1,\, a+\frac{3}{2},\, a+1 \\ a,\, a-\frac{1}{2},\, a-1 \end{matrix}\right] = 4y^{a-2}\,[-2y^{\frac{1}{2}} + \tfrac{2}{3}\, y^{\frac{3}{2}} + \tfrac{1}{3}\,(1-\log y)\,y],\quad |y| < 1.$$

5.10 $\Gamma(\tfrac{5}{2})\,\Gamma(\tfrac{8}{3})\,\Gamma(\tfrac{17}{6})\; G^{4,0}_{4,4}\left[y \,\middle|\, \begin{matrix} a,\, a+\frac{1}{3},\, a+\frac{2}{3},\, a \\ a-1,\, a-\frac{3}{2},\, a-2,\, a-\frac{5}{2} \end{matrix}\right]$

$$= y^{a-\frac{5}{2}}\,(1-y)^7 \int_0^1\!\!\int_0^1 t_2^{\frac{9}{2}}\, t_1^{\frac{11}{6}}\,(1-t_2)^{\frac{3}{2}}\,(1-t_1)^{\frac{5}{3}}$$

$$\times\,[1-t_2(1-y)]^{-2}\,[1-t_1t_2(1-y)]^{-\frac{13}{6}}\;{}_2F_1(\tfrac{4}{3};\, 1;\, \tfrac{17}{6};\, t_1t_2(1-y))\,dt_1dt_2,\quad |y|< 1.$$

5.11. $\int_0^1 (uv)^{-1}\, G^{a,0}_{a,a}\left[v \,\middle|\, \begin{matrix} \alpha_1+\beta_1,\ldots,\alpha_a+\beta_a \\ \alpha_1,\ldots,\alpha_a \end{matrix}\right] G^{b,0}_{0,b}\left[\tfrac{u}{v} \,\middle|\, \gamma_1,\ldots,\gamma_b\right] dv$

$$= G^{a+b,0}_{a,a+b}\left[u \,\middle|\, \begin{matrix} \alpha_1+\beta_1,\ldots,\alpha_a+\beta_a \\ \alpha_1,\ldots,\alpha_a,\, \gamma_1,\ldots,\gamma_b \end{matrix}\right],\quad |u| < 1 .$$

Problem 5.11 is obtained through a statistical technique. By specializing the parameters a number of results of this type can be obtained for the various elementary Special Functions.

CHAPTER VI

APPLICATION OF G-FUNCTION IN STATISTICS

During the past few years Meijer's G-function is extensively used in the theory of Statistical Distributions, Characterization of Distributions and in studying certain structural properties of statistical distributions. This chapter deals with the applications of G-function in the various fields of Statistics. Only the main results are discussed in the text. Some more results are given in the exercises at the end of this chapter and more can be found from the references. All the articles in the list of references do not make use of G-functions directly but the problems under consideration, in these articles, are such that atleast particular cases of G-function could be made use of. Some other applications in Physical Sciences, Engineering and related fields will be discussed in the next chapter.

6.1. EXACT DISTRIBUTIONS OF MULTIVARIATE TEST CRITERIA

In order to apply a test of a statistical hypothesis to a practical problem one needs the exact distribution of the test criterion in a form suitable for computing the exact percentage points. Several statistical tests associated with a multivariate normal distribution, along with the exact distributions in particular cases, of the likelihood ratio test criteria, are available in Anderson [16]. A method of obtaining approximate distributions in these cases is given by Box [50].

There are different techniques available to tackle these distribution problems. But none of them proved to be powerful enough to give the exact distributions of these test criteria in general cases. The methods of Fourier and Laplace transforms are useful in some of these problems. The method of Mellin transform is successfully used by Nair [215] to work out some exact distributions in particular cases. In a series of papers Consul ([68],[69],[70],[71],[72]) obtained the exact distributions in particular cases by using inverse Mellin transform technique and represented these in terms of Hypergeometric functions. These Hypergeometric functions are of the logarithmic type discussed in Section 5.4. The users of these results are often misled because it is not specifically mentioned in Consul's papers that he is getting the results in terms of Hypergeometric functions in the logarithmic cases.

In a number of recent papers on exact null and non-null distributions Pillai, Al-Ani and Jouris [231], Pillai and Jouris [234] and Pillai and Nagarsenker [235] expressed the exact distributions in G-functions. But unfortunately the problems under consideration in these papers are not solved because their problems are the cases where the poles of the integrand are not simple except in particular cases.

The exact null and non-null distributions of almost all the multivariate test criteria, in the general multinormal cases, are obtained in computable forms for the first time by Mathai ([173],[174],[179],[180],[181],[182]) and Mathai and Rathie ([186],[187],[189],[190],[191]) by using different techniques including the tecniques developed in Chapter V. The discussion in this section is mainly based on these articles. These problems were open problems since the 1930's and with the help of a number of techniques several authors have worked out particular cases. The expansions of Meijer's G-function given in Chapter V yielded the exact distributions of all these test criteria in computable forms and for the general cases. A detailed account of the different methods, applied to these problems so far, are available from Mathai [185].

Only the dens'ty functions are given in the following subsections.The distribution functions are available by term by term integration with the help of Theorem 6.1.1 and hence the discussion is omitted. The density functions are given in series forms, which are computable, but in particular cases they can be simplified in terms of elementary Special Functions, mostly in logarithmic cases. Some of these simplifications are given in the exercises.

All the density functions are represented in terms of series involving terms of the type $x^a(-\log x)^b$. Hence in order to obtain the distribution functions one needs the integral of the type given in Theorem 6.1.1.

Theorem 6.1.1. For $\alpha > 0$, k a positive integer, $0 < u < 1$.

$$\int_0^x u^{\alpha}(-\log u)^{k-1} du = x^{\alpha+1} \sum_{r=1}^{k} \frac{k(k-1)\ldots(k-r+1)}{k(\alpha+1)^r} (-\log x)^{k-r} . \tag{6.1.1}$$

This result follows from successive integration.

6.1.1. Testing linear Hypotheses on Regression Coefficients.

In this as well as in later sections we will only give the h-th moment of the likelihood ratio criterion of the problem under consideration and the exact density functions. The details of the tests and the methods of deriving these moments may be seen from any book on Multivariate Statistical Analysis or from Anderson [16] . In all these problems the (h-1)st moment is nothing but the Mellin transforms of the density functions. Hence the density functions are available from the inverse Mellin transforms. Since the density functions exist in all these cases, existence conditions are not stated separately.

In the problem of testing linear hypotheses on regression coefficients the h-th moment of the criterion U is as follows, [Anderson [16], pp. 192-194]:

$$E(U^h) = \prod_{j=1}^{p} \frac{\Gamma(\frac{n+1-j}{2} + h)\ \Gamma(\frac{n+q+1-j}{2})}{\Gamma(\frac{n+1-j}{2})\ \Gamma(\frac{n+q+1-j}{2} + h)} \tag{6.1.2}$$

where E denotes 'mathematical expectation' and p,q,n are all positive integers. From (6.1.2) it is easily seen that by taking the inverse Mellin transform the density function, denoted by f(u), is available as,

$$f(u) = \prod_{j=1}^{p} \frac{\Gamma(\frac{n+q+1-j}{2})}{\Gamma(\frac{n+q+j}{2})} u^{-1} G^{p,0}_{p,p}\left[u \middle| \begin{matrix} \frac{n+q+1-j}{2}, & j=1,\ldots,p \\ \frac{n+1-j}{2} & j=1,\ldots,p \end{matrix}\right],$$

$$0 < u < 1. \tag{6.1.3}$$

Evidently the G-function in (6.1.3) is in the logarithmic case but this can be evaluated by using the technique of Chapter V after identifying the poles. In order to identify the poles one has to consider four different cases, namely, Case I: p-even, q-even; Case II: p-odd; q-even; Case III: p-even; q-odd; Case IV: p-odd; q-odd. It is seen that in Cases I, II and III, $E(U^{s-1})$ can be written in the following form:

$$E(U^{s-1}) = C \prod_{j\epsilon a} (\alpha-j)^{-a_j} \prod_{j\epsilon b} (\alpha-\tfrac{1}{2}-j)^{-b_j}, \tag{6.1.4}$$

and for Case IV,

$$E(U^{s-1}) = C \frac{\Gamma(\alpha-\frac{1}{2})}{\Gamma(\alpha)} \prod_{j\epsilon a} (\alpha-j)^{-a_j} \prod_{j\epsilon b} (\alpha-\tfrac{1}{2}-j)^{-b_j}, \tag{6.1.5}$$

where

$$\alpha = s+\frac{n}{2}+\frac{q}{2}-1 \quad \text{and} \quad C = \prod_{j=1}^{p} \frac{\Gamma(\frac{n+q+1-j}{2})}{\Gamma(\frac{n+1-j}{2})}. \tag{6.1.6}$$

The quantities a,b, a_j and b_j are different for the different cases and these will be given below.

Case 1: p-even, q-even $(q \geq p)$

$$a_j = b_j = \begin{cases} j, j = 1,2,\ldots,\frac{p}{2}-1 . \\ \frac{p}{2}, j = \frac{p}{2}, \frac{p}{2}+1,\ldots,\frac{q}{2}, \\ \frac{p}{2}-i, j = \frac{q}{2}+i, i = 1,2,\ldots,\frac{p}{2}=1 ; \end{cases} \tag{6.1.7}$$

$$a = b = \{1,2,\ldots,\frac{p+q}{2}-1\}. \tag{6.1.8}$$

Case II: p-odd, q-even, $(q \geq p)$

$$a_j = \begin{cases} j, j = 1,2,\ldots,\frac{p-1}{2} . \\ \frac{p+1}{2}, j = \frac{p+1}{2}, \frac{p+1}{2}+1,\ldots,\frac{q}{2}, \\ \frac{p+1}{2}-i, j = \frac{q}{2}+i, i = 1,2,\ldots,\frac{p-1}{2} . \end{cases} \tag{6.1.9}$$

$$b_j = \begin{cases} j, j = 1,2,\ldots, \frac{p-3}{2}, \\ \frac{p-1}{2}, \; j = \frac{p-1}{2}, \frac{p-1}{2}+1,\ldots, \frac{q}{2}, \\ \frac{p-1}{2} - i, \; j = \frac{q}{2}+i, \; i = 1,2,\ldots, \frac{p-3}{2}; \end{cases} \tag{6.1.10}$$

$$a = \{1,2,\ldots, \frac{p+q-1}{2}\}, \quad b = \{1,2,\ldots, \frac{p+q-3}{2}\}. \tag{6.1.11}$$

Case III: p-even, q-odd ($q \geq p$)

$$a_j = \begin{cases} j, j = 1,2,\ldots, \frac{p}{2} - 1, \\ \frac{p}{2}, \; j = \frac{p}{2}, \frac{p}{2}+1,\ldots, \frac{q+1}{2}, \\ \frac{p}{2} - i, \; j = \frac{q+1}{2}+i, \; i = 1,2,\ldots, \frac{p}{2} - 1; \end{cases} \tag{6.1.12}$$

$$b_j = \begin{cases} j, j = 1,2,\ldots, \frac{p}{2} - 1 \\ \frac{p}{2}, \; j = \frac{p}{2}, \frac{p}{2}+1,\ldots, \frac{q-1}{2}, \\ \frac{p}{2} - i, j = \frac{q-1}{2}+i, \; i = 1,2,\ldots, \frac{p}{2} - 1; \end{cases} \tag{6.1.13}$$

$$a = \{1,2,\ldots, \frac{p+q-1}{2}\}, \quad b = \{1,2,\ldots, \frac{p+q-3}{2}\}. \tag{6.1.14}$$

Case IV: p-odd, q-odd ($q \geq p$)

$$a_j = \begin{cases} j-1, \; j = 2,3,\ldots, \frac{p-1}{2}, \\ \frac{p-1}{2}, \; j = \frac{p-1}{2}+1,\ldots, \frac{q+1}{2}, \\ \frac{p-1}{2} - i, \; j = \frac{q+1}{2}+i, \; i = 1,2,\ldots, \frac{p-3}{2}; \end{cases} \tag{6.1.15}$$

$$b_j = \begin{cases} j+1, \ j = 1,2,\ldots, \frac{p-1}{2}, \\ \frac{p-1}{2}+1, \ j = \frac{p-1}{2}+1,\ldots, \frac{q-1}{2}, \\ \frac{p-1}{2}, \quad j = \frac{q+1}{2}, \\ \frac{p-1}{2} - i, j = \frac{q+1}{2}+i, \ i = 1,2,\ldots, \frac{p-3}{2}; \end{cases} \tag{6.1.16}$$

$$a = \{2,3,\ldots, \frac{p+q-2}{2}\} \text{ and } b = \{1,2,\ldots, \frac{p+q-2}{2}\}. \tag{6.1.17}$$

For the cases I,II and III the poles are available by equating to zero the various factors of

$$\prod_{j\epsilon a} (\alpha - j)^{a_j} \prod_{j\epsilon b} (\alpha - \frac{1}{2} - j)^{b_j} \tag{6.1.18}$$

and for case IV the poles are available from,

$$\prod_{v=o}^{\infty} (\alpha - \frac{1}{2} + v) \prod_{j\epsilon a} (\alpha - j)^{a_j} \prod_{j\epsilon b} (\alpha - \frac{1}{2} - j)^{b_j}, \tag{6.1.19}$$

where the exponents denote the orders and the quantities a,b,a_j and b_j are available from (6.1.7) to (6.1.17). Now by using the results in Chapter V we can write down the density function as follows.

<u>Theorem 6.1.2.</u> For cases I, II and III, that is, when p and q are not both odd, the density function of U is given as,

$$f(u) = C \{ \sum_{j\epsilon a} \frac{u^{\frac{n}{2}+\frac{q}{2}-1-j}}{(a_j - 1)!} \sum_{v=o}^{a_j-1} \binom{a_j-1}{v} (-\log u)^{a_j-1-v} A_v V$$

$$+ \sum_{j\epsilon b} \frac{u^{\frac{n}{2}+\frac{q}{2}-\frac{3}{2}-j}}{(b_j - 1)!} \sum_{v=o}^{b_j-1} \binom{b_j-1}{v} (-\log u)^{b_j-1-v} B_v W\}$$

$0 < u < 1$, where (6.1.20)

$$V = \prod_{\substack{t\in a \\ t\neq j}} (j-t)^{-a_t} \prod_{t\in b} (j-\frac{1}{2}-t)^{-b_t}, \tag{6.1.21}$$

$$W = \prod_{t\in a} (j+\frac{1}{2}-t)^{-a_t} \prod_{\substack{t\in b \\ t\neq j}} (j-t)^{-b_t}, \tag{6.1.22}$$

$$A_v = \sum_{v_1=0}^{v-1} \binom{v-1}{v_1} A_0^{(v-1-v_1)} \sum_{v_2=0}^{v_1-1} \binom{v_1-1}{v_2} A_0^{(v_1-1-v_2)} \cdots \tag{6.1.23}$$

$$B_v = \sum_{v=0}^{v-1} \binom{v-1}{v_1} B_0^{(v-1-v_1)} \sum_{v_2=0}^{v_1-1} \binom{v_1-1}{v_2} B_0^{(v_1-1-v_2)} \cdots \tag{6.1.24}$$

$$A_0^{(r)} = (-1)^{r+1} r! \{ \sum_{\substack{t\in a \\ t\neq j}} [\frac{a_t}{(j-t)^{r+1}}]$$

$$+ \sum_{t\in b} [\frac{b_t}{(j-\frac{1}{2}-t)^{r+1}}] \}, \text{ for } r \geq 0 \text{ and} \tag{6.1.25}$$

$$B_0^{(r)} = (-1)^{r+1} r! \{ \sum_{t\in a} [\frac{a_t}{(\frac{1}{2}+j-t)^{r+1}}] + \sum_{\substack{t\in b \\ t\neq j}} [\frac{b_t}{(j-t)^{r+1}}] \}, \tag{6.1.26}$$

for $r \geq 0$ where C is given in (6.1.6).

<u>Theorem 6.1.3.</u> When p and q are both odd the density function of U is given by,

$$f(u) = C \{ \sum_{v=0}^{\infty} [\frac{(-1)^v u^{\frac{n}{2}+\frac{q}{2}-\frac{3}{2}+v}}{v!\,\Gamma(\frac{1}{2}-v)}] \prod_{t\in a} (\frac{1}{2}-v-t)^{-a_t}$$

$$\times \prod_{t\in b} (-v-t)^{-b_t}] + \sum_{j\in a} \frac{u^{\frac{n}{2}+\frac{q}{2}-1-j}}{(a_j-1)!} \sum_{v=o}^{a_j-1} \binom{a_j-1}{v}$$

$$\times (-\log u)^{a_j-1-v} A'_v V' + \sum_{t\in b} \frac{u^{\frac{n}{2}+\frac{q}{2}-\frac{3}{2}-j}}{(b_j-1)!} \sum_{v=o}^{b_j-1} \binom{b_j-1}{v}$$

$$\times (-\log u)^{b_j-1-v} B'_v W' \} , \quad 0 < u < 1, \tag{6.1.27}$$

where A'_v and B'_v have the same expressions in (6.1.23) and (6.1.24) with A_o and B_o replaced by A'_o and B'_o respectively and,

$$V' = \frac{\Gamma(j-\frac{1}{2})}{\Gamma(j)} V, \quad W' = \frac{\Gamma(j)}{\Gamma(\frac{1}{2}+j)} W , \tag{6.1.28}$$

$$A'_o = \psi(j-\tfrac{1}{2}) - \psi(j) + A_o \tag{6.1.29}$$

$$B'_o = \psi(j) - \psi(\tfrac{1}{2}+j) + B_o , \tag{6.1.30}$$

$$A'^{(r)}_o = (-1)^{r+1} r! \{\zeta(r+1, j-\tfrac{1}{2}) - \zeta(r+1,j) + A^{(r)}_o\} , \ r \geq 1, \tag{6.1.31}$$

$$B'^{(r)}_o = (-1)^{r+1} r!\{\zeta(r+1,j) - \zeta(r+1, \tfrac{1}{2}+j) + B^{(r)}\} , \ r \geq 1, \tag{6.1.32}$$

where $\psi(\cdot)$ and $\zeta(.,.)$ are defined in Chapter V. The cummulative distribution can be easily worked out by using Theorem 6.1.1 and hence the discussion is omitted.

Exact percentage points are computed by using Theorems 6.1.1, 6.1.2 and 6.1.3. These are available in Mathai [180] . Also the exact percentage points for a number of likelihood ratio test criteria are computed for the first time by using the expansions given in Chapter V. Only a few of them will be given here. A number of others are available from the references cited at the end of this chapter.

6.1.2. The problem of Testing Independence.

The h-th moment of a criterion for testing independence of sub-vectors in a multinormal case is available in Anderson ([16], p.235) as,

$$E(V^h) = C_1 \frac{\prod_{j=p_o+1}^{p} \Gamma(\frac{n+1-j}{2} + h)}{\prod_{i=1}^{q} \prod_{j=1}^{p_i} \Gamma(\frac{n+1-j}{2} + h)} \tag{6.1.33}$$

where

$$C_1 = \frac{\prod_{i=1}^{q} \prod_{j=1}^{p_i} \Gamma(\frac{n+1-j}{2})}{\prod_{j=p_o+1}^{p} \Gamma(\frac{n+1-j}{2})},$$

$p = p_o + p_1 + \ldots + p_q$, $n, p_o, \ldots, p_q$ are all non-negative integers. Evidently the density function of V is available from the inverse Mellin transform as,

$$f_1(v) = C_1 v^{-1} G^{p-p_o,0}_{p-p_o,p-p_o}\left[v \,\Big|\, {n/2,(n-1)/2,\ldots,(n-p_1+1)/2,\ldots,n/2,(n-1)/2,\ldots(n-p_q+1)/2 \atop (n-p_o)/2,(n-p_o-1)/2,\ldots,(n-p+1)/2} \right]$$

$$0 < v < 1 . \tag{6.1.34}$$

In order to use the results of Chapter V and to put (6.1.34) in computable forms one has to identify the poles of the integrand in (6.1.34). In this connection one has to consider three different cases. Case I: $p_o \geq p_1 \geq \cdots \geq p_r$ all even and p_{r+1} odd such that p_{r+1} in magnitude is in between p_t and p_{t+1} for some $t \leq r$ and $p_o, \ldots, p_r, p_{r+1}$ exhaust all the $p_o, p_1, \ldots, p_q$;, Case II: $p_o \geq p_1 \geq \cdots \geq p_r$ all even and $p_{r+1} \geq p_{r+2} \geq \cdots \geq p_q$ all odd with $q-r = 2m$, $m = 0,1,\ldots,$; Case III: $p_o \geq p_1 \geq \cdots \geq p_r$ all even and $p_{r+1} \geq \cdots \geq p_q$ all odd with $q-r=2m+1$, $m = 0,1,\ldots$. The ordering of the p's is done without loss of any generality.

The simplification of the gammas in (6.1.33) is a lengthy process and hence further discussion is omitted. The details of simplification and the exact density $f_1(v)$, in computable form, are available from Mathai and Rathie [190].

Remark: In the problems discussed in Sections 6.1.1 and 6.1.2 it may be observed that the gammas cancel out leaving linear factors in the denominator of the moment expressions in (6.1.2) and (6.1.33) for a number of cases. Hence the distributions can also be worked out by using a generalized partial fraction technique developed by Mathai and Rathie [189].

6.1.3. Testing the Hypothesis that the Covariance Matrix is Diagonal.

This test is described in Anderson ([16], p.262) and the h-th moment of a criterion W_1 for testing this hypothesis is given as,

$$E(W_1^h) = \frac{\Gamma^p(\frac{n}{2})}{\Gamma^p(\frac{n}{2}+h)} \prod_{j=1}^{p} \frac{\Gamma(\frac{n+1-j}{2}+h)}{\Gamma(\frac{n+1-j}{2})}, \quad 0 < \omega_1 < 1, \qquad (6.1.35)$$

where n and p are positive integers and $\Gamma^p(\cdot) = \{\Gamma(\cdot)\}^p$. Therefore the density function of W_1, denoted by $f_2(\omega_1)$, is as follows:

$$f_2(\omega_1) = C_2\, \omega_1^{-1}\, G^{p-1,0}_{p-1,p-1}\left[\omega_1 \middle| \begin{matrix} n/2, n/2, \ldots, n/2 \\ (n-1)/2,\ (n-2)/2, \ldots, (n+1-p)/2 \end{matrix}\right]$$

$$0 < \omega_1 < 1, \qquad (6.1.36)$$

where

$$C_2 = \frac{\Gamma^p(\frac{n}{2})}{\prod_{j=1}^{p} \Gamma(\frac{n+1-j}{2})}$$

Case I, p-even: When p is even the poles of the integrand in (6.1.36) are available from,

$$\prod_{j=1}^{\frac{p-2}{2}} (\alpha - \frac{p}{2} + j)^{j} \prod_{j=1}^{\infty} (\alpha - \frac{1}{2} - \frac{p}{2} + j)^{b_j} , \qquad (6.1.37)$$

where

$$b_j = \begin{cases} j, j = 1,2,\ldots, \frac{p}{2} - 1 , \\ \frac{p}{2}, j \geq \frac{p}{2} , \end{cases}$$

by equating to zero the various factors in (6.1.37). Hence when p is even, $f_2(\omega_1)$ reduces to the following form.

$$f_2(\omega_1) = C_2\omega_1^{-1} \Sigma_j \{\omega_1^{\frac{n}{2} - \frac{p}{2} + j} M_{a_j}(\omega_1) + \omega_1^{\frac{n}{2} - \frac{1}{2} - \frac{p}{2} + j} M_{b_j}(\omega_1)\} ,$$

$$0 < \omega_1 < 1. \qquad (6.1.38)$$

where for example $M_{a_j}(\omega_1)$ stands for the expression,

$$M_{a_j}(\omega_1) = \frac{1}{(a_j - 1)!} \sum_{j_1=o}^{a_j-1} \binom{a_j-1}{j_1} (-\log \omega_1)^{a_j-1-j_1} [\sum_{j_2=o}^{j_1-1} \binom{j_1-1}{j_2} A_{a_j}^{(j_1-1-j_2)}$$

$$\times \sum_{j_3=o}^{j_2-1} \binom{j_2-1}{j_3} A_{a_j}^{(j_2-1-j_3)} \cdots] B_{a_j} \qquad (6.1.39)$$

a_j for (6.1.38) is $a_j = j, j = 1,\ldots, \frac{p}{2} - 1$ and b_j for (6.1.38) is given in (6.1.37). The notation in (6.1.39) will be retained throughout the remaining subsections. In the various problems to be discussed below $a_j, b_j, A_{a_j}, A_{b_j}, B_{b_j}$ are different for different problems. Further $A_{a_j}^{(t)}$ is available from B_{a_j} by the following technique. Introduce a dummy variable y in each factor of B_{a_j}. Then evaluate the logarithmic derivative of B_{a_j} at y = 0. That is,

$$A_{a_j}^{(t)} = \frac{\partial^{t+1}}{\partial y^{t+1}} \log B_{a_j} \quad \text{at } y = 0 . \tag{6.1.40}$$

Similarly $A_{b_j}^{(t)}$ is available from B_{b_j}. Hence in the following sections only a_j, b_j, B_{a_j} and B_{b_j} will be listed. The method of obtaining $A_{a_j}^{(t)}$ from B_{a_j} for (6.1.38) will be illustrated here. In this case

$$B_{a_j} = \frac{\Gamma(\frac{p}{2} - j - \frac{1}{2})\, \Gamma(\frac{p}{2} - j - \frac{3}{2}) \ldots \Gamma(\frac{p}{2} - j - \frac{p-1}{2})}{\Gamma^{\frac{p}{2}}(\frac{p}{2} - j)\, \{(-j+1)(-j+2)^2 \ldots (-1)^{j-1}\, 1^{j+1} 2^{j+2} \ldots (\frac{p}{2} - j - 1)^{\frac{p}{2}-1}\}} . \tag{6.1.41}$$

Introduce a variable y in each factor and write $B_{a_j}(y)$ as

$$B_{a_j}(y) = \frac{\Gamma(\frac{p}{2} - j - \frac{1}{2} + y)\, \Gamma(\frac{p}{2} - j - \frac{3}{2} + y) \ldots \Gamma(\frac{p}{2} - j - \frac{p-1}{2} + y)}{\Gamma^{\frac{p}{2}}(\frac{p}{2} - j + y)\, \{(-j+1+y)(-j+2+y)^2 \ldots (-1+y)^{j-1} \ldots (\frac{p}{2} - j - 1 + y)^{\frac{p}{2}-1}\}} . \tag{6.1.42}$$

Now take the logarithmic derivative of (6.1.42) and evaluate at y = 0 then one gets,

$$A^{(r)},\ r \geq 1 = (-1)^{r+1}\, r!\, \Big\{ \sum_{k=o}^{\frac{p}{2}-1} \zeta(r+1, \tfrac{p}{2} - j - \tfrac{1}{2} - k)$$

$$- \frac{p}{2} \zeta(r+1, \tfrac{p}{2} - j) + \Big[\frac{1}{(-j+1)^{r+1}} + \frac{2}{(-j+2)^{r+1}} + \ldots + \frac{j-1}{(-1)^{r+1}}$$

$$+ \frac{(j+1)}{1^{r+1}} + \frac{(j+2)}{2^{r+1}} + \ldots + \frac{\frac{p}{2}-1}{(\frac{p}{2} - j - 1)^{r+1}} \Big]\Big\}; \tag{6.1.43}$$

$A^{(0)}$ is obtained from (6.1.43) by putting r = 0 and replacing $\zeta(r+1,.)$ by $-\psi(.)$, where ψ and zeta functions are defined in (5.3.2) and (5.3.12) respectively. Also in (6.1.38), B_{b_j} is as follows:

$$B_{b_j} = \frac{\Gamma(\frac{p}{2}-j)\,\Gamma(\frac{p}{2}-j-1)\ldots\Gamma(2)}{(-1)^{j-1}(-2)^{j-2}\ldots(-j+1)\,[\Gamma^{\frac{p}{2}}(\frac{1}{2}+\frac{p}{2}-j)]\,[(\frac{3}{2}-j)(\frac{5}{2}-j)^2\ldots(\frac{p}{2}-\frac{1}{2}-j)^{\frac{p}{2}-1}]}$$

for $j = 1,2.,\ldots,\frac{p}{2}-1$; (6.1.44)

$$= [(-1)^{\frac{p}{2}}(-2)^{\frac{p}{2}}\ldots(\frac{p}{2}-j)^{\frac{p}{2}}(\frac{p}{2}-j-1)^{\frac{p}{2}-1}\ldots(-j+1)][\Gamma^{\frac{p}{2}}(\frac{1}{2}+\frac{p}{2}-j)$$

$$[(\frac{3}{2}-j)(\frac{5}{2}-j)^2\ldots(\frac{p}{2}-\frac{1}{2}-j)^{\frac{p}{2}-1}]]^{-1} \quad \text{for } j \geq \frac{p}{2}\,. \qquad (6.1.45)$$

Case II, p-odd: When p is odd the density function is of the form

$$f_2(\omega_1) = C_2\,\omega_1^{-1}\sum_j[\omega_1^{\frac{n}{2}-\frac{p+1}{2}+j}M_{a_j}(\omega_1)+\omega_1^{\frac{n}{2}-\frac{1}{2}-\frac{p-1}{2}+j}M_{b_j}(\omega_1)]$$

(6.1.46)

where $0 < \omega_1 < 1$, a_j and b_j have the same forms as in (6.1.38) with p replaced by p+1 and p-1 respectively. B_{a_j} and B_{b_j} are given below.

$$B_{a_j} = \frac{\Gamma(\frac{p+1}{2}-j-\frac{1}{2})\,\Gamma(\frac{p+1}{2}-j-\frac{3}{2})\ldots\Gamma(\frac{p+1}{2}-j-\frac{p-2}{2})}{\Gamma^{\frac{p-1}{2}}(\frac{p+1}{2}-j)\,[(-j+1)(-j+2)^2\ldots(-1)^{j-1}(1)^{j+1}\ldots(\frac{p-1}{2}-j)^{\frac{p-1}{2}}]}\,.$$

for $j = 1,2,\ldots,\frac{p-1}{2}$. (6.1.47)

$$B_{b_j} = \frac{\Gamma(\frac{p}{2}-\frac{1}{2}-j)\,\Gamma(\frac{p}{2}-\frac{3}{2}-j)\ldots\Gamma(2)}{\Gamma^{\frac{p-1}{2}}(\frac{p}{2}-j)[(-1)^{j-1}(-2)^{j-2}\ldots(-j+1)][(-j+\frac{1}{2})(-j+\frac{3}{2})^2\ldots(-j+\frac{p-2}{2})^{\frac{p-1}{2}}]}$$

for $j = 1,2,\ldots,\frac{p-1}{2}-1$;

$$= \{[\Gamma^{\frac{p-1}{2}}(\tfrac{p}{2}-j)][(-1)^{\frac{p-1}{2}}(-2)^{\frac{p-1}{2}}\cdots(\tfrac{p}{2}-\tfrac{1}{2}-j)^{\frac{p-1}{2}}(\tfrac{p}{2}-j-\tfrac{3}{2})^{\frac{p-1}{2}-1}\cdots$$

$$\cdots(-j+1)][(-j+\tfrac{1}{2})(-j+\tfrac{3}{2})^{2}\cdots(-j+\tfrac{p-2}{2})^{\frac{p-1}{2}}]\}^{-1}$$

$$\text{for } j \geq \frac{p-1}{2}. \tag{6.1.48}$$

Again $A_{a_j}^{(r)}$ and $A_{b_j}^{(r)}$ are available by the procedure described in (6.1.43)

6.1.4. Testing Equality of Diagonal Elements.

In the multinormal case the h-th moment of a criterion W_2 for testing the hypothesis that the diagonal elements are equal given that the covariance matrix is diagonal, is given in Anderson ([16], p.262) as,

$$E(W_2^h) = \frac{p^{hp}\,\Gamma^p(\frac{n}{2}+h)\,\Gamma(\frac{np}{2})}{\Gamma^p(\frac{n}{2})\,\Gamma(\frac{np}{2}+ph)}. \tag{6.1.49}$$

The density of W_2 can be written in terms of an H-function given in Section 5.8 of Chapter V or in terms of a G-function after simplifying the Gammas in (6.1.49) with the help of Gauss-Legendre multiplication formula (1.2.6). By using the technique of Chapter V the density can be represented in computable form. These different forms are given here.

$$f_3(\omega_2) = C_3'\,\omega_2^{-1}\,H_{1,0}^{p,0}\left[\frac{\omega_2}{p^p}\,\middle|\,\begin{matrix}(\frac{np}{2},p)\\(\frac{n}{2},1),(\frac{n}{2},1),\ldots,(\frac{n}{2},1),(\frac{n}{2},1),\ldots,(\frac{n}{2},1)\end{matrix}\right]$$

$$= C_3' \, w_2^{-1} \, G^{p-1,0}_{p-1,p-1}\left[w_2 \,\middle|\, \begin{matrix} \frac{n}{2}+\frac{1}{p}, \frac{n}{2}+\frac{2}{2}, \ldots, \frac{n}{2}+\frac{p-1}{2} \\ \frac{n}{2}, \frac{n}{2}, \ldots, \frac{n}{2} \end{matrix} \right]$$

$$= C_3 \, w_2^{-1} \sum_{\nu=0}^{\infty} w_2^{\frac{n}{2}+\nu} M_{a_j}(w_2), \quad 0 < w_2 < 1 . \qquad (6.1.50)$$

where $M_{a_j}(w_2)$ has the same form as in (6.1.39) with $a_j = p-1$,

$$C_3' = \frac{\Gamma(\frac{np}{2})}{\Gamma^p(\frac{n}{2})}, \quad C_3 = \frac{\Gamma(\frac{np}{2})}{p^{(np-1)/2} \, (2\pi)^{\frac{1-p}{2}} \, \Gamma^p(\frac{n}{2})}, \qquad (6.1.51)$$

$$B_{a_j} = \frac{(-1)^{\nu(p-1)}}{[(1)(2)\ldots(\nu)]^{p-1}} \prod_{r=1}^{p-1} \Gamma(-\nu+\tfrac{r}{2}), \qquad (6.1.52)$$

n and p are positive integers and

$$A_{a_j}^{(r)}, \; r \geq 1, = (-1)^{r+1} \, r! \, \{(p-1)\, \zeta(r+1,1) - \sum_{j=1}^{p-1} \zeta(r+1,-\nu+\tfrac{j}{p})$$

$$+ (p-1) \left[\frac{1}{(-1)^{r+1}} + \ldots + \frac{1}{(-\nu)^{r+1}} \right] \} . \qquad (6.1.53)$$

Put $r = 0$ in (6.1.53) and replace $\zeta(r+1,.)$ by $-\psi(\cdot)$ to obtain $A_{a_j}^{(0)}$, where ψ and ζ-functions are discussed in (5.3.2) and (5.3.12) respectively.

6.1.5. The Sphericity Test.

The hypothesis that the covariance matrix Σ is of the form $\sigma^2 I$ where σ^2 is an unknown scalar and I is an identity matrix, is often known as the sphericity test. This test is a combination of the tests given in 6.1.3 and 6.1.4. The h-th moment of a criterion W for testing sphericity in the multinormal case is given in Anderson [16] as,

$$E(W^h) = \frac{p^{hp}\,\Gamma(\frac{np}{2})}{\Gamma(\frac{np}{2}+ph)} \prod_{j=1}^{p} \frac{\Gamma(\frac{n+1-j}{2}+h)}{\Gamma(\frac{n+1-j}{2})} . \tag{6.1.54}$$

As indicated in Section 6.1.4 the density function of W can be put in terms of a H-function, a G-function and in computable elementary functions as follows:

$$f_4(\omega) = C_4' \,\omega^{-1} H^{p,0}_{1,p}\left[\frac{\omega}{p^p} \middle| \begin{matrix} (\frac{np}{2}, p) \\ (\frac{n-2}{2},1), (\frac{n-3}{2},1), \ldots, (\frac{n-p-1}{2},1) \end{matrix} \right]$$

$$= C_4\, G^{p-1,0}_{p-1,p-1}\left[\omega \middle| \begin{matrix} \frac{n-2}{2}+\frac{j}{p},\ j=1,2,\ldots,p-1 \\ \frac{n-2}{2}-\frac{j}{2},\ j=1,2,\ldots,p-1 \end{matrix} \right],$$

$$= C_4 \left[\sum_{j=1}^{\infty} \omega^{\frac{n}{2}-\frac{p}{2}-1+j} M_{a_j}(\omega) + \sum_{j=1}^{\infty} \omega^{\frac{n}{2}-\frac{p}{2}-\frac{3}{2}+j} M_{b_j}(\omega) \right], \tag{6.1.55}$$

$0 < \omega < 1$, $M_{a_j}(\omega)$ is given in (6.1.39), n and p are positive integers,

$$C'_4 = \frac{\Gamma(\frac{np}{2})}{p^p \prod_{j=1}^{p} \Gamma(\frac{n+1-j}{2})}, \quad C_4 = \frac{\Gamma(\frac{np}{2})}{(2\pi)^{\frac{1-p}{2}}\, p^{\frac{np-1}{2}} \prod_{j=1}^{p} \Gamma(\frac{n+1-j}{2})}, \tag{6.1.56}$$

a_j, b_j, B_{a_j} and B_{b_j} are given below. $A^{(r)}_{a_j}$ and $A^{(r)}_{b_j}$ occurring in (6.1.55) are available by the procedure described in (6.1.43).

Case I, p-odd ($p \geq 3$):

$$a_j = b_j = \begin{cases} j, j = 1,2,\ldots, \frac{p-3}{2} , \\ \frac{p-1}{2} , \quad j = \frac{p-1}{2}, \frac{p+1}{2} ,\ldots \end{cases}$$

$$B_{a_j} = \frac{\prod_{r=2}^{\frac{p-1}{2}} \Gamma(r) \prod_{r=1}^{\frac{p-1}{2}} \Gamma(-j-\frac{1}{2}+r)}{\prod_{r=1}^{j-1} (-r)^{j-r} \prod_{r=1}^{p-1} \Gamma(-j+\frac{p}{2}+\frac{r}{p})} , \quad j = 1,2,\ldots, \frac{p-3}{2};$$

$$= \frac{\prod_{r=1}^{\frac{p-1}{2}} \Gamma(-j-\frac{1}{2}+r) \left[\prod_{r=1}^{p-1} \Gamma(-j+\frac{p}{2}+\frac{r}{p})\right]^{-1}}{[(-1)(-2)\ldots(-j+\frac{p-1}{2})]^{\frac{p-1}{2}} [(-j+\frac{p-1}{2}-1)^{\frac{p-1}{2}-1} \ldots (-j+1)]}$$

for $j = \frac{p-1}{2} , \frac{p+1}{2} ,\ldots ;$

$$B_{b_j} = \frac{\prod_{r=2}^{\frac{p-1}{2}-j} \Gamma(r) \prod_{r=1}^{\frac{p-1}{2}} \Gamma(-j+\frac{1}{2}+r)}{\prod_{r=1}^{j-1} (-r)^{j-r} \prod_{r=1}^{p-1} \Gamma(-j+\frac{1}{2}+\frac{p}{2}+\frac{r}{p})} , \quad j = 1,2,\ldots, \frac{p-3}{2} ;$$

$$= \frac{\prod_{r=1}^{\frac{p-1}{2}} \Gamma(-j+\frac{1}{2}+r) \left[\prod_{r=1}^{p-1} \Gamma(-j+\frac{1}{2}+\frac{p}{2}+\frac{r}{p})\right]^{-1}}{[(-1)(-2)\ldots(-j+\frac{p-1}{2})]^{\frac{p-1}{2}} [(-j+\frac{p-1}{2}-1)^{\frac{p-1}{2}-1} \ldots (-j+1)]} , \qquad (6.1.57)$$

$j = \frac{p-}{2} , \frac{p+1}{2} , \ldots$

Case II, p-even ($p \geq 4$):

$$a_j = \begin{cases} j, & j = 1,2,\ldots,\frac{p}{2}-2 \\ \frac{p}{2}-1, & j = \frac{p}{2}-1, \frac{p}{2}, \ldots \end{cases}$$

$$b_j = \begin{cases} j, & j = 1,2,\ldots,\frac{p}{2}, \\ \frac{p}{2}-1, & j = \frac{p}{2}+1, \frac{p}{2}+2, \ldots \end{cases}$$

$$B_{a_j} = \frac{\prod_{r=2}^{\frac{p}{2}-1-j} \Gamma(r) \prod_{r=1}^{\frac{p-2}{2}} \Gamma(-j-\frac{1}{2}+r)}{[\prod_{r=1}^{j-1} (-r)^{j-r}]^{\frac{p}{2}-2} (-j+\frac{p-1}{2}) \prod_{\substack{r=1 \\ r \neq p/2}}^{p-1} \Gamma(-j+\frac{p}{2}+\frac{r}{p})}, \quad j = 1,2,\ldots,\frac{p}{2}-2 ;$$

$$= \frac{\prod_{r=1}^{\frac{p-2}{2}} \Gamma(-j-\frac{1}{2}+r)}{[\prod_{r=1}^{j-\frac{p}{2}+1} (-r)]^{\frac{p}{2}-1} (-j+\frac{p-1}{2})^{\frac{p}{2}-2} \prod_{r=1}^{\frac{p}{2}-2} (-j+r)^r \prod_{\substack{r=1 \\ r \neq p/2}}^{p-1} \Gamma(-j+\frac{p}{2}+\frac{r}{p})},$$

for $j \geq \frac{p}{2}-1$.

$$B_{b_j} = \frac{\prod_{r=2}^{\frac{p}{2}-1-j} \Gamma(r) \prod_{r=1}^{\frac{p}{2}-1} \Gamma(-j+\frac{1}{2}+r)}{[\prod_{r=1}^{j-1} (-r)^{j-r}] (-j+\frac{p}{2}) \prod_{\substack{r=1 \\ r \neq p/2}}^{p-1} \Gamma(-j+\frac{1}{2}+\frac{p}{2}+\frac{r}{p})}, \quad j = 1,2,\ldots,\frac{p}{2}-2 ;$$

$$= \frac{\prod_{r=1}^{\frac{p}{2}-1} \Gamma(-j+\frac{1}{2}+r)}{[\prod_{r=1}^{j-\frac{p}{2}+1} (-r)]^{\frac{p}{2}-1} \prod_{r=1}^{\frac{p}{2}-2} (-j+r)^r] (-j+\frac{p}{2}) \prod_{\substack{r=1 \\ r \neq p/2}}^{p-1} \Gamma(-j+\frac{1}{2}+\frac{p}{2}+\frac{r}{p})}, \qquad (6.1.58)$$

for $j \geq \frac{p}{2} - 1$, with the restriction that when $j = p/2$ the factor $(-j+\frac{p}{2})$ is to be deleted from the denominator.

There are many other test criteria described in books on Multivariate Analysis. The exact distributions of the likelihood ratio criteria for several other tests are given in Mathai [172]. Some of these fall under the category of H-functions but these are reducable to G-functions and hence the techniques of Chapter V are applicable. There are some other test criteria for which the distributions do not fall in the category of G-functions and the exact distributions for these are not yet available for the general cases. One such test is the one for testing that the mean vector equals a given vector and the covariance matrix equals a given matrix in a multivariate normal distribution. Only approximations are available for the distribution of the likelihood ratio criterion for testing this hypothesis. A description of the test is available in Anderson [16].

6.2. THE EXACT NON-NULL DISTRIBUTIONS OF MULTIVARIATE TEST CRITERIA

This is another topic where the theory of G-function is applicable. Non-null distributions are the distributions of statistical test criteria when the null hypothesis is not assumed to hold. These distributions and their representations in computable forms are needed for comparison of tests and for studying various other properties of tests. The exact non-null distributions of the likelihood ratio criteria for testing hypotheses on multinormal populations, were not available for most of the problems till the 1960's. A breakthrough is achieved in this direction with the help of the theory of Zonal Polynomials and Hypergeometric Functions of matrix arguments developed by several authors of which the main articles are James ([122], [123], [124], [125]), Herz [117] and Constantine [67]. Mathai and Saxena [199] defined and developed the theory of the G-function with a matrix argument.

Exact non-null distributions of various test criteria are represented in terms of G-functions by several authors which include Khatri and Srivastava [136], Pillai, Al-Ani and Jouris [231], Pillai and Jouris [234] and Pillai and Nagarsenker [235]. In these articles, as remarked earlier, the problems under consideration are not solved because the problems under consideration are the logarithmic cases and hence their representations do not give computable forms. The non-central distribution of the determinant of a Wishart matrix and the non-null distributions of a collection of multivariate criteria are obtained for the general cases and for the first time in Mathai ([183], [184]). This section is mainly based on Mathai ([183]).

In order to introduce the non-null distributions one needs a Hypergeometric function of a matrix argument which is in turn defined in terms of Zonal Polynomials. Hence a brief description of Zonal Polynomial is given here.

6.2.1. Zonal Polynomials.

Let A be a positive definite, symmetric $m \times m$ matrix and $\emptyset(A)$ a polynomial in the elements of A. Consider the transformation,

$$\emptyset(A) \rightarrow \emptyset(L^{-1}AL'^{-1}), \quad L \in GL(m) \tag{6.2.1}$$

which defines a representation of the real linear group GL(m) in the vector space of all polynomials in A. L' denotes the transpose of L. Under the transformation (6.2.1) the space V_k of homogeneous polynomials of degree k is invariant and V_k decomposes into the direct sum of irreducible subspaces,

$$V_k = \sum_K (+) V_{k,K} \tag{6.2.2}$$

where

$$K = (k_1,\ldots,k_m),\quad k_1 \geq k_2 \geq \ldots \geq k_m,\quad k_1+k_2+\ldots+k_m = k, \qquad (6.2.3)$$

and the summation is over all partitions of k into not more than m parts. Each $V_{k,K}$ contains a unique one dimensional subspace invariant under the orthogonal group $O(m)$. These subspaces are generated by the Zonal Polynomials, $Z_K(A)$ which is invariant under the orthogonal group. That is,

$$Z_K(H'AH) = Z_K(A), \quad H \in O(m) . \qquad (6.2.4)$$

The Zonal Polynomials are homogeneous symmetric polynomials in the eigenvalues of A. Instead of $Z_K(A)$, Constantine [67] uses a normalized polynomial $C_K(A)$ where

$$C_K(A) = c(K)\, Z_K(A)/[(1)(2)\ldots(2k-1)], \qquad (6.2.5)$$

where $c(K)$ is the degree of the representation [2K] of the symmetric group on 2k symbols.

6.2.2. Hypergeometric Functions with Matrix Arguments.

Let Z be a complex symmetric $m \times m$ matrix. The Hypergeometric functions with argument Z is defined as

$${}_pF_q\,(a_1,\ldots,a_p;\, b_1,\ldots,b_q;\, Z) = \sum_{k=o}^{\infty} \sum_{K} \frac{(a_1)_K\cdots(a_p)_K\; C_K(Z)}{(b_1)_K\cdots(b_q)_K \quad k!}, \qquad (6.2.6)$$

with $p < q+1$ or $p = q+1$ and $\|Z\| < 1$, where $\|Z\|$ is a suitable "norm", for example the maximum of the absolute values of the characteristic roots of Z,

$$(a)_K = \prod_{i=1}^{m} (a- \frac{i-1}{2})_{k_i}, \quad K = (k_1,\ldots,k_m),\ k_1 \geq \ldots k_m \geq 0,$$

$$k = k_1 + k_2 +\ldots+ k_m,$$

$$(a)_n = (a)(a+1)\ \ldots\ (a+n-1). \qquad (6.2.7)$$

The parameters a_i and b_j are all complex numbers and none of the b_j is an integer or a half integer $\leq \frac{m-1}{2}$. When any a_i is a negative integer the series terminates. By using the above concepts the non-null distributions can be worked out in terms of computable series. For a detailed discussion of the various problems in this direction see Mathai [172]. For the purpose of illustration we will discuss one problem here.

6.2.3 A Test Statistic and Its Non-null Distribution.

Consider the independent matrix variates $X(p \times n_1)$ and $Y(p \times n_2)$, $p \leq n_i$, $i = 1,2$, with the columns of X and Y distributed according to multinormal distributions $N_p(0,\Sigma_1)$ and $N_p(0,\Sigma_2)$ respectively, where $N_p(\mu,\Sigma)$ stands for a p-variate multinormal density with mean vector μ and covariance matrix Σ. Then $S_1 = XX'$ and $S_2 = YY'$ are independent and have the Wishart distributions $W_p(n_i,\Sigma_i)$, $i = 1,2,$. Let $0 < f_1 < \dots < f_p < \infty$ be the eigenvalues of,

$$|S_1 - f\, S_2| = 0, \tag{6.2.8}$$

and $0 < \lambda_1 \leq \lambda_2 \leq \dots \leq \lambda_p < \infty$ be the eigenvalues of,

$$|\Sigma_1 - \lambda\, \Sigma_2| = 0, \tag{6.2.9}$$

where $W_p(n,\Sigma)$ stands for the density of a Wishart distribution on p variates, n degrees of freedom and covariance matrix Σ. Then a criterion W for testing the hypothesis

$$\delta\, \Lambda = I_p, \quad \delta > 0, \text{ known}$$

is given by

$$W = \prod_{i=1}^{p} (1-e_i) = |I_p - E_1| \tag{6.2.10}$$

where I_p is an identity matrix of order p, $\Lambda = \text{diag}\,(\lambda_1,\dots,\lambda_p)$, $E_1 = \text{diag}\,(e_1,\dots,e_p)$, $e_i = \delta f_i/(1+\delta f_i)$, $i = 1,2,\dots,p$ and $|(\cdot)|$ stands for the determinant of $(\cdot)$. The h-th moment of W is available in Pillai, Al-Ani and

Jouris ([231]) as,

$$E(W^h) = \frac{\Gamma_p(\frac{n}{2})\,\Gamma_p(\frac{n_2}{2}+h)\,|\delta\,|^{-\frac{n_1}{2}}\;{}_2F_1(\frac{n}{2},\frac{n_1}{2};\frac{n}{2}+h,\,I_p-(\delta\Lambda)^{-1})}{\Gamma_p(\frac{n_2}{2})\,\Gamma_p(\frac{n}{2}+h)} \qquad (6.2.11)$$

where E denotes 'mathematical expectation' , $n = n_1 + n_2$ and

$$\Gamma_p(u) = \pi^{p(p-1)/4} \prod_{i=1}^{p} \Gamma(u-\frac{i-1}{2}) \,. \qquad (6.2.12)$$

By taking the inverse Mellin transform of (6.2.11) the density function can be written as,

$$g(\omega) = A_K \quad \omega^{\frac{n_2}{2}-1} \; G^{p,0}_{p,p}\left[\,\omega\middle|\begin{matrix}\frac{n_1}{2}+k_i-\frac{i-1}{2},\; i=1,\ldots,p\\ -\frac{(i-1)}{2},\; i=1,2,\ldots,p\end{matrix}\right] \qquad (6.2.13)$$

$0 < \omega_1 < 1$ and where A_K stands for the expression

$$A_K = \frac{\Gamma_p(\frac{n}{2})\,|\,\delta\Lambda|^{-\frac{n_1}{2}}}{\Gamma_p(\frac{n_2}{2})} \sum_{k=o}^{\infty}\sum_{K} \frac{(\frac{n}{2})_K\,(\frac{n_1}{2})_K\,C_K(I_p-(\delta\Lambda)^{-1}}{k\,!} \,. \qquad (6.2.14)$$

The density function in (6.2.13) can be put into computable form by expressing the G-function in (6.2.13) in computable form. By using the technique of Chapter V and by using the notations in (6.1.39) to (6.1.43) we can write $g_1(\omega)$ in the following form where for convenience $g(\omega)$ in (6.2.13) is written as $g(\omega) = A_K\,\omega^{\frac{n_2}{2}-1}\;g_1(\omega)$ and in $g_1(\omega)$ we will write n_1 as q for convenience. For Cases I,II and III to be discussed below,

$$g_1(\omega) = \sum_{j\epsilon a} \omega^{-\frac{p}{2}+j} M_{a_j}(\omega) + \sum_{j\epsilon b} \omega^{-\frac{p}{2}-\frac{1}{2}+j} M_{b_j}(\omega) \qquad (6.2.15)$$

$0 < \omega < 1$, where a_j, b_j, B_{a_j} and B_{b_j} are listed below for the different values of p_1 and $q = n_1$.

Case I: p-even, q-even ($p \leq q$):

$$a_j = \begin{cases} j, & j = 1,2,\ldots,\frac{p}{2}-1 \\ \frac{p}{2}, & j = \frac{p}{2},\ldots,\frac{q}{2}+k_{p-1}, \\ \frac{p}{2}-i, & j = \frac{q}{2}+k_{p-(2i-1)}+i,\ldots,\frac{q}{2}+k_{p-(2i+1)}+i, \end{cases}$$

$$i = 1,2,\ldots,\frac{p}{2}-1;\ j \in a,\ a = \{1,2,\ldots,\frac{q}{2}+\frac{p}{2}+k_1-1\};$$

$$b_j = \begin{cases} j, & j = 1,2,\ldots,\frac{p}{2}-1 \\ \frac{p}{2}, & j = \frac{p}{2},\ldots,\frac{q}{2}+k_p, \\ \frac{p}{2}-i, & j = \frac{q}{2}+k_{p-(2i-2)}+i,\ldots,\frac{q}{2}+k_{p-2i}+i, \end{cases}$$

$$i = 1,2,\ldots,\frac{p}{2}-1,$$

$$j \in b \text{ and } b = \{1,2,\ldots,\frac{q}{2}+\frac{p}{2}+k_2-1\}. \tag{6.2.16}$$

Case II: p-odd, q-even ($p < q$):

$$a_j = \begin{cases} j, & j = 1,2,\ldots,\frac{p-1}{2}-1 \\ \frac{p-1}{2}, & j = \frac{p-1}{2},\ldots,\frac{q}{2}+k_{p-1}, \\ \frac{p-1}{2}-i, & j = \frac{q}{2}+k_{p-(2i-1)}+i,\ldots,\frac{q}{2}+k_{p-(2i+1)}+i, \end{cases}$$

$$j \in a,\ a = \{1,2,\ldots,\frac{q}{2}+\frac{p-1}{2}+k_2-1\};$$

$$b_j = \begin{cases} j, & j = 1,2,\ldots,\frac{p+1}{2}-1, \\ \frac{p+1}{2}, & j = \frac{p+1}{2},\ldots,\frac{q}{2}+k_p, \\ \frac{p+1}{2}-i, & j = \frac{q}{2}+k_{p-(2i-2)}+i,\ldots,\frac{q}{2}+k_{p-2i}+i, \end{cases}$$

$$i = 1,2,\ldots,\frac{p+1}{2}-1,$$

$$j \in b \text{ and } b = \{1,2,\ldots,\frac{q}{2}+\frac{p+1}{2}+k_1-1\}. \tag{6.2.17}$$

Case III: p-even, q-odd (p < q):

$$a_j = \begin{cases} j, & j = 1,2,\ldots,\frac{p}{2}-1 \\ \frac{p}{2}, & j = \frac{p}{2},\ldots,\frac{q-1}{2}+k_p, \\ \frac{p}{2}-i, & j = \frac{q-1}{2}+k_{p-(2i-2)}+i,\ldots,\frac{q-1}{2}+k_{p-2i}+i \end{cases}$$

$$j \in a, \quad a = \{1,2,\ldots,\frac{q-1}{2}+\frac{p}{2}+k_2-1\};$$

$$b_j = \begin{cases} j, & j = 1,2,\ldots,\frac{p}{2}-1, \\ \frac{p}{2}, & j = \frac{p}{2},\ldots,\frac{q+1}{2}+k_{p-1}, \\ \frac{p}{2}-i, & j = \frac{q+1}{2}+k_{p-(2i-1)}+i,\ldots,\frac{q+1}{2}+k_{p-(2i+1)}+i, \end{cases}$$

$$i = 1,2,\ldots,\frac{p}{2}-1,$$

$$j \in b \text{ and } b = \{1,2,\ldots,\frac{q+1}{2}+\frac{p}{2}+k_1-1\}. \qquad (6.2.18)$$

Case IV: p-odd, q-odd, $p \leq q$:

$$a_j = \begin{cases} j, & j = 1,2,\ldots,\frac{p-1}{2}-1, \\ \frac{p-1}{2}, & j = \frac{p-1}{2},\ldots,\frac{q-1}{2}+k_p, \\ \frac{p-1}{2}-i, & j = \frac{q-1}{2}+k_{p-(2i-2)}+i,\ldots,\frac{q-1}{2}+k_{p-2i}+i, \end{cases}$$

$$i = 1,2,\ldots,\frac{p-1}{2}-1,$$

$$j \in a, \quad a = \{1,2,\ldots,\frac{q-1}{2}+\frac{p-1}{2}+k_3-1\};$$

$$b_j = \begin{cases} j, \ j = 1,2,\dots,\frac{p-1}{2}, \\ \frac{p+1}{2}, \quad j = \frac{p+1}{2},\dots,\frac{q+1}{2}+k_{p-1}, \\ \frac{p+1}{2} - i, \ j = \frac{q+1}{2}+k_{p-(2i-1)}+i,\dots,\frac{q+1}{2}+k_{p-(2i+1)}+i \end{cases}$$

$$i = 1,2,\dots,\frac{p-1}{2} - 1,$$

$$j \in b \quad \text{and } b = \{1,2,\dots,\frac{q+1}{2}+\frac{p-1}{2}+k_2-1\}. \tag{6.2.19}$$

For Cases I, II and III,

$$B_{a_j} = \prod_{\substack{t\in a\\ t\neq j}} (-j+t)^{-a_t} \prod_{t\in b} (-j-\tfrac{1}{2}+t)^{-b_t}; \tag{6.2.20}$$

$$B_{b_j} = \prod_{t\in a} (\tfrac{1}{2}-j+t)^{-a_t} \prod_{\substack{t\in b\\ t\neq j}} (-j+t)^{-b_t}. \tag{6.2.21}$$

For Case IV,

$$g_1(\omega) = \sum_{j\in a} \omega^{-\frac{p}{2}+j} M_{a_j}(\omega) + \sum_{j\in b} \omega^{-\frac{p}{2}-\frac{1}{2}+j} M_{b_j}(\omega) + \sum_{\nu=0}^{\infty} R_\nu,$$

$$0 < \omega < 1, \tag{6.2.22}$$

where B_{a_j} and B_{b_j} of Case IV are the B_{a_j} and B_{b_j} in (6.2.20) and (6.2.21) multipliplied by$\{\Gamma(\frac{p}{2}-j+\frac{q-1}{2}+k_2)/\Gamma(\frac{p}{2}-j+\frac{q}{2}+k_1)\}$ and $\{\Gamma(\frac{p}{2}+\frac{1}{2}-j+\frac{q-1}{2}+k_2)/\Gamma(\frac{p}{2}+\frac{1}{2}-j+\frac{q}{2}+k_1)\}$ respectively and

$$R_\nu = \frac{(-1)^\nu\,\Gamma(1)\,\omega^{\frac{q+1}{2}+k_2+\nu}}{\nu!\,\Gamma(\frac{1}{2}+k_1-k_2-\nu)} \prod_{t\in a} (-\tfrac{q}{2}+\tfrac{1}{2}-k_2-\nu-\tfrac{p}{2}+t)^{-a_t}$$

$$\times \prod_{t\in b} (-\tfrac{q}{2}-\tfrac{p}{2}-k_2-\nu+t)^{-b_t} \tag{6.2.23}$$

Now the density in (6.2.13) is available in computable form from (6.2.13) to (6.2.23).

Remark 1: There are several test criteria associated with the test criterion discussed in this section. Two of them are the likelihood ratio test criteria associated with Wilks criterion for testing regression and for testing independence when there are only two subvectors. The exact distributions of these are available from Mathai [183].

Remark 2: The exact non-null distributions of most of the test criteria mentioned in Section 6.1 are still open problems. From the discussion in Section 6.2 it should be remarked that once the non-null moments of these criteria are available in Hypergeometric functions of matrix arguments then the exact distributions are easily available by using the procedure discussed in this section.

Remark 3: In a recent series of papers Davis has employed the method of differential equations in deriving the exact distributions and he obtained some particular cases. For a paper of this category see Davis [77]. It should be remarked that whenever the density functions are representable in terms of G-functions the differential equations can be written down directly from (1.5.1) which may be noticed from the different problems considered in Sections 6.1 and 6.2. The method of differential equation, for getting exact distributions of multivariate test criteria, was employed for the first time by Nair [215] and from this paper it can be noticed that the method is not a powerful one. The most powerful technique, available so far, for tackling these types of problems, is the one given in Chapter V and Sections 6.1 and 6.2.

6.3. CHARACTERIZATIONS OF PROBABILITY LAWS.

Characterization of probability laws is a fast developing branch of Statistics and Probability Theory. A forthcoming book by C.R. Rao to be published by Wiley and Sons, New York contains an up to date literature in this field. Some applications of characterization theorems may be seen from Csörgo and Seshadri [75].

There is another type of characterization problem in statistics, namely, the characterization of fundamental concepts or different measures which are used in Statistics and Information Theory. This is a method of giving axiomatic definitions to the various concepts in Statistics and Information Theory. Recent development in this direction are available from the report Mathai and Rathie [192].

Characterization of a Probability law means that a particular Probability distribution is shown to be the only distribution enjoying a given set of properties. As it is evident that a G-function, due to its generality, may not be useful to show that a particular property is the characteristic property of a Probability law. But a G-function is useful to show the converse that a particular property is not a unique property of a certain distribution and thus a G-function helps to get counter examples in such problems. Some such uses will be indicated here.

6.3.1. The Gamma Property

Consider a real Gamma Variate with the density function,

$$h(x) = \frac{\beta^{\alpha}}{\Gamma(\alpha)} x^{\alpha-1} e^{-\alpha x}, \alpha > 0, \beta > 0, 0 < x < \infty . \qquad (6.3.1)$$

The characteristic function of X is given by,

$$\phi_x(t) = E(e^{itX}) = \frac{\beta^{\alpha}}{\Gamma(\alpha)} \int_0^{\infty} x^{\alpha-1} e^{-\beta x} e^{itx} dx$$

$$= (1 - \frac{it}{\beta})^{-\alpha} \qquad (6.3.2)$$

$i = \sqrt{-1}$, t is a real arbitrary constant and E denotes 'mathematical expectation'. If $X_1,\ldots,X_n$ is a simple random sample from this gamma population, that is, a set of independently and identically distributed gamma variates, then the characteristic function of the sample mean,

$$\bar{X} = (X_1 + \ldots + X_n)/n$$

is given by

$$\phi_{\bar{X}}(t) = [\phi_x(\tfrac{t}{n})]^n = (1 - \frac{it}{n\beta})^{-n\alpha}. \tag{6.3.3}$$

Since the characteristic function uniquely determines the distribution, which follows from the properties of Fourier transforms, by comparing (6.3.2) and (6.3.3) it is seen that the sample mean $\bar{X}$ again has a gamma distribution with parameters α and β scaled by the sample size n. That is, the new parameters are $n\alpha$ and $n\beta$. It is natural to ask whether this property is a unique property of the gamma distribution or not. By using some properties of Special Functions it can be shown that this is not a characteristic property of the gamma distribution. We will illustrate it by taking a density function associated with a Bessel function.

Theorem 6.3.1. If a real stochastic variable Y has the probability law,

$$h_1(y,\nu,p,a) = \frac{\nu e^{-py} I_\nu(ay)}{y a^\nu [p+(p^2-a^2)^{1/2}]^{-\nu}} \tag{6.3.4}$$

for $y > 0$, $\nu > 0$, $p > 0$ and $h_1(y,\nu,p,a) = 0$ elsewhere, then the sample mean has the probability law $h_1(y,n\nu,np,na)$, where n is the sample size and $I_\nu(\cdot)$ is the modified Bessel function of the first kind which is defined in Chapter II.

Proof: From Erdélyi, A. et al ([86], I, p.195), we have

$$\int_0^\infty e^{-pt} I_\nu(at) \frac{dt}{t} = \frac{a^\nu}{\nu} [p + (p^2-a^2)^{1/2}]^{-\nu}. \tag{6.3.5}$$

Now the characteristic function of Y, denotes by $\phi(t,\nu,p,a)$ is available from (6.3.5). That is,

$$\phi(t,\nu,p,a) = \frac{\{(p-it) + [(p-it)^2 - a^2]^{1/2}\}^{-\nu}}{[p + (p^2 - a^2)^{1/2}]^{-\nu}}, \tag{6.3.6}$$

Therefore,

$$[\phi(\frac{t}{n}, \nu,p,n)]^n = \phi(t,n\nu,np,na) . \qquad (6.3.7)$$

This compleres the proof.

Remark: By using a G-function one can construct more general classes of distributions enjoying the same property. Similar properties are investigated by Mathai and Saxena [195]. Due to the presence of a number of parameters, the various functions appearing in this and in the remaining sections are all assumed to be non-negative without loss of any generality.

6.3.2. The Ratio Property.

It is a well known fact that the ratio of two independent standard normal variables is distributed according to a Cauchy probability distribution. It is natural to ask the converse question that if the ratio of two independent variates is distributed according to a Cauchy distribution, are the individual variables standard normal? Several authors have given counter examples, see Laha [145]. There are also two other ratio distributions frequently used in Statistical Inference. They are the Student-t and F-statistics. Mathai and Saxena [197] developed a general technique for giving counter examples for such problems. These counter examples belong to very general classes of distributions and wider classes can be constructed by using G-functions. We will state one such theorem here.

Theorem 6.3.2. If two independent stochastic variables X_1 and X_2 have the density functions

$$M_{a_{i},\nu}(x) = \frac{\Gamma(a_i+1+\nu)x^{\nu}}{2^{\nu}\Gamma(\nu+1)\Gamma(a_i)}\ {}_2F_1\left(\frac{a_i+\nu+1}{2},\frac{a_i+\nu+2}{2};\ \nu+1;\ -x^2\right), \qquad (6.3.8)$$

for $x > 0$, $i = 1,2$, then X_1/X_2 has the F-distribution for all values of ν such that $M_{a_{i},\nu}(x)$ exists.

Proof: Note: A variable X is said to have an F-distribution if its density function is of the form,

$$h_2(x) = \frac{\Gamma(\beta_1+\beta_2)}{\Gamma(\beta_1)\Gamma(\beta_2)} \frac{x^{\beta_1-1}}{(1+x)^{\beta_1+\beta_2}}, \; x > 0, \beta_1 > 0, \beta_2 > 0. \tag{6.3.9}$$

This is a slightly modified form of the density. It is easy to see that (6.3.8) is a density, that is, $M_{a_{i,\nu}}(x) \geq 0$ for all x and $\int_0^\infty M_{a_{i,\nu}}(x)dx = 1$ for $i = 1,2,.$ This follows from Erdélyi, A.et.al ([86], I.P. 336 (3)). Now the theorem is proved if we can show that

$$\int_0^\infty u\, M_{a_{1,\nu}}(ux)M_{a_{2,\nu}}(u)du = \frac{\Gamma(a_1+a_2)}{\Gamma(a_1)\,\Gamma(a_2)} \frac{x^{a_2-1}}{(1+x)^{a_1+a_2}}, \tag{6.3.10}$$

because when X_1 and X_2 are independently distributed according to (6.3.8) the ratio X_1/X_2 should satisfy the L.H.S. of (6.3.10). From Erdélyi, A. et.al([86], II, p.29) we have,

$$\int_0^\infty \frac{t\,J_\nu(ut)}{\Gamma(a_i)} t^{a_i-1} e^{-t} dt = M_{a_{i,\nu}}(u), \tag{6.3.11}$$

$R(a_i + \nu) > -1$, $u > 0$ and

$$\int_0^\infty \frac{t\,J_\nu(ut)}{\Gamma(a_i)} t^{a_i-1} e^{-t/a}dt = M_{a_{i,\nu}}(au)a^{a_i+1}. \tag{6.3.12}$$

Now by applying the Parseval property of the Hankel transform we have

$$\int_0^\infty u\, M_{a_{1,\nu}}(au)\, M_{a_{2,\nu}}(u)du = \int_0^\infty \left[\frac{t\,t^{a_1-1}e^{-t/a}}{\Gamma(a)a^{a_1+1}}\right] \left[\frac{t^{a_2-1}e^{-t}}{\Gamma(a_2)}\right]dt$$

$$= \frac{\Gamma(a_1+a_2)}{\Gamma(a_1)\Gamma(a_2)} \frac{a^{a_2-1}}{(1+a)^{a_1+a_2}}, \text{ for all } \nu. \tag{6.3.13}$$

This complètes the proof

Note: For convenience the Parseval property of the Hankel transform will be stated here. If,

$$\int_0^\infty t\, J_\nu(pt)\, \rho_1(t)dt = \psi_1(p)$$

and

$$\int_0^\infty t\, J_\nu(pt)\, \rho_2(t)dt = \psi_2(p)$$

then

$$\int_0^\infty t\, \rho_1(t)\, \rho_2(t)dt = \int_0^\infty p\, \psi_1(p)\, \psi_2(p)dp,$$

where $J_\nu(\cdot)$ is the Bessel function defined in Chapter II.

Remark: From the nature of Theorem 6.3.2 it can be seen that wider classes can be constructed by using G-functions. Several theorems similar to the one stated above and further references to the literature in this topic are given in Mathai and Saxena [197].

6.4. PRIOR AND POSTERIOR DISTRIBUTIONS

In one branch of statistical Inference known as Bayesian Inference statistical distributions involving random parameters play an important role. A number of authors have investigated the posterior distributions of random variables when one or more parameters have prior distributions. Several such practical problems can be found in standard books on Bayesian Statistical Inference and on Probability Theory, for example see Karlin ([133], p.21). Instead of dealing with particular conditional and unconditional distributions one can use generalized Special Functions to derive wider classes of unconditional distributions from a very general class of conditional distributions. Several such results are obtained by Kaufman, Mathai and Saxena [134]. One such theorem will be given here.

Theorem 6.4.1. Let the conditonal distribution of a real stochastic variable X (discrete or continuous), given the parameter p, be a generalized gamma with the density function

$$f(x|p) = \frac{\beta\, p^{\alpha/\beta}}{\Gamma(\alpha/\beta)} x^{\alpha-1} e^{-px^{\beta}}, \quad x > 0,\ \alpha,\beta,p > 0, \tag{6.4.1}$$

and $f(x|p) = 0$ elsewhere, where α and β are additional parameters. Let the marginal distribution of p be given by a density g(p) where,

$$g(p) = \frac{\prod_{j=k+1}^{s} \Gamma(-b_j) \prod_{j=\nu+1}^{r} \Gamma(1+a_j)}{\prod_{j=1}^{k} \Gamma(1+b_j) \prod_{j=1}^{\nu} \Gamma(-a_j)} \lambda\, G_{r,s}^{k,\nu}\left[\lambda p \,\middle|\, \begin{matrix} a_1,\ldots,a_r \\ b_1,\ldots,b_s \end{matrix}\right] \tag{6.4.2}$$

for $p > 0$. The parameters in the G-function are assumed to be such that the G-function under consideration exists and is real and non-negative. The unconditional distribution of X is given by the density

$$h(x) = \frac{\lambda\beta \prod_{j=k+1}^{s} \Gamma(-b_j) \prod_{j=\nu+1}^{r} \Gamma(1+a_j)}{\prod_{j=1}^{k} \Gamma(1+b_j) \prod_{j=1}^{\nu} \Gamma(-a_j)\, \Gamma(\frac{\alpha}{\beta})} x^{-(\beta+\alpha)}$$

$$\times G_{r+1,s}^{k,\nu+1}\left[\frac{\lambda}{x^{\beta}} \,\middle|\, \begin{matrix} -\frac{\alpha}{\beta},\ a_1,\ldots,a_r \\ b_1,\ldots,b_s \end{matrix}\right]$$

where $k+\nu > (\frac{r+s}{2})$, $R(b_j + 1 + \frac{\alpha}{\beta}) > 0$, $j = 1,\ldots, k$. (6.4.3)

Proof: The unconditional distribution is given by

$$h(x) = \int_0^{\infty} f(x|p)\, g(p)dp\ . \tag{6.4.4}$$

Now (6.4.3) follows from (6.4.4) and Erdélyi, A. et. al ([86], II, p.419 (5)).

Some useful particular cases of Theorem 6.4.1 are given below for different $g(p)$ but with the same $f(x|p)$ as in (6.4.1)

SOME PRIOR AND POSTERIOR DISTRIBUTIONS

$g(p)$	$h_1(x) = \int_0^\infty g(x\|p)\, g(p)dp$
(a) Exponential	
$\lambda e^{-p\lambda} = \lambda G^{1,0}_{0,1}[\lambda p\|0]$	$\lambda \alpha x^{\alpha-1} (\lambda+x^{\beta})^{-\frac{\alpha}{\beta}-1}$
(b) Gamma	
$\dfrac{p^{\alpha-1} \lambda^{\alpha} e^{-p\lambda}}{\Gamma(\alpha)}$	$\dfrac{\lambda^{\alpha} \beta\, \Gamma(\alpha+\frac{\alpha}{\beta})\, x^{\alpha-1}(\lambda+x^{\beta})^{-\alpha(1+\frac{1}{\beta})}}{\Gamma(\alpha)\, \Gamma(\frac{\alpha}{\beta})}$
(c) Beta	
$\dfrac{\Gamma(\alpha+\beta)\, p^{\alpha-1}(1-p)^{\beta-1}}{\Gamma(\alpha)\,\Gamma(\beta)}$	$\dfrac{\beta\, \Gamma(\alpha+\beta)\, \Gamma(\alpha+\frac{\alpha}{\beta})}{\Gamma(\alpha)\, \Gamma(\frac{\alpha}{\beta})\, \Gamma(\alpha+\beta+\frac{\alpha}{\beta})}\ {}_1F_1(\alpha+\frac{\alpha}{\beta}; \alpha+\beta+\frac{\alpha}{\beta}; -x^{\beta})x^{\alpha-1}$

6.5. GENERALIZED PROBABILITY DISTRIBUTIONS

Wells, Anderson and Cell [354] considered the distribution of the product of two independent stochastic variables one of which is a central Raleigh variable and the other is a non-central Raleigh variable. Their problem came from a problem in Physical Sciences. There are several papers on the ratio, product and linear combinations of stochastic variables and all these problems come from Physical, Natural and Biological Sciences and from Economics. The literature in this direction is vast and a short list of references is available from Miller [213]. In all these problems the different authors have considered combinations of stochastic variables having some specified distributions. Mathai and Saxena [196] considered the distribution of a product of two independent stochastic variables where the density of each one is defined in terms of a product of two H-functions. After analysing the structural set up of densities the authors have noticed that almost all the ratio and product distributions which are used in statistical literature will come as special cases from the results obtained in Mathai and Saxena [196]. A large number of other product and ratio distributions are also available from the results in this article. The technique can also be extended to the product and ratios of several independent variables.

In the field of the distribution of linear combinations of independent stochastic variables there is a large number of papers. Mathai and Saxena [198] considered the linear combination of k independent stochastic variables where each one has a density defined in terms of H-functions with different parameters and obtained the exact distribution. The results of this article cover almost all distributions of linear combinations of stochastic variables used in statistical literature and many more in this category. Since the H-function is the most generalized Special Function available so far, by specialysing the parameters, a large number of results are available from the above mentioned articles. In Section 6.5, we are not giving any specific result in order to save space. These results are readily available from the references cited at the end of this chapter.

From Mathai and Saxena ([196],[197],[198]) it is evident that G and H-functions and the techniques of Special Functions are useful and powerful in tackling problems of statistical distributions. While solving statistical problems the authors have come across some statistical techniques of deriving new results in the theory of Special Functions. This may be seen from Mathai and Saxena ([200],[201]) and articles of the author's in this line are given in the list of references.

From the discussions in this chapter it is evident that G-functions are applicable in a number of statistical problems. It is quite likely that in the next few years G-functions will be applied to many other topics in Statistics as well. Several authors have started applying G-function to statistical problems. These may be seen from the recent articles by Pillai and Jouris [234], Pillai and Nagarsenker [235] and Khatri and Srivastava [136]. Two more articles in these lines by Bagai, O.P. are scheduled to appear in Sankhya Series A.

Some applications of G-function in other fields are given in Chapter VII.

EXERCISES

Note: For a discussion of the Hypergeometric function in the logarithmic case see Section 5.4 of Chapter V. These appear in some of the following problems.

6.1. Show that f(u) in (6.1.3) reduces to the following forms for particular cases.

(i) $p = 3, q = 3$: $C\, u^{\frac{n}{2}-2}\,[(1-u)^{\frac{3}{2}} - 3u^{\frac{1}{2}} \sin^{-1}(1-u)^{\frac{1}{2}} + 3u \log\{u^{-\frac{1}{2}} + u^{-\frac{1}{2}}(1-u)^{\frac{1}{2}}\}]$

$$C = \Gamma(n+2)\,\Gamma(\tfrac{n+1}{2})/[\,\Gamma(n-1)\,\Gamma(\tfrac{n}{2}-1)3\pi^{\frac{1}{2}}\,];$$

(ii) $p = 3, q = 4$: $C\, u^{\frac{n}{2}-2}\,[1-u^2 + 8u^{\frac{1}{2}}(1-u) - 6u \log u\,]$,

$$C = \Gamma(n+3)\,\Gamma(\tfrac{n}{2}+1)/[24\,\Gamma(n-1)\,\Gamma(\tfrac{n}{2}-1)]\ ;$$

(iii) $p = 3, q = 5$: $C[u^{\frac{n}{2}-2}\,(1-u)^{\frac{5}{2}} - 45u(1-u)^{\frac{1}{2}} + (30\,u^{\frac{3}{2}} - \frac{15}{2}\,u^{\frac{1}{2}})\sin^{-1}(1-u)^{\frac{1}{2}}$

$$+ (30u - \tfrac{15}{2}\,u^2)\log\,(u^{-\frac{1}{2}} + u^{-\frac{1}{2}}(1-u)^{\frac{1}{2}})\ ,$$

$$C = \Gamma(n+4)\,\Gamma(\tfrac{n+3}{2})/[90\,\pi^{\frac{1}{2}}\,\Gamma(n-1)\,\Gamma(\tfrac{n}{2}-1)]\ ;$$

(iv) $p = 3, q = 6$: $C\, u^{\frac{n}{2}-2} [1 - 16\, u^{\frac{1}{2}} - 65\, u + 16\, u^{\frac{3}{2}} - 65\, u^2 - 16\, u^{\frac{5}{2}} + u^3$

$$- 30\, u(1-u) \log u)],$$

$$C = \Gamma(n+5)\, \Gamma(\tfrac{n}{2} + 2)/[1440\, \Gamma(n-1)\, \Gamma(\tfrac{n}{2} - 1)];$$

(v) $p = 3, q = 7$: $C\, u^{\frac{n}{2}-2} [(1-u)^{\frac{7}{2}} - (\frac{1855}{8}) u(1-u)^{\frac{3}{2}} - (\frac{105}{8})\, (1 - 20\, u + 8u^2)$

$$\times u^{\frac{1}{2}} \sin^{-1}(1-u)^{\frac{1}{2}} + (\tfrac{105}{8})(8 - 20\, u + u^2) u \log(\{1+(1-u)^{\frac{1}{2}}\} u^{-\frac{1}{2}}\ ],$$

$$C = \Gamma(n+6)\, \Gamma(\tfrac{n+5}{2})/[2\, \Gamma(7)\, \Gamma(\tfrac{9}{2})\, \Gamma(n-1)\, \Gamma(\tfrac{n}{2} - 1)];$$

(vi) $p = 3, q = 8$: $C\, u^{\frac{n}{2}-2} [1 - (\frac{128}{5}) u^{\frac{1}{2}}(1-u^3) - (\frac{1428}{5})\, u(1-u^2) + 896\, u^{\frac{3}{2}}(1-u)$

$$-u^4 - 84u(1 - 5u + u^2) \log u]\ ,$$

$$C = \Gamma\,(n+7)\ \Gamma(\tfrac{n}{2} + 3)/[2\, \Gamma(8)\, \Gamma(5)\, \Gamma(n-1)\ (\tfrac{n}{2} - 1)];$$

(vii) $p = 4, q = 4$: $C\, u^{\frac{n-5}{2}} [1 - 15\, u^{\frac{1}{2}} - 80u + 80u^{\frac{3}{2}} + 15u^2 - u^{\frac{5}{2}}$

$$-30(u + u^{\frac{3}{2}}) \log u]\ ,$$

$$C = \Gamma(n+1)\ \Gamma(n+3)/\Gamma\ (n-3);$$

(viii) $p = 4, q = 5$: $C\, u^{\frac{n-5}{2}} [1 - 24\, u^{\frac{1}{2}} - 375u + 375u^2 + 24u^{\frac{5}{2}} - u^3$

$$- 30(3u + 8u^{\frac{3}{2}} + 3u^2) \log u]\ ,$$

$$C = \Gamma(n+2)\ \Gamma(n+4)/[(2)4!\ 6!\ \Gamma(n-3)\ \Gamma(n-1)];$$

(ix) $p = 4, q = 6$: $C\, u^{\frac{n-5}{2}} [1 - 35u^{\frac{1}{2}} - 1099u - 1575u^{\frac{3}{2}} + 1575u^2$

$$+ 1099u^{\frac{5}{2}} + 35u^3 - u^{\frac{7}{2}} - 210(u + 5u^{\frac{3}{2}} + 5u^2 + u^{\frac{5}{2}}) \log u],$$

$$C = \Gamma(n+3)\ \Gamma(n+5)/[(2)5!\ 7!\ \Gamma(n-3)\ \Gamma(n-1)];$$

(x) $p = 4, q = 7$: $C\,u^{\frac{n-5}{2}}\,[1-u^4-48u^{\frac{1}{2}}(1-u^3)-2548u(1-u^2)$

$-8624\,u^{\frac{3}{2}}\,(1-u)-420(u + 8u^{\frac{3}{2}} + 15u^2 + 8u^{\frac{5}{2}} + u^3)\log u]$,

$C = \Gamma(n+4)\,\Gamma(n+6)/[(2)6!\;8!\;\Gamma(n-3)\,\Gamma(n-1)]$;

(xi) $p = 4, q = 8$: $C\,u^{\frac{n-5}{2}}\,[1-u^{\frac{9}{2}} - 63(u^{\frac{1}{2}} - u^4)-(5104-\frac{4}{5})(u-u^{\frac{7}{2}})$

$-\ 29988\,(u^{\frac{3}{2}} - u)-28244(u^2-u^{\frac{5}{2}}) - 252(3u$

$+\ 35u^{\frac{3}{2}} + 105u^2 + 105u^{\frac{5}{2}} + 35u^3 + 3u^{\frac{7}{2}})\log u]$,

$C = \Gamma(n+5)\,\Gamma(n+7)/[(2)7!\;9!\;\Gamma(n-3)\,\Gamma(n-1)]$;

(Consul, 1966, [68])

6.2. Show that the density function $f_1(v)$ in (6.1.34) reduces to the following forms in particular cases. [Most of these are worked out by Consul (1967,[69]).

(i) $q = 1$, all p_1, $p_o = 1$:

$C\,v^{(n-p_1-2)/2}\,(1-v)^{\frac{p_1}{2}-1}$,

$C = \Gamma(n/2)[\,\Gamma(\frac{p_1}{2})\,\Gamma(\frac{n-p_1}{2})\,]$;

(ii) $q = 1$, all p_1, $p_o = 2$:

$C\,v^{(n-p_1-3)/2}\,(1-v^{1/2})^{p_1-1}$

$C = \Gamma(n-1)/[2\,\Gamma(n-p_1-1)\,\Gamma(p_1)]$

(iii)	$q = 1$	$C\, v^{(n-p_1-4)/2} (1-v)^{p_1/2} \sum_{i=o}^{p_1-1} \binom{p_1-1}{i} (-v^{1/2})^i$
	all p_1	$\times\, {}_2F_1(\frac{p_1}{2}, \frac{i}{2}; \frac{p_1}{2} + 1; 1-v)$
	$p_o = 3$	$C = \Gamma(n-1)\, \Gamma(\frac{n}{2} -1)/[2\Gamma(n-p_1-1)\, \Gamma(\frac{n-p_1}{2} -1)\, \Gamma(p_1)\, \Gamma(\frac{p_1}{2} + 1)]$
(iv)	$q = 1$	$C\, v^{(n-p_1-5)/2} (1-v^{1/2})^{2p_1-1}\, {}_2F_1(p_1-2, p_1; 2p_1; 1-v^{1/2})$
	all p_1	
	$p_o = 4$	$C = \Gamma(n-3)\, \Gamma(n-1)/[2\, \Gamma(2p_1)\, \Gamma(n-p_1-1)\, \Gamma(n-p_1-3)]$
(v)	$q = 2$	$C\, v^{(n-p_2-3)} (1-v)^{p_2-1/2}\, {}_2F_1(\frac{p_2}{2}, \frac{p_2}{2}; p_2+1/2; 1-v)$
	all p_2	
	$p_o = p_1 = 1$	$C = \Gamma^2(\frac{n}{2})/[\Gamma(\frac{n}{2} - \frac{p_2}{2})\, \Gamma(\frac{n-p_2-1}{2})\, \Gamma(p_2+1/2)]$
(vi)	$q = 2$	$C\, v^{(n-p_2-3)/2} (1-v)^{\frac{p_2}{2} + 1} \sum_{r=o}^{p_2-1} \binom{p_2-1}{r} (-v^{1/2})^r$
	all p_2	$\times\, {}_2F_1(\frac{p_2}{2} + 1, \frac{r}{2} + \frac{3}{2}; \frac{p_2}{2} + 2; 1-v)$
	$p_o = 1,\ p_1 = 2$	$C = \Gamma(n-1)\, \Gamma(\frac{n}{2})/[2\, \Gamma(n-p_2-1)\, \Gamma(\frac{n}{2} - \frac{p_2}{2} -1)\, \Gamma(p_2)\, \Gamma(\frac{p_2}{2} +1)]$
(vii)	$q = 2$	$C\, v^{(n-p_2-5)/2} (1-v^{1/2})^{2p_2+1}\, {}_2F_1(p_2, p_2; 2p_2+2; 1-v^{1/2})$
	all p_2	$V = \Gamma^2(n-1)/[2\, \Gamma(n-p_2-1)\, \Gamma(n-p_2-3)\, \Gamma(2p_2+2)]$
	$p_o = p_1 = 2$	

6.3. Show that for p = 3 the density function $f_2(\omega_1)$ in (6.1.36) can be put in terms of a hypergeometric function in the logarithmic case as follows:

$$C\,\omega_1^{\frac{n}{2}-2}(1-\omega_1)^{\frac{1}{2}}\;{}_2F_1(\tfrac{1}{2},\tfrac{1}{2};\tfrac{3}{2};1-\omega_1)$$

(Mathai 1971, [181]).

6.4. Show that for p = 3 the density in (6.1.50) reduces to the form

$$C\,\omega_2^{\frac{n}{2}-1}\;{}_2F_1(\tfrac{2}{3},\tfrac{1}{3};1;1-\omega_2)\;,$$

$$C = \Gamma(\tfrac{3n}{2})/[3^{(3n-1)/2}(2\pi)^{-1}\Gamma^3(\tfrac{n}{2})]$$

(Mathai, 1971, [182]).

6.5. Show that the density function in (6.1.55) reduces to the following form for particular cases:

(i) p = 3: $C\,\omega^{\frac{n}{2}-2}(1-\omega)^{\frac{3}{2}}\;{}_2F_1(\tfrac{5}{6},\tfrac{7}{6};\tfrac{5}{2};1-\omega),$

$$C = 2^{n+1}\Gamma(\tfrac{3n}{2})/[\Gamma(n-1)\Gamma(\tfrac{n}{2}-1)3^{(3n+1)/2}]\;;$$

(ii) p = 4: $C\,\omega^{\frac{n-5}{2}}(1-\omega^2)^{\frac{1}{2}\frac{7}{2}}\;{}_2F_1(1,\tfrac{3}{2};\tfrac{9}{2};1-\omega^{\frac{1}{2}})\;,$

$$C = (n-1)\Gamma(n+\tfrac{1}{2})/[7\Gamma(n-3)\Gamma(\tfrac{7}{2})]\;.$$

(Consul, 1967, [70])

6.6. Obtain the exact densities of (det S), W_2 and W_3 from the following moment expressions.

(i) $$E[(\det S)^t] = \frac{\Gamma_m(t+\frac{n}{2})}{\Gamma_m(\frac{n}{2})}(\det 2\Sigma)^t \exp(\operatorname{tr}\Omega)\;{}_1F_1(1+\tfrac{n}{2};\tfrac{n}{2};\Omega),$$

where Σ, Ω, S are positive definite matrices, n,m are positive integers, (det S) denotes the determinant of S and tr(·) denotes the trace (·) = the sum of the leading diagonal elements of (·);

(ii) $$E(W_2^h) = \frac{\Gamma_p(h+\frac{t}{2})\,\Gamma_p(\frac{n}{2})}{\Gamma_p(\frac{t}{2})\,\Gamma_p(h+\frac{n}{2})}\;{}_1F_1(h;h+\frac{n}{2};-\Omega),$$

where Ω is a positive definite matrix and n and p are positive integers, $0 < W_2 < 1$;

(iii) $$E(W_3^h) = \frac{\Gamma_p(\frac{n}{2})\,\Gamma_p(\frac{n-q}{2}+h)}{\Gamma_p(\frac{n-q}{2})\,\Gamma_p(\frac{n}{2}+h)}\,|I_p-P^2|^{\frac{n}{2}}\;{}_2F_1(\frac{n}{2},\frac{n}{2};\frac{n}{2}+h;P^2),$$

where P is a positive definite matrix, n,p,q are positive integers, $n \le q$, $p+q \le n$, $0 < W_3 < 1$ and $|(\;)|$ denotes the determinant of (·)

(Mathai, 1972, 1971, [184],[183]).

6.7. Show that the following densities have the gamma property discussed in Section 6.3.1 .

(i) $$\frac{(\alpha-\beta)^{\frac{1}{2}-\nu}\, e^{-px}\, x^{\nu-\frac{1}{2}}\, e^{-(\alpha+\beta)\frac{\alpha}{2}}}{\Gamma(\alpha)\;.(p+\alpha)^{-\nu}\;(p+\beta^{-\nu}}\; I_{\nu-\frac{1}{2}}[\frac{(\alpha-\beta)x)}{2}]\;, x > 0,$$

$$\alpha \ge \beta,\; \nu > 0\;;$$

(ii) $$\frac{\nu(\alpha-\beta)^{-\nu}\, e^{-px}\, x^{-1}\, e^{-(\alpha+\beta)\frac{x}{2}}}{[(p+\alpha)^{\frac{1}{2}} + (p+\beta)^{\frac{1}{2}}]^{-2\nu}}\; I_\nu\,[\frac{(\alpha+\beta)x)}{2}]\;,\; x > 0, \nu > 0,\; \alpha \ge \beta;$$

(iii) $$\frac{e^{-px}\, 2^{2\nu-2m-1}\, e^{\alpha m}}{\Gamma(2\nu-2m)(p-\alpha)^{2m}(p-\beta)^{-2\nu}}\;{}_1F_1(2\nu,2\nu-2m;(\beta-\alpha)x),\; x > 0,\; (\nu-m) > 0;$$

(iv) $$\frac{e^{-px}\, x^{2m+2\nu-1}}{\Gamma(2m+2\nu)\, p^{-2m}(p^2+\alpha^2)^{-\nu}}\; {}_1F_2(\nu;\, m+\nu,\, m+\nu+\frac{1}{2};\, \frac{-\alpha^2x^2}{4}),$$

$$x > 0, \quad m+\nu > 0 .$$

(Mathai and Saxena, 1968, [195]).

6.8 Show that the ratio of any two independent stochastic variables having the density functions in the class of $N_\nu(x)$ has a Cauchy probability law, where,

$$N_\nu(x) = \frac{\Gamma(\frac{\nu}{2}+1)\, x^\nu\, e^{-\frac{x^2}{2}}}{\Gamma(\nu+1)\,(2\pi)^{1/2}\, 2^{\nu/2}}\; {}_1F_1(\frac{\nu}{2};\, \nu+1;\, \frac{x^2}{2}),\ x > 0 .$$

6.9. Evaluate the exact density of $p_1X_1 + \ldots + p_nX_n$ if $X_1,\ldots,X_n$ are independent variates, $p_1,\ldots,p_n$ are constants and the density function of X_i being,

(i) $C_1\, x_i^{\alpha_i-1}\, e^{-a_ix_i^{\beta_i}}$, $x_i > 0,\ \alpha_i > 0,\ a_i > 0,\ \beta_i > 0$;

(ii) $C_2\, x_i^{\alpha_i-1}\, e^{-a_ix_i^{\alpha_i}}$, $x_i > 0,\ \alpha_i > 0,\ a_i > 0$;

(iii) $C_3\, x_i^{\alpha_i-1}\, e^{-a_ix_i}$, $x_i > 0,\ \alpha_i > 0,\ a_i > 0$;

(iv) $C_4\, e^{-a_ix_i}$, $x_i > 0,\ a_i > 0$,

where C_1, C_2, C_3 and C_4 are the normalizing constants.

CHAPTER VII

OTHER APPLICATIONS OF THE G-FUNCTION

In this chapter we consider three main problems, namely the production of heat in a cylinder, dual integral equations and hard limiting of sinusoidal signals, where the theory of G-function is applicable. In these problems we give only results which have physical interpretations. Other papers which are more of theoretical interest and which make use of G-functions extensively are mentioned in the text. There are other problems in other fields where G-functions are applicable eventhough no one has written articles in this direction. As an example one can examine the structure of the gamma factors appearing in the problem of functional equations considered by Bochner [45]. From this structure one can see that G and H-functions can come in handy in such problems in the theory of Functional Equations. Our discussion in this chapter is confined to some simple problems in Physical and Engineering Sciences.

7.1. PRODUCTION OF HEAT IN A CYLINDER.

From ([60], pp.12-13) it is seen that the problem in which heat is produced in the solid are important in technical applications. Heat is produced by the passage of an electrical current, dielectric or induction heating, radio active dicay, absorption from radiation, mechanical generation in viscous or plastic flow, chemical reaction, hydration of cement etc.

In this section we consider the application of a special case of G-function, namely, Jacobi polynomial, in the study of production of heat in a cylinder. From the following discussion it may be noticed that a general G-function can be used instead of a Jacobi polynomial but the physical interpretation may become difficult. Hence we limit our attention to the work of Bhonsle [42]. Some generalizations are given in the exercises. For further results in this line involving G-functions see Kalla and Kushwaha [129] and Bajpai [23].

Consider the difussion of heat in a cylinder of radius a when there are sources of heat within it which lead to an axially symmetrical temperature distribution. It is well-known that the fundamental differential equation for this problem is of the form, ([305], p.202),

$$\frac{\partial u}{\partial t} = \frac{k}{r}\frac{\partial}{\partial r}\left(r\frac{\partial u}{\partial r}\right) + \theta(r,t) . \tag{7.1.1}$$

This is the case if it is assumed that the rate of generation of heat is independent of temperature and that the cylinder is infintely long. Suppose that the initial distribution of temperature is zero and that the temperature at the surface $r = a$ is kept at zero. Let,

$$\theta(r,t) = \frac{k}{K} f(r)\, g(t), \tag{7.1.2}$$

where k is the diffusivity and K is the conductivity of the material. Let,

$$f(r) = \left(1-\frac{r^2}{a^2}\right) P_n^{(0,\beta)}\left(1-\frac{2r^2}{a^2}\right) \tag{7.1.3}$$

where $P_n^{(\alpha,\beta)}(x)$ is the Jacobi polynomial defined in Chapter II section 2.5.4. Throughout this section β will be assumed to be non-negative.

It may be noticed that f(r) can represent both sources and sinks embedded in the system. Whenever f(r) g(t) gives a negative value it should be treated as a sink. Heat sources will be characterized by the function g(t).

7.1.1 A Finite Hankel Transform.

The finite Hankel transform of f(r) is defined as, ([305], P.83),

$$J[f(r)] = \int_0^a t\, f(t)\, J_o(t\,\xi_i)dt, \tag{7.1.4}$$

where $J_\nu(\cdot)$ is a Bessel function defined in Chapter II. From Bhonsle ([40],P.189) it can be seen that,

$$J\left[\left(1-\frac{r^2}{a^2}\right)^\beta P_n^{(0,\beta)}\left(1-\frac{2r^2}{a^r}\right)\right] = \frac{a^{1-\beta}2^\beta\,\Gamma(\beta+n+1)\, J_{\beta+2n+1}(a\,\xi_i)}{n!\;\xi_i^{\beta+1}}, \tag{7.1.5}$$

where $R(\beta) > -1$ and ξ_i is the root of the equation

$$J_o(a\,\xi_i) = 0\,. \tag{7.1.6}$$

Also it is well-known ([305], P.83), that

$$\left(\frac{1-r^2}{a^2}\right)^{\beta} P_n^{(0,\beta)}\left(1-\frac{2r^2}{a^2}\right) = \frac{2^{\beta+1}\,\Gamma(\beta+n+1)}{n!\;a^{1+\beta}} \sum_j \frac{J_{\beta+2n+1}(a\,\xi_i)\,J_o(r\,\xi_i)}{\xi_i^{\beta+1}\,[J_1(a\,\xi_i)]^2} \tag{7.1.7}$$

where the sum is taken over all positive roots of equation (7.1.6).

Before discussing the solution of (7.1.1) we will discuss the behaviour of $f(r)$. This will help in interpreting the problem. We have,

$$f(r) = \left(1-\frac{r^2}{a^2}\right)^{\beta} P_n^{(0,\beta)}\left(1-\frac{2r^2}{a^2}\right) = \left(1-\frac{r^2}{a^2}\right)^{\beta}\ {}_2F_1\left(-n, n+\beta+1; 1; \frac{r^2}{a^2}\right). \tag{7.1.8}$$

When $n = 1$,

$$f(r) = \left(1-\frac{r^2}{a^2}\right)^{\beta}\left[\,1-(\beta+2)\,\frac{r^2}{a^2}\right]. \tag{7.1.9}$$

That is,

$$f(r) = 0 \quad \text{when } r = a\,(\beta+2)^{-\frac{1}{2}} \quad \text{and } r = a. \tag{7.1.10}$$

As β increases r decreases and

$$\begin{aligned} f(r) &> 0 \quad \text{when} \quad 0 \le r < a(\beta+2)^{-\frac{1}{2}} \\ &< 0 \quad \text{when} \quad a(\beta+2)^{-\frac{1}{2}} < r < a\,. \end{aligned} \tag{7.1.11}$$

If $g(t) > 0$ then we shall interpret that inner circular cylinder will enclose sources, while the volume between two concentric cylinders will contain sinks. If $g(t) < 0$ then sources and sinks will interchange their roles. The value of β determines the radius of the inner cylinder. When $n = 0$, $f(r) = \left(1-\frac{r^2}{a^2}\right)^{\beta}$ and when $n = 0$, $\beta = 0$ then evidently $f(r) = 1$.

7.1.2. Solution of the Differential Equation Concerning the Problem.

By applying finite Hankel transforms one can write down the solution of (7.1.1) by using ([305], P.203) as,

$$u(r,t) = \frac{2^{\beta+1} k\, \Gamma(\beta+n+1)}{n!\ a^{\beta+1}\ K} \sum_j \frac{J_o(r\,\xi_j) J_{\beta+2n+1}(a\,\xi_j)\ h(\xi_j,t)}{[J_1(a\,\xi_j)]^2\ \xi_j^{\beta+1}} \tag{7.1.12}$$

where the sum is over all the positive roots of the equation (7.1.6) and

$$h(\xi_j,t) = \int_o^t g(\tau)\, e^{-k\,\xi_j^2 (t-\tau)}\, d\tau\ . \tag{7.1.13}$$

By applying convolution theorem of the Laplace transformation one can evaluate (7.1.13). In order to apply this theorem $g(t)$ should be of order $e^{\gamma t}$ as t tends to infinity where $\gamma > 0$ and $g(t)$ must be sectionally continuous on each interval $0 \leq t \leq T$. Hence we will examine the following three particular cases of $g(t)$ which can be given physical interpretations.

(i) Heat sources acting for a finite interval of time;

(ii) Heat source of exponential character;

(iii) Heat source of Legendre polynomial character.

Case (i): Let

$$\begin{aligned} g(t) &= g_o\ , \quad 0 < t < b \\ &= 0, \quad t > b \end{aligned} \tag{7.1.14}$$

then the Laplace transform,

$$\begin{aligned} L[g(t)] &= \frac{g_o}{p}\ , \quad 0 < t < b \\ &= \frac{g_o}{p}(1-e^{-bp}), \quad t > b\ , \end{aligned} \tag{7.1.15}$$

and

$$L[h(\xi_i,t)] = \frac{g_o}{p(p+k\,\xi_i^2)}\ , \quad 0 < t < b$$

$$= \frac{g_o(1-e^{-bp})}{p(p+k\,\xi_i^2)}, \quad t > b. \tag{7.1.16}$$

Therefore,

$$h(\xi_i,t) = \frac{g_o(1-e^{-kt\,\xi_i^2})}{k\,\xi_i^2}, \quad 0 \le t \le b$$

$$= \frac{g_o}{k\,\xi_i^2}[1-e^{-k\,\xi_i^2\,t} + e^{-k\,\xi_i^2\,(t-b)}], \quad t \ge b, \tag{7.1.17}$$

and thus,

$$u(r,t) = \frac{g_o\, 2^\beta\,\Gamma\,(\beta+n+1)}{n!\;K\;a^{\beta+1}} \sum_j \frac{J_o(r\,\xi_j)\,J_{\beta+2n+1}(a\,\xi_j)\,(1-e^{-kt\,\xi_j^2})}{\xi_i^{\beta+3}\;[J_1(a\,\xi_j)]^2},$$

$$0 \le t \le b$$

$$= \frac{g_o\, 2^\beta\,\Gamma(\beta+n+1)}{n!\;K\;a^{\beta+1}} \sum_j \frac{J_o(r\,\xi_j)\,J_{\beta+2n+1}(a\,\xi_j)[1-e^{-k\xi_j^2 t}+e^{-k\xi_j^2(t-b)}]}{[J_1(a\,\xi_j)]^2\,\xi_j^{\beta+3}}$$

$$t \ge b. \tag{7.1.18}$$

Evidently $u(r,0) = 0$

Case (ii) Let

$$g(t) = g_o\,t^{\nu-1}\,e^{-\alpha t}, \quad \nu > 0, \quad \alpha > 0 \tag{7.1.19}$$

then

$$L[g(t)] = \frac{g_o\,\Gamma(\nu)}{(p+\alpha)^\nu} \tag{7.1.20}$$

and

$$L\,[h(\xi_i,t)] = g_o\,\Gamma(\nu) \sum_{r=o}^{\infty} \frac{(\alpha - k\,\xi_i^2)^r}{(p+\alpha)^{\nu+r+1}}. \tag{7.1.21}$$

The term by term inversion will give

$$h(\xi_i,t) = \frac{g_o\,t^\nu\,e^{-\alpha t}}{\nu}\;{}_1F_1\,[1;\,\nu+1;\,(\alpha-k\,\xi_i^2)t] \tag{7.1.22}$$

and thus

$$u(r,t) = \frac{2^{\beta}\ \Gamma(\alpha+n+1)\ g_o\ k\ t^{\nu}\ e^{-\alpha t}}{n!\ a^{\beta+1}\ \nu\ K}$$

$$\times \sum_j \frac{J_o(r\ \xi_j)\ J_{\beta+2n+1}(a\ \xi_j)}{[J_1(a\ \xi_j)]^2\ \xi_j^{\beta+1}}\ {}_1F_1\ [1;\ \nu+1;\ (\alpha-k\ \xi_j^2)t]. \quad (7.1.23)$$

Case (iii). Let

$$g(t) = g_o\ P_n(e^{-\alpha t}),\quad \alpha > 0 \quad (7.1.24)$$

where $P_n(x)$ is the Legendre polynomial defined in Chapter II.

$$L[g(t)] = \frac{g_o(p-\alpha)(p-2\alpha)\ldots[p-(n-1)\alpha]}{(p+n\alpha)\ [p+(n-2)\alpha]\ \ldots\ [p-(n-2)\alpha]} \quad (7.1.25)$$

and

$$L[h(\xi_i,t)] = \frac{g_o\ Q(p)}{(p+k\ \xi_i^2)\ P(p)}$$

where $Q(p) = (p-\alpha)(p-2\alpha)\ \ldots\ [p-(n-1)\alpha\ ]$

and $P(p) = (p-\beta_1)\ (p-\beta_2)\ \ldots\ (p-\beta_n)$

where $\beta_r = \alpha(\ -n+2r-2)\ ,\quad r = 1,\ldots,n\ .$

Therefore,

$$h(\xi_i,t) = \frac{g_o\ Q(-k\ \xi_i^2)\ e^{-k\ \xi_i^2\ t}}{P(-k\ \xi_i^2)} + \sum_{r=1}^{n} \frac{Q(\beta_r)\ e^{t\beta_r}}{P_r(\beta_r)(\beta_r+k\ \xi_i^2)} \quad (7.1.27)$$

and

$$P_r(\beta_r) = \left[\frac{P(p)}{p-\beta_r}\right]_{p=\beta_r}$$

It has been verified that $h(\xi_i,0) = 0$ for $n = 1,2,3$ and 4.

7.1.3. Verification of the Solution.

Verfication is easily done by substituting the following expressions in (7.1.1)

$$\frac{k}{r}\ \frac{\partial}{\partial r}\left(r\ \frac{\partial u}{\partial r}\right) = -\ \frac{2^{\beta+1} k\Gamma(\beta+n+1)}{n!\ K\ a^{1+\beta}}\ \sum_j\ \frac{\xi_j^2\ J_o(r\ \xi_j)\ J_{p+2n+1}(a\ \xi_j)}{[J_1(a\ \xi_j)]^2\ \xi_j^{\beta+1}}$$

$$* \int_o^t g(\tau)\ e^{-k\ \xi_j^2\ (t-\tau)}\ d\tau\ . \qquad (7.1.28)$$

$$\theta(r,t) = \frac{k}{K}\ \frac{2^{\beta+1}\ \Gamma(\beta+n+1)}{n!\ \ a^{1+\beta}}\ \sum_j\ \frac{J_{\beta+2n+1}(a\ \xi_j)\ J_o(r\ \xi_j)\ g(t)}{\xi_j^{\beta+1}\ [J_1(a\ \xi_j)]^2}\ . \qquad (7.1.29)$$

$$\frac{\partial u}{\partial t} = \frac{2^{\beta+1} k\ \Gamma(\beta+n+1)}{n!\ a^{\beta+1}\ K}\ \sum_j\ \frac{J_o(r\ \xi_j)\ J_{\beta+2n+1}(a\ \xi_j)}{[J_1(a\ \xi_j)]^2\ \ \xi_j^{\beta+1}}$$

$$* \left[g(t) - k\ \xi_j^2\ \int_o^t g(\tau)\ e^{-k\ \xi_j^2\ (t-\tau)}\ d\tau\ \right]. \qquad (7.1.30)$$

The boundary condition $u(a,t) = 0$ is satisfied because $J_o(a\ \xi_i)$ which is present in every term of $u(a,t)$ is zero. The initial condition is satisfied because $h(\xi_i,0) = 0$.

For a discussion of the justification of the differentiation of $u(r,t)$ with respect to t and r, see Bhonsle [42].

Remark 1: A circular section obtained by the intersection of the plane perpendicular to the axis of the cylinder considered in this section will have three sets of points S_1, S_2 and S_3. S_1 contains source points, S_2 contains sink points and S_3 contains the boundary points. We have discussed the temperature distribution in the region $S_1 \cup S_2 \cup S_3$.

Remark 2: S_1 and S_2 interchange their roles whenever g(t) changes sign. g(t) may be selected such that it fluctuates a large number of times in a finite interval of time. This will illustrate the situation when S_1 and S_2 interchange their roles a large number of times within a finite interval of time.

Remark 3: It may be pointed out that the two sets of values considered in ([305], p.204) become the subsets of the set considered in this section. By virtue of the expansion property of the orthogonal polynomials the set containing source of Legendre polynomial character may include several cases of interest.

7.2. DUAL INTEGRAL EQUATION

7.2.1. Introduction.

In the analysis of mixed boundary value problems, we often encounter pairs of dual integral equations which can be written in the form:

$$\int_0^\infty u^\lambda J_\mu(ux)\, f(u)\, du = g(x), \quad 0 < x < 1 , \tag{7.2.1}$$

and

$$\int_0^\infty u^\sigma J_\nu(ux)\, f(u)\, du = h(x), \quad x > 1 , \tag{7.2.2}$$

where $J_\alpha(x)$ is the ordinary Bessel function, g(x) and h(x) are given and f(x) is to be determined.

The interesting case when $\lambda = \mu = \nu = 0$, $\sigma = 1$, g(x) is equal to a constant and h(x) = 0 occurs in the problem of finding the electrostatic field arising from a flat circular disk charged to a constant potential. This problem seem to have been solved first by Weber in 1873 and since then a number of methods are developed by Busbridge [55], Erdélyi and Sneddon [87] Noble [221], Peters [230] and others to deal with the more general pair of integral equations than (7.2.1) and (7.2.2) above. A systematic treatment of this subject is given by Fox [97].

In the present section we give a general method based on the theory of Fractional Integration operators to obtain the solution of dual integral equations associated with any Special Function which has a Mellin-Barnes type integral representation. This discussion here is based on the work of Saxena [271]. We consider the most general case in which the dual integral equations contain general H-functions as kernels . It has been shown that the given dual integral equations can be reduced into two others having the same kernel by the application of Fractional Integration operators. Then by an appeal to the theory of generalized Fourier kernels the solution is immediately obtained. The final solution is obtained in an elegant form which is most suitable for practial applications in various problems of Mathematical Physics. The results obtained in this section may be regarded as the key formulas for the solution of dual integral equations associated with any Special Function governing the problem under consideration.

For a discussion of dual integral equations associated with a Symmetrical Fourier kernel due to Fox [94], see Saxena [272].

7.2.2. The H-function.

We define the H-function by means of the Mellin-Barnes integral as

$$H^{m,n}_{p+n,q+m}\left(x\middle|\begin{matrix}(a_j,A_j)\\(b_k,B_k)\end{matrix}\right)$$

$$= H^{m,n}_{p+n,q+m}\left(x\middle|\begin{matrix}(a_1,A_1),\ldots,(a_{n+p},A_{n+p})\\(b_1,B_1),\ldots,(b_{m+q},B_{m+q})\end{matrix}\right)$$

$$= \frac{1}{2\pi i}\int_L \chi_1(s)\, x^{-s} ds \qquad (7.2.3)$$

where

$$\chi_1(s) = \frac{\prod_{j=1}^{m}\Gamma(b_j+s\,B_j)\prod_{j=1}^{n}\Gamma(a_j-s\,A_j)}{\prod_{j=1}^{q}\Gamma(b_{m+j}-s\,B_{m+j})\prod_{j=1}^{p}\Gamma(a_{n+j}+s\,A_{n+j})}. \qquad (7.2.4)$$

An empty product is to be interpreted as unity and the following simplified assumptions are made.

(i) a_j, A_j, b_k and B_k are all real for $j = 1,2,\ldots,(p+n)$ and $k = 1,2,\ldots,(q+m)$.

(ii) A_j's and B_k's are positive for $j = 1,2,\ldots,(p+n)$ and $k = 1,2,\ldots,(q+m)$.

(iii) All the poles of the integrand in (7.2.3) are simple.

(iv) Let $s = \sigma+it$, σ and t being real. Then the contour L is a straight line parallel to the imaginary axis in the complex s-plane whose equation is $\sigma = \sigma_o$, where σ_o is a constant, and is such that all the poles of $\Gamma(b_k + B_k s)$ for $k = 1,2,\ldots,m$ lie to the left and those of $\Gamma(a_j - A_j s)$ for $j = 1,2,\ldots,n$, to the right of it

(v) $$\sum_{j=1}^{p} A_j + \sum_{j=1}^{m} B_j = \sum_{j=1}^{p} A_{n+j} + \sum_{j=1}^{q} B_{m+j} .$$

(vi) $$\lambda = \sum_{j=1}^{n} A_j + \sum_{j=1}^{p} A_{n+j} - \sum_{j=1}^{m} B_j - \sum_{j=1}^{q} B_{m+j} .$$

(vii) $$\mu = \sum_{j=1}^{n} a_j + \sum_{j=1}^{m} b_j - \sum_{j=1}^{p} a_{n+j} - \sum_{j=1}^{q} b_{m+j} + \frac{(p+q-m-n)}{2} .$$

(viii) $x > 0$.

(ix) $\lambda\,\sigma_o + \mu + 1 < 0$.

It can be easily seen from the asymptotic representation of the gamma function ([85], Vol. I. p.47)

$$\lim_{|t|\to\infty} |\Gamma(\sigma+it)|\;|t|^{\frac{1}{2}-\sigma} \exp\left(\frac{\lambda|t|}{2}\right) = (2\pi)^{\frac{1}{2}}$$

on taking $s = \sigma_o + it$, $x = Re^{i\emptyset}$ ($R > 0$, $\emptyset$ real) that the absolute value of the integrand of (7.2.3) is comparable with

$$\left(\frac{\rho}{R}\right)^{\sigma_o} |t|^{\lambda\sigma_o+\mu} e^{\phi t}$$

where

$$\rho = \prod_{j=1}^{n} (A_j)^{A_j} \prod_{j=1}^{m} (B_j)^{-B_j} \prod_{j=1}^{p} (A_{n+j})^{A_{n+j}} \prod_{j=1}^{q} (B_{n+j})^{-B_{n+j}} ,$$

when $|t|$ is large. Hence the integral (1.7.3) converges absolutely if the conditions (viii) and (ix) are satisfied.

The solution of the following dual integral equations will be developed here.

$$\int_0^\infty H_{p+n,q+m}^{m,n}\left[xu \,\middle|\, \begin{matrix}(a_j,A_j)\\(b_k,B_k)\end{matrix}\right] f(u)du = \phi(x), \quad 0 < x < 1 \tag{7.2.5}$$

and

$$\int_0^\infty H_{p+n,q+m}^{m,n}\left[xu \,\middle|\, \begin{matrix}(c_j,A_j)\\(d_k,B_k)\end{matrix}\right] f(u)du = \psi(x), \quad x > 1 , \tag{7.2.6}$$

where $\phi(x)$ and $\psi(x)$ are given and $f(x)$ is to be determined. It is assumed that the H-function of (7.2.6) satisfies all the conditions given above with a_j replaced by c_j and b_k replaced by d_k for $j = 1,2,\ldots,(n+p)$ and $k = 1,2,\ldots,(m+q)$. Further it is assumed that a common value of σ_o can be found for both the H-functions.

7.2.3. Mellin Transform.

The following results will be required in the analysis that follows.

The notations $M[f(u)]$ or $F(s)$ will be used to denote the integral

$$\int_0^\infty u^{s-1} f(u)du \tag{7.2.7}$$

and $M^{-1}\{F(s)\}$ to represent the inverse Mellin transform, namely

$$M^{-1}[F(s)] = f(u) = \frac{1}{2\pi i} \int_L F(s)u^{-s} ds . \tag{7.2.8}$$

For the conditions of the validity of (7.2.7) and (7.2.8), the reader is referred to the book by Titchmarsh [329, sections 1.29 and 1.37].

Fox ([97], p.391)has restated the Parseval theorem of Mellin transform in the following form:

Lemma 7.2.1. If

$$M[h(u)] = H(s)$$

and

$$M[f(xu)] = x^{-s}F(s),$$

then

$$\int_0^\infty h(xu)\, f(u)du = \frac{1}{2\pi i} \int_L x^{-s} H(s)\, F(1-s)ds \; . \tag{7.2.9}$$

We also make use of the generalized Fourier transformation which consists of the reciprocity relations,

$$\emptyset(x) = \int_0^\infty p(ux)\, f(u)du \tag{7.2.10}$$

and

$$f(x) = \int_0^\infty q(ux)\, \emptyset(u)du \, . \tag{7.2.11}$$

The functions p(x) and q(x) are known as the kernels of the transformation. The transformation is said to be symmetrical if p(x) and q(x) are equal and unsymmetrical otherwise .

Among the conditions required for the validity of (7.2.10) and (7.2.11) is that the functional equation

$$P(s)\; Q(1-s) = 1 \tag{7.2.12}$$

is satisfied, where $M[p(u)] = P(s)$ and $M[q(u)] = Q(s)$.

Remark: The conditions under which the G and H-functions form a pair of symmetrical Fourier kernels and unsymmetrical Fourier kernels are respectively given by Fox [94] and Narain [217].

By the application of (7.2.9) to (7.2.10) and (7.2.11) the following Lemma is obtained by Fox ([97], p.395).

Lemma 7.2.2.

$$\emptyset(x) = \frac{1}{2\pi i} \int_C P(s)x^{-s}\, F(1-s)ds \tag{7.2.13}$$

and

$$f(x) = \frac{1}{2\pi i} \int_C \frac{x^{-s}\,\phi(1-s)ds}{P(1-s)}, \qquad (7.2.14)$$

where $M[p(u)] = P(s)$; $\phi(x)$ and $f(x)$ are connected by the relations (7.2.10) and (7.2.11).

From (7.2.3) and (7.2.8), it follows that

$$M[\, H^{m,n}_{p+n,q+m}\left(x \,\middle|\, \begin{matrix}(a_j, A_j)\\(b_k, B_k)\end{matrix}\right)] = \chi_1(s)\,. \qquad (7.2.15)$$

On using $M[f(u)] = F(s)$ and applying (7.2.9) to (7.2.5) and (7.2.6) it is seen that

$$\frac{1}{2\pi i} \int_L \chi_1(s) x^{-s}\, F(1-s)ds = \phi(x) \qquad (7.2.16)$$

where $0 < x < 1$, $\chi_1(s)$ is defined in (7.2.4) and

$$\frac{1}{2\pi i} \int_L \chi_2(s) x^{-s}\, F(1-s)ds = \psi(x), \qquad (7.2.17)$$

where $x > 1$ and $\chi_2(s)$ is obtained by $\chi_1(s)$ by replacing the a_j's by c_j's and b_j's and d_j's .

7.2.4. Fractional Integration Operators and Their Applications.

In this section we will transform the equations (7.2.16) and (7.2.17) into two others with the same kernel by the application of fractional integration operators. The main technique is to transform

$$\frac{\prod_{j=1}^{n} \Gamma(a_j - s\,A_j)}{\prod_{j=1}^{q} \Gamma(b_{m+j} - s\,B_{m+j})} \text{ of (7.2.16) into } \frac{\prod_{j=1}^{n} \Gamma(c_j - s\,A_j)}{\prod_{j=1}^{q} \Gamma(d_{m+j} - s\,B_{m+j})}, \qquad (7.2.18)$$

of (7.2.17) and

$$\frac{\prod_{j=1}^{m} \Gamma(d_j + s\,B_j)}{\prod_{j=1}^{p} \Gamma(c_{n+j} + s\,A_{n+j})} \text{ of (7.2.17) into } \frac{\prod_{j=1}^{m} \Gamma(b_j + s\,B_j)}{\prod_{j=1}^{p} \Gamma(a_{n+j} + s\,A_{n+j})}, \qquad (7.2.19)$$

of (7.2.16).

In making these transformations, we will make use of the fractional integration operators due to Erdélyi [84].

$$\tau[\alpha,\beta:r:\, w(x)] = \{\frac{r}{\Gamma(\alpha)}\}\, x^{-r\alpha + r-\beta-1} \times \int_0^x (x^r - v^r)^{\alpha-1} v^{\beta}\, w(v)dv \,. \qquad (7.2.20)$$

$$R[\alpha,\beta:r:\, w(x)] = \frac{r\,x^{\beta}}{\Gamma(\alpha)} \int_x^{\infty} (v^r - x^r)^{\alpha-1} v^{-\beta-r\alpha+r-1} w(v)dv \,. \qquad (7.2.21)$$

For $r = 1$, these operators reduce to the ones studied by Kober [138]. Extensions of these operators are given by Kalla and Saxena [130].

It is convenient to employ the following contracted notations.

$$\tau[(a_j - c_j),\ c_j A_j^{-1} - 1:\ A_j^{-1}:\, w(x)] = \tau_j[w(x)] \,. \qquad (7.2.22)$$

$$\tau[(d_{m+k} - b_{m+k}),\ (b_{m+k} B_{m+k}^{-1} - 1):\ B_{m+k}^{-1} :\, w(x)] = \tau_K^{*}[w(x)] \,. \qquad (7.2.23)$$

$$R[(d_\rho - b_\rho),\ b_\rho B_\rho^{-1} : B_\rho^{-1} :\, w(x)] = R_\rho[w(x)] \,. \qquad (7.2.24)$$

$$R[(a_{n+h} - c_{n+h}),\ c_{n+h} A_{n+h}^{-1} : A_{n+h}^{-1} :\, w(a)] = R_h^{*}[w(x)] \,. \qquad (7.2.25)$$

To effect the first transformation, let us replace x by v, multiply by $v^{c_n e_n - 1} (x^{e_n} - v^{e_n})^{a_n - c_n - 1}$ where $e_n = A_n^{-1}$, in (7.2.16), integrate with respect to v from 0 to x, $0 < x < 1$ and apply the well known Beta function formula, we then obtain

$$\frac{1}{2\pi i}\int_L \frac{\prod_{j=1}^{m}\Gamma(b_j+sB_j)\prod_{j=1}^{n-1}\Gamma(a_j-A_js)\,\Gamma(c_n-sA_n)}{\prod_{j=1}^{q}\Gamma(b_{m+j}-sB_{m+j})\prod_{j=1}^{p}\Gamma(a_{n+j}+sA_{n+j})}\,x^{-s}F(1-s)ds$$

$$= \frac{e_n x^{e_n-a_ne_n}}{\Gamma(a_n-c_n)}\int_0^x v^{e_nc_n-1}(x^{e_n}-v^{e_n})^{a_n-c_n-1}\phi(v)dv$$

$$= \tau_n[\phi(v)], \qquad (7.2.26)$$

where $0 < x < 1$, and $e_n = A_n^{-1}$, on using (7.2.20)and (7.2.22).

Again transforming (7.2.26) sucessively for $k = n-1, n-2, \ldots, 3,2,1$ by the application of the operators τ_k and by τ_j^* for $j = q, q-1, \ldots, 3,2,1$, we finally get

$$\frac{1}{2\pi i}\int_L \frac{\prod_{j=1}^{m}\Gamma(b_j+sB_j)\prod_{j=1}^{n}\Gamma(c_j-sA_j)\,x^{-s}F(1-s)ds}{\prod_{j=1}^{q}\Gamma(d_{m+j}-sB_{m+j})\prod_{j=1}^{p}\Gamma(a_{n+j}+sA_{n+j})}$$

$$= \tau_1^*[\tau_2^* \ldots \tau_q^*\ \tau_1 \ldots \tau_n[\phi(x)]\ldots], \qquad (7.2.27)$$

where $0 < x < 1$.

The second transformation can be effected by using the operators R_ρ and R_h^* given in (7.2.24) and (7.2.25) respectively for $\rho = m, m-1, \ldots, 3,2,1$ and $h = p, p-1, \ldots, 3,2,1$ to transform (7.2.17) into the desired form

$$\frac{1}{2\pi i}\int_L \frac{\prod_{j=1}^{m}\Gamma(b_j+sB_j)\prod_{j=1}^{n}\Gamma(c_j-sA_j)\,x^{-s}F(1-s)ds}{\prod_{j=1}^{q}\Gamma(d_{m+j}-sB_{m+j})\prod_{j=1}^{p}\Gamma(a_{n+j}+sA_{n+j})}$$

$$= R_1^*[R_2^* \ldots R_p^*\ R_1 \ldots R_m[\psi(x)]\ldots], \qquad (7.2.28)$$

where $x > 1$.

If we set

$$t(x) = \begin{cases} \tau_1^* [\tau_2^* \ldots \tau_q^* \tau_1 \tau_2 \ldots \tau_n [\phi(x)] \ldots], & 0 < x < 1 \\ R_1^* [R_2^* \ldots R_p^* R_1 R_2 \ldots R_m [\psi(x)] \ldots], & x > 1 \end{cases} \tag{7.2.29}$$

the equations (7.2.27) and (7.2.28) can be put into a compact form

$$\frac{1}{2\pi i} \int_L \frac{\prod_{j=1}^{m} \Gamma(b_j + s B_j) \prod_{j=1}^{n} \Gamma(c_j - s A_j) x^{-s} F(1-s) ds}{\prod_{j=1}^{q} \Gamma(d_{m+j} - s B_{m+j}) \prod_{j=1}^{p} \Gamma(a_{n+j} + s A_{n+j})} = t(x) . \tag{7.2.30}$$

Here (7.2.30) is the reduction of (7.2.16) and (7.2.17) to two equations with a common kernel on applying Lemma 7.2.2, we find that (7.2.30) can be written as

$$f(x) = \frac{1}{2\pi i} \int_L \frac{\prod_{j=1}^{q} \Gamma(d_{m+j} + s B_{m+j} - B_{m+j}) \prod_{j=1}^{p} \Gamma(a_{n+j} + A_{n+j} - s A_{n+j})}{\prod_{j=1}^{m} \Gamma(b_j + B_j - s B_j) \prod_{j=1}^{n} \Gamma(c_j - A_j + s A_j)} x^{-s} T(1-s) ds, \tag{7.2.31}$$

where $M\, t(x) = T(x)$.

Applying (7.2.19) to (7.2.31) , it becomes $f(x) = \int_0^\infty t(u)$

$$\times H^{q,p}_{p+n,q+m}\left(xu \,\middle|\, \begin{matrix} (a_{n+j} + A_{n+j}, A_{n+j}), (c_j - A_j, A_j) \\ (d_{m+k} - B_{m+k}, B_{m+k}), (b_k + B_k, B_k) \end{matrix}\right) du . \tag{7.2.32}$$

It is clear from the nature of this solution that it can be written down by inspection from (7.2.5) and (7.2.6).

The solution (7.2.32) can be verified by reversing all the steps from (7.2.5) (7.2.6) to (7.2.32) by means of the operators τ^{-1} and R^{-1} defined by Fox ([97], p.397), namely,

$$\tau^{-1}[\alpha,\beta:r: f(x)] = \tau[-\alpha, \beta+r\alpha :r: f(x)] \tag{7.2.33}$$

$$R^{-1}[\alpha,\beta:r: f(x)] = R[-\alpha,\beta+r\alpha :r: f(x)] , \tag{7.2.34}$$

and establishing conditions which justify these reversals. The H-function of (7.2.32) exists if $\lambda\sigma_o + \mu' + \lambda + 1 < 0$, where μ' is obtained from μ by interchanging a_j's by $-c_j$'s, $j = 1,\ldots,n$; b_j's by $-b_j$'s for $j = 1,2,\ldots,m$; a_{n+j} by $-a_{n+j}$; b_{m+j} by $-d_{m+j}$ for $j = 1,2,\ldots,q$ and $\frac{1}{2}(p+q-m-n)$ by $\frac{1}{2}(m+n-p-q)$.

7.2.5. Interesting Particular Cases.

(i) Fox's result ([97], p.395) can be derived from (7.2.32) by taking $n = p = 0$, $q = m$, $B_{m+j} = B_j = a_j$, $b_j = \alpha_j$, $b_{m+j} = \beta_j$, $c_j = \lambda_j$, $d_{m+j} = \mu_j$, for $j = 1,2,\ldots m$.

(ii) If we take $A_j = B_k = 1$, for $j = 1,2,\ldots, (n+p)$ and $k = 1,2,\ldots, (m+q)$, H-functions then reduce to G-functions and consequently the function

$$f(x) = \int_o^1 G^{q,p}_{p+n,q+m}\left(xu \,\middle|\, \begin{matrix} -a_{n+1},\ldots,-a_{n+p}, -c_1,\ldots,-c_n \\ -d_{m+1},\ldots,-d_{m+q}, -b_1,\ldots,-b_m \end{matrix}\right)$$

$$\times \tau_1^* [\tau_2^* \ldots \tau_1 \ldots \tau_n [\phi(u)]\ldots]du$$

$$+ \int_1^\infty G^{q,p}_{p+n,q+m}\left(xu \,\middle|\, \begin{matrix} -a_{n+1},\ldots,-a_{n+p}, -c_1,\ldots,-c_n \\ -d_{m+1},\ldots,-d_{m+q}, - b_1,\ldots,-b_m \end{matrix}\right)$$

$$\times R_1^* [R_2^* \ldots R_p^* R_1 R_2 \ldots R_m [\psi(u)]\ldots] du , \tag{7.2.35}$$

satisfies the dual integral equations

$$\int_o^\infty G^{m,n}_{p+n,q+m}\left(xu \,\middle|\, \begin{matrix} a_1,\ldots,a_{n+p} \\ b_1,\ldots,b_{m+q} \end{matrix}\right) f(u)du = \phi(x) , \quad 0 < x < 1 ,$$

and

$$\int_0^\infty G^{m,n}_{p+n,q+m}\left(xu \,\Big|\, \begin{matrix} c_1,\ldots,c_{n+p} \\ d_1,\ldots,d_{m+q} \end{matrix}\right) f(u)du = \psi(x) \ , \quad x > 1.$$

Here the operators are to be taken with a_j replaced by $1-a_j$, $j = 1,\ldots,n$; b_{m+j} by $1-b_{m+j}$, $j = 1,2,\ldots,q$, c_j by $1-c_j$, $j = 1,2,\ldots,n$; d_{m+j} by $1-d_{m+j}$, $j = 1,2,\ldots,q$,

$$B_j = A_k = 1 \quad \text{for } j = 1,2,\ldots, (m+q) \quad \text{and } k = 1,2,\ldots, ,(n+p) \ .$$

Under these conditions the operators become Kober operators.

For further discussion on dual integral equations associated with H-function, see the work of Saxena [272].

Remark 1: The importance of the results obtained in this section lies in the fact that the solution of dual integral equations involving nearly all the special functions occurring in Applied Mathematics can be derived as their particular cases.

Remark 2: Solution of an integral equation whose kernel has a Mellin-Barnes type integral representation is obtained by Saxena [270] by the application of fractional integral operators. The result is important from the point of view that the inversion formula for all the important integral transforms given in the monograph by Erdélyi, A. et al [86] and the transform introduced by Bhise [36] can be obtained as its particular cases.

7.3 HARD LIMITING OF SEVERAL SINUSOIDAL SIGNALS

This is a problem of interest to engineers and communication technologists. In this section we consider only the mathematical aspect of the problem for its engineering and other aspects see Jones [126], Sevy [277], Shaft [278] and Sollfrey [309]. These are only a few of the several papers on this topic.

The problem of limiting several sinusoidal signals of arbitrary amplitudes can be solved in terms of generalized functions of several variables. But these functions, except for some particular cases, can not be put into computable forms. In the following discussion we will use the standard notations, usually used in this field. Consider the input,

$$e_{in} = a\cos(\omega_1 t + e_1) + b[\cos(\omega_2 t + e_2) + \cos(\omega_3 t + e_3)], \qquad (7.3.1)$$

where the frequencies ω_1, ω_2 and ω_3 are assumed incommensurable. Let the limiter characteristic be,

$$\begin{aligned} e_{out} &= 1 \quad , \quad e_{in} > 0 \\ &= 0 \quad , \quad e_{in} = 0 \\ &= -1 \quad , \quad e_{in} < 0 \, . \end{aligned} \qquad (7.3.2)$$

This characteristic can be represented as

$$e_{out} = \frac{2}{\pi} \int_0^\infty \sin(x\, e_{in}) \frac{dx}{x} \, . \qquad (7.3.3)$$

Sollfrey [309] has given a detailed derivation of the output signal components, which is reproduced here for completeness. The output signal components can be written as,

$$e_s = c \cos r_1 + d\,(\cos r_2 + \cos r_3) \qquad (7.3.4)$$

where for example, $r_1 = \omega_1 t + e_1$ and

$$c = \frac{4}{\pi} \int_0^\infty J_1(ax)\, [J_o(bx)]^2 \frac{dx}{x} \, , \qquad (7.3.5)$$

and

$$d = \frac{4}{\pi} \int_0^\infty J_o(ax)\, J_o(bx)\, J_1(bx) \frac{dx}{x} \, , \qquad (7.3.6)$$

where $J_\nu(\cdot)$ is a Bessel function. In order to evaluate (7.3.5) and (7.3.6) Sollfrey [309] used the integral representations of the product of two Bessel functions, introduced one more Bessel function through a complicated procedure and then evaluated the integrals by using Calculus of residues. But these results are available as particular cases of a known result. By making the substitutions,

$\rho = 0$, $\lambda = \mu$, $\nu = 1$ and replacing a by b and b by a/2, formula (7) of Erdélyi, A.et.al ([86], Vol. 2, p.350) reduces to

$$_3F_2\left(\frac{1}{2}, \frac{1}{2}, -\frac{1}{2}; 1,1; \frac{4b^2}{a^2}\right), \quad 2b < a . \tag{7.3.7}$$

This is Sollfrey's result (16). By making the substitution $\rho = 0$, $\lambda = 0$, $\nu = 0$, $\mu = 1$ and replacing a by b and b by a/2 the same formula Erdélyi, A. et. al ([86], Vol.2, 350 (7)) reduces to

$$\frac{b}{2a}\ {}_3F_2\left(\frac{3}{2}, \frac{1}{2}, \frac{1}{2}; 2,2; \frac{4b^2}{a^2}\right), \quad ab < a . \tag{7.3.8}$$

This is Sollfrey's result (20) which also has a misprint, namely, $\Gamma(n + \frac{1}{2})$ should be replaced by $\Gamma^2(n + \frac{1}{2})$. Also in Erdélyi, A. et. al ([86], Vol.2, P.350 (7)), $2^{\lambda+\mu}$ should be $2^{\lambda+\mu+1}$ and the condition $R(\rho) < 5/2$ is missing. When the amplitudes are different, integrals corresponding to (7.3.5) and (7.3.6) can be evaluated by using Erdélyi, A.et. al ([86], Vol.2, 351 (9)) in terms of Appell's hypergeometric function of two variables F_4. When there are only three Bessel functions involved this F_4 can be put into a form which can be handled by a digital computer.

In order to solve the problem of limiting four sinusoidal signals involving two different-amplitudes one has to evaluate the integrals,

$$c' = \frac{4}{\pi} \int_0^\infty J_o(ax)\ J_1(ax)\ [J_o(bx)]^2\ \frac{dx}{x} , \tag{7.3.9}$$

and

$$d' = \frac{4}{\pi} \int_0^\infty [\ J_o(ax)]^2\ \ J_o(bx)\ J_1(bx)\ \frac{dx}{x} , \tag{7.3.10}$$

where a and b are constant. In this section we give a simple alternate method of evaluating (7.3.9) and (7.3.10). The following discussion is based on Mathai [177]. From Erdélyi, A. et. al ([86], Vol. 1, P.331(33)), we have,

$$\int_0^\infty x^{s-1}\ J_o(ax)\ J_1(ax)dx = \frac{a^{-s}\ \Gamma(1-\frac{s}{2})\ \Gamma(\frac{1}{2}+\frac{s}{2})}{2\sqrt{\pi}\ \ \Gamma^2(\frac{3}{2}-\frac{s}{2})} , \tag{7.3.11}$$

$$-1 < R(s) < 1, \quad a > 0 \ \text{ and}$$

$$\int_0^\infty x^{s-1}\,[J_o(bx)]^2\,dx = \frac{b^{-s}\,\Gamma(\frac{1}{2}-\frac{s}{2})\,\Gamma(\frac{s}{2})}{2\sqrt{\pi}\,\Gamma^2(1-\frac{s}{2})} \tag{7.3.12}$$

$$0 < R(s) < 1, \quad b > 0.$$

From (7.3.11) and (7.3.12) one can write,

$$\int_0^\infty x^{-1}\; J_o(ayx)\, J_1\,(ayx)\, J_o(bx)\, J_o(bx)dx$$

$$= \frac{1}{2\pi}\, G^{2,2}_{4,4}\left[\frac{a^2y^2}{b^2}\,\middle|\,\begin{matrix}0,1,1,1\\ \frac{1}{2},\frac{1}{2},-\frac{1}{2},-\frac{1}{2}\end{matrix}\right), \quad \frac{a^2y^2}{b^2} < 1. \tag{7.3.13}$$

Now (7.3.13) can be evaluated by using the technique in Chapter V.

7.3.1. Limiting of Five Sinusoidal Signals.

Proceeding as in Sollfrey [309] one can easily see that the essential problem in this case is the evaluation of an integral involving product of five Bessel functions of the following type.

$$I_1 = \int_0^\infty x^{-1}\, J_1(ax)\, J_o(ax)\, J_o(2bx)\, [J_o(cyx)]^2 dx, \tag{7.3.14}$$

$a,b,c,y > 0$. By using Erdélyi, A. et.al ([86], Vol.2, p.350 (7) Vol. 1,p.308 (13)), (7.3.14) can be written as,

$$I_1 = \frac{1}{2\pi}\sum_{r=o}^{\infty}\frac{(\frac{a^2}{b^2})^r\;\Gamma(\frac{3}{2}+r)}{r!\;\Gamma^2\;(2+r)}\; G^{1,2}_{3,3}\left(\frac{c^2y^2}{b^2}\,\middle|\,\begin{matrix}\frac{1}{2}-r,\,\frac{1}{2}-r,\,\frac{1}{2}\\ 0,\;0,\;0\end{matrix}\right) \tag{7.3.15}$$

$c^2y^2/b^2 < 1$, $0 < a < b$. Now from Erdélyi, A. et. al ([85],Vol.1, p.208 (5)) we have,

$$G^{1,2}_{3,3}\left(\frac{c^2y^2}{b^2}\,\middle|\,\begin{matrix}\frac{1}{2}-r,\,\frac{1}{2}-r,\,\frac{1}{2}\\ 0,\;0,\;0\end{matrix}\right] = \frac{\Gamma^2(\frac{1}{2}+r)}{\Gamma(\frac{1}{2})}$$

$$\times\; {}_3F_2(\tfrac{1}{2}+r,\,\tfrac{1}{2}+r,\,\tfrac{1}{2};\,1,1;\,\frac{c^2y^2}{b^2})\;. \tag{7.3.16}$$

If $(c^2y^2/b^2) > 1$ then by using (1.2.8) we have,

$$G^{1,2}_{3,3}\left(\frac{c^2y^2}{b^2}\middle|\begin{matrix}\frac{1}{2}-r,\ \frac{1}{2}-r,\ \frac{1}{2}\\ 0,\ 0,\ 0\end{matrix}\right] = G^{2,1}_{3,3}\left[\frac{b^2}{c^2y^2}\middle|\begin{matrix}1,\ 1,\ 1\\ \frac{1}{2}+r,\ \frac{1}{2}+r,\ \frac{1}{2}\end{matrix}\right)$$

$$= \sum_{\nu=0}^{\infty} \frac{\Gamma(\frac{1}{2}+r+\nu)\left(\frac{b^2}{c^2y^2}\right)^{r+\nu+\frac{1}{2}}}{(\nu!)^2\,\Gamma(1+r+\nu)\,\Gamma^2(\frac{1}{2}-r-\nu)}\,[-\log(b^2/c^2y^2)$$

$$+ 2\,\psi(\nu+1) + \psi(\tfrac{1}{2}+r+\nu) - \psi(1+r+\nu) - 2\,\psi(\tfrac{1}{2}-r-\nu)], \quad (7.3.17)$$

by using the technique in Chapter V where the ψ-function is defined in (5.3.2).

7.3.2. Limiting of Six Sinusoidal Signals.

In this case the essential problem is the evaluation of the integral,

$$I_2 = \int_0^\infty x^{-1}\,J_1(ayx)\,J_o(ayx)[J_o(cx)]^2\,[J_o(dx)]^2 dx,\quad a,c,d,y > 0\,. \qquad (7.3.18)$$

From Erdélyi, A.et.al ([85],Vol.1,p.217(20); [86],Vol.2,p.422(14) [85],Vol.1, p.209(8); and [85], Vol.1, p.208 (5)), we have,

$$\int_0^\infty x^{s-1}\,[J_o(cx)]^2\,[J_o(dx)]^2 dx = \frac{(c^2)^{-\frac{s}{2}}}{2\pi}\,G^{2,2}_{4,4}\left[\frac{d^2}{c^2}\middle|\begin{matrix}-\frac{s}{2}+1,\frac{1}{2},-\frac{s}{2}+1,-\frac{s}{2}+1\\ \frac{1}{2}-\frac{s}{2},\ 0,\ 0,\ 0\end{matrix}\right],$$

$$0 < R(s) < 2,\quad (d^2/c^2) < 1, \qquad (7.3.19)$$

$$= \frac{(c^2)^{-\frac{s}{2}}}{2\pi}\left\{\frac{\Gamma(-\frac{1}{2}+s)\,\Gamma(\frac{1}{2})\,\Gamma(1-\frac{s}{2})\left(\frac{d^2}{c^2}\right)^{\frac{1}{2}-\frac{s}{2}}}{\Gamma^2(\frac{3}{2}-\frac{s}{2})\,\Gamma^2(\frac{1}{2})}\;{}_4F_3\left(\tfrac{1}{2},1-\tfrac{s}{2},\tfrac{1}{2},\tfrac{1}{2};\right.\right.$$

$$\left.\tfrac{3}{2}-\tfrac{s}{2},\ \tfrac{3}{2}-\tfrac{s}{2},\ \tfrac{3}{2}-\tfrac{s}{2}\ ;\ \frac{d^2}{c^2}\right) + \frac{\Gamma(\frac{1}{2}-\frac{s}{2})\,\Gamma(\frac{s}{2})\,\Gamma(\frac{1}{2})}{\Gamma^2(-\frac{s}{2}+1)}$$

$$\left.\times\,{}_4F_3\left(\tfrac{s}{2},\ \tfrac{1}{2},\ \tfrac{s}{2},\ \tfrac{s}{2};\tfrac{1}{2}+\tfrac{s}{2},1,1;\ \frac{d^2}{c^2}\right)\right\} = g_2(s),\quad \text{say}. \qquad (7.3.20)$$

Then I_2 in (7.3.18) is the inverse Mellin transform of $g_1(s)\ g_2(-s)$ where $g_1(s)$ is the R.H.S. of (7.3.11). Thus I_2 can be seen to be the following.

$$I_2 = \frac{1}{2\pi^{3/2}} \left\{ \frac{-(d^2/c^2)^{1/2}}{\Gamma^4(\frac{1}{2})} \sum_{r=o}^{\infty} \frac{(d^2/c^2)^r\, \Gamma^3(\frac{1}{2}+r)}{r!} \right.$$

$$\times G^{3,2}_{5,5}\left(\frac{a^2y^2}{d^2} \,\middle|\, \begin{matrix} \frac{1}{2}, 0, \frac{3}{2}+r, \frac{3}{2}+r, \frac{3}{2}+r \\ \frac{1}{2}, \frac{1}{2}, 1+r, -\frac{1}{2}, -\frac{1}{2} \end{matrix}\right) + \sum_{r=o}^{\infty} \frac{(d^2/c^2)^r\, \Gamma(\frac{1}{2}+r)}{r!\, \Gamma^2(1+r)}$$

$$\left. \times G^{2,5}_{7,7}\left(\frac{a^2y^2}{c^2} \,\middle|\, \begin{matrix} \frac{1}{2}, 0, 1-r, 1-r, 1-r, 1, 1 \\ \frac{1}{2}, \frac{1}{2}, -\frac{1}{2}, -\frac{1}{2}, 1, 1, \frac{1}{2}-r \end{matrix}\right) \right\}, \quad a^2y^2 < d^2 < c^2. \qquad (7.3.21)$$

By using the technique of Chapter V we can write down the G-functions in (7.3.21) as follows. For $(a^2y^2/c^2) < 1$,

$$G^{2,5}_{7,7}\left(\frac{a^2y^2}{c^2} \,\middle|\, \begin{matrix} \frac{1}{2}, 0, 1-r, 1-r, 1-r, 1, 1 \\ \frac{1}{2}, \frac{1}{2}, -\frac{1}{2}, \frac{1}{2}, 1, 1, \frac{1}{2}-r \end{matrix}\right) = \sum_{\nu=o}^{\infty} \frac{\Gamma(\frac{3}{2}+\nu)\, \Gamma^3(\frac{1}{2}+r+\nu)}{(\nu\,!)\, \Gamma^3(2+\nu)}$$

$$\frac{}{\Gamma^2(\frac{1}{2}+\nu)\, \Gamma(1+r+\nu)\Gamma^2(\frac{1}{2}-\nu)} \left[-\log(a^2y^2/c^2) + 3\,\psi(\nu+1) + \psi(\tfrac{3}{2}+\nu)\right.$$

$$\left. + 3\,\psi(\tfrac{1}{2}+r+\nu) - 3\,\psi(2+\nu) - 2\,\psi(\tfrac{1}{2}+\nu) - \psi(1+r+\nu) - 2\,\psi(\tfrac{1}{2}-\nu)\right].$$

(7.3.22)

For $(a^2y^2/c^2) > 1$,

$$G^{2,5}_{7,7}\left(\frac{a^2y^2}{c^2} \,\middle|\, \begin{matrix} \frac{1}{2}, 0, 1-r, 1-r, 1-r, 1, 1 \\ \frac{1}{2}, \frac{1}{2}, -\frac{1}{2}, -\frac{1}{2}, 1, 1, \frac{1}{2}-r \end{matrix}\right) = G^{5,2}_{7,7}\left[\frac{c^2}{a^2y^2} \,\middle|\, \begin{matrix} \frac{1}{2}, \frac{1}{2}, \frac{3}{2}, \frac{3}{2}, 0, 0, \frac{1}{2}+r \\ \frac{1}{2}, 1, r, r, r, 0, 0 \end{matrix}\right].$$

(7.3.23)

For $(a^2y^2/d^2) > 1$,

$$G^{3,2}_{5,5}\left(\frac{a^2y^2}{d^2}\,\middle|\,\begin{matrix}\frac{1}{2},\ 0,\ \frac{3}{2}+r,\ \frac{3}{2}+r,\ \frac{3}{2}+r\\ \frac{1}{2},\ \frac{1}{2},\ 1+r,\ -\frac{1}{2},\ -\frac{1}{2}\end{matrix}\right]$$

$$= G^{2,3}_{5,5}\left(\frac{d^2}{a^2y^2}\,\middle|\,\begin{matrix}\frac{1}{2},\ \frac{1}{2},\ -r,\ \frac{3}{2},\ \frac{3}{2}\\ \frac{1}{2},\ 1,-\frac{1}{2}-r,\ -\frac{1}{2}-r,-\frac{1}{2}-r\end{matrix}\right]$$

$$= \frac{\Gamma(-\frac{1}{2})\,\Gamma(\frac{3}{2}+r)}{\Gamma^3\ (2+r)}\ {}_5F_4(1,1,\ \tfrac{3}{2}+r,0,0;\ \tfrac{1}{2},\ 2+r,2+r,2+r;\frac{d^2}{a^2y^2})$$

$$+ \frac{\Gamma(\frac{3}{2})\,\Gamma(2+r)}{2\,\Gamma^3(\frac{5}{2}+r)}\ {}_4F_3\,(\tfrac{3}{2},\ 2+r,\ \tfrac{1}{2},\tfrac{1}{2};\ \tfrac{5}{2}+r,\ \tfrac{5}{2}+r,\ \tfrac{5}{2}+r;\ \frac{d^2}{a^2y^2})\,. \tag{7.3.24}$$

For $(a^2y^2/d^2) < 1$,

$$G^{3,2}_{5,5}\left(\frac{a^2y^2}{d^2}\,\middle|\,\begin{matrix}\frac{1}{2},\ 0,\ \frac{3}{2}+r,\ \frac{3}{2}+r,\ \frac{3}{2}+r\\ \frac{1}{2},\ \frac{1}{2},\ 1+r,\ -\frac{1}{2},\ -\frac{1}{2}\end{matrix}\right)$$

$$= \sum_{\nu=o}^{\infty} \frac{(-1)^{\nu}\ \Gamma^2(-\frac{1}{2}-r-\nu)\ \Gamma(\frac{3}{2}+r+\nu)\Gamma(2+r+\nu)}{\nu!\ \Gamma^2(\frac{5}{2}+r+\nu)\ \Gamma^3(\frac{1}{2}-\nu)} \left(\frac{a^2y^2}{d^2}\right)^{1+r+\nu}$$

$$+ \sum_{\nu=1}^{r+1} \frac{\Gamma(-\frac{1}{2}+\nu)\ \Gamma(2+r-\nu)\ \ \Gamma(\frac{5}{2}+r-\nu)}{\Gamma^3(3+r-\nu)\ \ \Gamma(\nu)\ \prod\limits_{\substack{t=1\\ t\neq\nu}}^{r+1}(\nu-t)^2} \left(\frac{a^2y^2}{d^2}\right)^{\frac{3}{2}+r-\nu}$$

$$\times\,[-\log\,(a^2y^2/d^2) + \psi(-\tfrac{1}{2}+\nu) + \psi(2+r-\nu) + \psi(\tfrac{5}{2}+r-\nu)$$

$$-\ 3\psi(3+r-\nu\ ) - \psi(\nu) - 2\sum_{\substack{t=1\\ t\neq\nu}}^{r+1}(\nu-t)^{-1}\ ]. \tag{7.3.25}$$

Now (7.3.21) to (7.3.25) solve the problem of six sinusoidal signals.

Remark 1: By using the technique of Section 7.3.2 one can evaluate and represent in computable forms, an integral of the type,

$$\int_0^\infty x^{-r} J_{\nu 1}(ayx) J_{\nu 2}(ayx) J_{\nu 3}(cx) J_{\nu 4}(cx) J_{\nu 5}(dx) J_{\nu 6}(dx) dx,$$

under the conditions of its existence. Also the same technique can be extended to the case of many sinusoidal signals but the final results will be in multiple series which may not be suitable for computational purposes.

Another problem in this category is the problem of input to a limiter consisting of a desired sinusoidal signal, Gaussian noise and n interfering sinusoids. In this problem the integral to be evaluated is the following:

$$Q_k(a) = \int_0^\infty e^{-\frac{\sigma^2 u^2}{2}} \prod_{j=1}^{k} J_o(s_j u) J_1(au) \frac{du}{u}, \qquad (7.3.26)$$

where σ^2, a, $s_1, \ldots, s_k$ are all constants. As in the case of the previous problem here also only approximations and particular cases are available so far, see Campbell [57]. Computable representations of (7.3.26) and the general integral corresponding to (7.3.19) are of immense value to engineers and communication technologists.

Remark 2: For the integrals involving products of Bessel functions see the papers by Saxena (Proc. Glasgow Math. Assoc. 6, 1964, 130-132 and Monatsh. Math.70,1966, 161-163).

EXERCISES

7.1. Establish the integral

$$\int_0^1 x^{\gamma-1}(1-x)^{\rho-1} e^{-zx} {}_2F_1(\alpha,\beta;\gamma;x)$$

$$\times G^{m,n}_{p,q}\left(\zeta(1-x)^{\delta/t}\,\Big|\,\begin{matrix}a_1,\dots,a_p\\ b_1,\dots,b_q\end{matrix}\right)dx$$

$$= (2\pi)^{c^*(1-t)}\, t^{\sum_{j=1}^{q} b_j - \sum_{j=1}^{p} a_j + \frac{p}{2} - \frac{q}{2} + 1}\, \delta^{-\gamma} e^{-z}\,\Gamma(\gamma)$$

$$\times \sum_{r=0}^{\infty} \frac{z^r}{r!}\, G^{tm,\,tn+2\delta}_{tp+2\delta,\,tq+2\delta}\left(\zeta^t\, t^{t(p-q)}\,\big|\,\eta\right)$$

where η stands for the set of parameters,

$$\eta \equiv \begin{matrix}\Delta(\delta,1-\rho-r),\ \Delta(\delta,1+\alpha+\beta-\gamma-\rho-r),\ \Delta(t,a_1),\dots,\ \Delta(t,a_p)\\ \Delta(t,b_1),\dots,\Delta(t,b_q),\ \Delta(\delta,\ 1+\alpha-\gamma-\rho-r),\ \Delta(\delta,1+\beta-\gamma-\rho-r)\end{matrix},$$

$c^* = m + n - \frac{p}{2} - \frac{q}{2} > 0$, $|\arg\zeta| < c^*\pi$,

$R(\gamma) > 0$, $R(\rho + \frac{\delta\, b_j}{t}) > 0$, $R(\gamma+\rho-\alpha-\beta+\frac{\delta b_j}{t}) > 0$, $(j = 1,\dots,m)$, and δ and t are positive integers (Bajpai, 1968, p.1049(1-3),[23]) .

7.2 By taking

$$f(r) = r^{\rho-2}\left(1-\frac{r}{a}\right)^{\gamma-1} e^{-z(1-\frac{r}{a})}\, {}_2F_1\left(\alpha,\beta;\gamma;\ 1-\frac{r}{a}\right)$$

and following the procedure adopted in Section 7.1, show that the solution of the differential equation

$$\frac{\partial u}{\partial t} = \frac{k}{r}\frac{\partial}{\partial r}\left(r\,\frac{\partial u}{\partial r}\right) + \frac{k}{K}\, f(r)\, g(t)$$

where k is the diffusivity and K the conductivity of the material, under the assumption that the surface r = a is maintained at zero temperature and the initial distribution of the temperature is also zero, are given by

$$u(r,t) = 2^{1-\gamma} a^{\rho-2} e^{-z} \Gamma(r) \frac{k}{K} \sum_{i} \sum_{s=o}^{\infty} \frac{z^s}{s!}$$

$$\times \frac{\Gamma(\rho+s)\, \Gamma(\gamma+\rho-\alpha-\beta+s)}{\Gamma(\gamma+\rho-\alpha+s)\, \Gamma(\gamma+\rho-\beta+s)} \frac{J_o(r\,\xi_i)}{[J_1(a\,\xi_i)]^2} h(\xi_i,t)$$

$$\times {}_4F_5\left[\begin{matrix}\Delta(2,\rho+s),\ \Delta(2,\ \gamma+\rho-\alpha-\beta+s) \\ 1,\ \Delta(2,\gamma+\rho-\alpha+s),\ \Delta(2,\gamma+\rho-\beta+s)\end{matrix} ;\ \frac{-a^2\xi_i^2}{4}\right]$$

and

$$u(r,t) = 2^{1-\gamma} a^{\rho-2} \frac{\Gamma(\gamma)\, \Gamma(\rho)\, \Gamma(\gamma+\rho-\alpha-\beta)\ \ J_o(r\,\xi_i)\ h(\xi_i,t)}{\Gamma(\gamma+\rho-\alpha\,)\ \Gamma(\gamma+\rho-\beta)\ \ [J_1(a\,\xi_i)]^2}$$

$$\times \frac{k}{K} \sum_{i} {}_4F_5\left[\begin{matrix}\Delta(2,\rho),\ \Delta(2,\gamma+\rho-\alpha\,-\beta) \\ 1,\Delta\ (2,\gamma+\rho-\alpha\,),\ \Delta(2,\gamma+\rho-\beta)\end{matrix} ;\ \frac{-a\ \xi_i^2}{4}\right]$$

where the sum is taken over all the positive roots of $J_o(a\,\xi_i) = 0$ and

$$h(\xi_i,t) = \int_o^t g(r)\exp\,\{-k\,\xi_i^2(t-\tau)\}\,d\tau\ .$$

Also verify the solution.

7.3. (i) By taking the heat source of hypergeometric character, namely

$$g(\tau) = e^{-\tau z}\ \tau^{\nu-1}\ (t-\tau)^{h-1}\ {}_2F_1(\sigma,\mu;\ \nu;\ \tfrac{\tau}{t}),$$

show that

$$h(\xi_i,t) = g_o \frac{\Gamma(\nu)\ \Gamma(h)\ \Gamma(h+\nu-\sigma-\mu)}{\Gamma(\nu+h-\sigma)\ \Gamma(\nu+h-\mu)}$$

$$\times e^{-zt}\, t^{\nu+h-1}\; {}_2F_1\left(\begin{matrix} h, \nu+h-\sigma-\mu \\ \nu+h-\sigma,\ \nu+h-\mu \end{matrix}\; ;\ (z-k\,\xi_i^{\,2})t\right)$$

where $R(\nu) > 0$, $R(h) > 0$, $R(\nu+h-\sigma-\mu) > 0$.

Hence obtain the value of $u(r,t)$ from the previous example.

(ii) Show that if we take

$$g(\tau) = g_o\, e^{-\tau z}\, \tau^{\sigma}(t-\tau)^{h-1}\; \frac{m!}{(1+\sigma)m}\; P_m^{(\sigma,\mu)}\left(1-\frac{2\tau}{t}\right)$$

where $P_m^{(\sigma,\mu)}$ is the Jacobi polynomial, then the value of $h(\xi_i,t)$ and $u(r,t)$ are given by

$$h(\xi_i,t) = g_o\, \frac{\Gamma(1+\sigma)\,\Gamma(h)\,\Gamma(h-\mu)}{\Gamma(1+\sigma+h+m)\,\Gamma(h-\mu-m)}\; e^{-zt}\, t^{\sigma+h}$$

$$\times\, {}_2F_2\left(\begin{matrix} h,\ h-\mu \\ 1+\sigma+h+m,\ h-\mu-m \end{matrix}\; ;\ (z-k\xi_i^{\,2})t\right)$$

and

$$u(r,t) = g_o\, \frac{\Gamma(1+\sigma)\,\Gamma(h)\,\Gamma(h-\mu)\,\Gamma(\mu)\,\Gamma(\rho)\,\Gamma(\gamma+\rho-\alpha-\beta)}{\Gamma(1+\sigma+h+m)\,\Gamma(h-\mu-m)\,\Gamma(\gamma+\rho-\alpha)\,\Gamma(\gamma+\rho-\beta)}$$

$$\times\, e^{-zt}\, t^{\sigma+h}\, 2^{1-\gamma}\, a^{\rho-2}\, \frac{k}{K}\; \sum_i \frac{J_o(r\,\xi_i)}{[J_1(a\,\xi_i)]^2}$$

$$\times\, {}_4F_5\left[\begin{matrix} \Delta(2,\rho),\ \Delta(2,\gamma+\rho-\alpha-\beta) \\ 1,\ \Delta(2,\gamma+\rho-\alpha),\ \Delta(2,\gamma+\rho-\beta) \end{matrix}\; ;\ \frac{-a^2\,\xi_i^{\,2}}{4}\right]$$

$$\times\, {}_2F_2\left(\begin{matrix} h,\ \nu+h-\sigma-\mu \\ \nu+\sigma-\sigma,\ \nu+h-\mu \end{matrix}\; ;\ (z-k\,\xi_i^{\,2})t\right)$$

respectively (Bajpai, 1968, pp. 1052, 1053, [23]).

7.4. Show that the formal solutions of the dual integral equations

$$\int_0^\infty H^{q,p+m}_{2p+m,2q+m}(xu)\, f(u)du = \emptyset(x), \quad 0 < x < 1$$

and

$$\int_0^\infty H^{q+n,p}_{2p+n,\, 2q+n}(xu)\, f(u)du = \psi(x), \quad x > 1$$

where $\emptyset(x)$ and $\psi(x)$ are given and $f(x)$ is to be determined, is given by

$$f(x) = \int_0^\infty g(u)\, H^{q,p}_{2p,2q}(xu)du = \int_0^1 T_1[T_2 \ldots T_m[\emptyset(u)]\ldots]\, H^{q,p}_{2p,2q}(xu)du$$

$$+ \int_0^\infty R_1\, [R_2 \ldots R_n\, [\psi(u)]\ldots]\, H^{q,p}_{2p,2q}(xu)du$$

where $H^{q,p}_{2p,2q}(xu)$ is the symmetrical Fourier kernel due to Fox ([94],p.408).
The definitions of the various H-functions are as follows:

$$H^{q,p+m}_{2p+m,2q+m}(x) = \frac{1}{2\pi i} \int_L \chi_{p,q,m}(s)x^{-s}ds ,$$

$$\chi_{p,q,m}(s) = \chi(p,q) \prod_{j=1}^{m} \frac{\Gamma(\alpha_j - s\tau_j)}{\Gamma(\beta_j - s\tau_j)} ;$$

$$H^{q+n,p}_{2p+n,2q+n}(x) = \frac{1}{2\pi i} \int \chi_{p,q,n}(s)x^{-s}\, ds$$

where

$$\chi_{p,q,n} = \chi(p,q) \prod_{j=1}^{n} \frac{\Gamma(\lambda_j + s\xi_j)}{\Gamma(\mu_j + s\xi_j)}$$

and

$$H^{q,p}_{2p,2q}(x) = \frac{1}{2\pi i} \int \chi_{p,q}(s)\, x^{-s}\, ds ,$$

where

$$\chi_{p,q}(s) = \prod_{j=1}^{q} \frac{\Gamma(b_j + s\, B_j)}{\Gamma(b_j+B_j-s\, B_j)} \prod_{j=1}^{p} \frac{\Gamma(a_j - s\, A_j)}{\Gamma(a_j - A_j + s\, A_j)} .$$

The operators T's and R's are defined by the following relations

$$T[(\alpha_k - \beta_k), (\beta_k \tau_k^{-1} - k); \tau_k^{-1}: w(x)] = T_k[w(x)]$$

$$R[(\lambda_\rho - \mu_\rho), \mu_\rho \xi_\rho^{-1}; \xi_\rho^{-1}: w(x)] = R_\rho[w(x)]$$

resembling the equations (7.2.20) and (7.2.21) (Saxena [272] 1967).

BIBLIOGRAPHY

ABIODUN, R.F. and SHARMA, B.L.

[1] Summation of series involving generalized hypergeometric functions. Glasnik Mat. 6(26)2, 253-264, 1971.

ABRAMOWITZ, M and STEGUN, I. (eds.)

[2] Handbook of Mathematical Functions with Formulas, Graphs and Mathematical Tables. Appl. Math. Ser. 55, U.S. Govt. Printing Office, Washington, D.C., 1964.

AGARWAL, B.M.

[3] Transformation of a certain series involving the solutions of F-equations and confluent hypergeometric functions. Proc. Nat. Acad. Sci.India, Sect.A, 36, 469-472, 1966.

[4] On generalized Meijer's H-functions satisfying the Truesdell F-equations. Proc. Nat. Acad. Sci. India, Sect. A, 38, 259-264. 1968.

[5] Application of Δ and E operators to evaluate certain integrals. Proc. Cambridge Philos. Soc. 64, 99-104, 1968.

[6] Rodrigues and Schlafli formulae for the solution of F-equation J. Indian Math. Soc. 32(3-4), 173-177, 1968.

AGARWAL, I and SAXENA, R.K.

[7] Integrals involving Bessel functions. Univ. Nac. Tucuman Rev. Ser.A 19, 245-254, 1969.

AGARWAL, Km. NIRMALA

[8] A q-analogue of MacRobert's generalized E-function. Ganita, 11, 49-63, 1960.

AGARWAL, R.P. (AGARWAL, RATAN PRAKASH).

[9] A basic analogue of MacRobert's E-function. Proc. Glasgow Math. Assoc. 5, 4-7, 1961.

[10] Some further properties of a q-analogue of MacRobert's E-function. Proc. Glasgow Math. Assoc. 6, 34-38, 1963.

[11] Generalized Hypergeometric Series. Asia Publishing House, Bombay, New York, 1963.

[12] A note on evaluation of the integral $\int_0^\infty e^{-kt} I_0^n(t)dt$. Proc. Edinburgh Math.Soc.(2)14, 85-86, 1964-65.

[13] An extension of Meijer's G-function. Proc. Nat. Inst. Sci. India, Part A, 31, 536-546, 1965.

[14] A note on a bilateral q-series transform. Ganita 19, 1-4, 1968.

[15] On certain transformation formulae and Meijer's G-function of two variables. Indian J. Pure Appl.Math. 1, No.4, 537-551, 1970.

AGARWAL, R.P. and VERMA, A.

[16] Generalized basic hypergeometric series with unconnected bases. Proc. Cambridge Philos. Soc. 63, 727-734, 1967.

[17] Generalized Basic Hypergeometric series with unconnected bases, II, Quart. J. Math. Oxford Ser. (2)18, 181-192, 1967.

AL-SALAM, NADHLA, A.

[18] Orthogonal polynomials of hypergeometric type. Duke Math.J. 33,109-121, 1966.

[19] A class of hypergeometric polynomials. Ann. Mat. Pura. Appl.(4)75,95-120, 1967.

Al-SALAM, W.A.(AL-SALAM,WALEED A; ALSO SEE CARLITZ,L).

[20] The Bessel polynomials. Duke Math. J. 24,529-545, 1957.

[21] Some fractional q-integrals and q-derivatives. Proc. Edinburgh Math. Soc. (2)15, 135-140, 1966-67.

ANANDANI, P.(ANANDANI, PARMANAND)

[22] Some expansion formulae for H-function II, Ganita 18, 89-101,1967.

[23] Some integrals involving products of Meijer's G-function and H-function. Proc. Indian Acad. Sci. Sect. A.67, 312-321,1968.

[24] Summation of some series of products of H-functions. Proc.Nat. Inst.Sci. India, Part A, 34, 216-223, 1968.

[25] Fourier series for H-function. Proc. Indian Acad. Sci. Sect.A, 68, 291-295, 1968.

[26] On some recurrence formulae for the H-function. Ann. Polon. Math. 21, 113-117, 1969.

[27] On finite summation, recurrence relations and identities of H-functions. Ann. Polon. Math. 21, 125-137, 1969.

[28] Some integrals involving H-functions. Labdev J. Sci. Tech. Part A, 7, 62-66, 1969.

[29] On some identities of H-function. Proc. Indian Acad. Sci. Sect. A,70, 89-91, 1969.

[30] Some expansion formulae for the H-function III, Proc. Nat. Acad. Sci. India. Sect. A, 39, 23-24,1969.

[31] Some integrals involving generalized associated Legendre's functions and the H-function. Proc. Nat. Acad.Sci.India, Sect.A, 39, 127-136, 1969.

[32] Some integrals involving products of generalized Legendre's associated functions and the H-function. J. Sci. Eng. Res.13, 274-279, 1969.

[33] On some integrals involving generalized Legendre's associated functions and H-functions. Proc. Nat. Acad. Sci. India, Sect. A, 39, 341-348, 1969.

[34] Some infinite series of H-function I. Math. Student, 37, 117-123, 1969.

[35] Some infinite series of H-function II, Vijnana Parishad Anusandhan Patrika, 13, 57-66, 1970.

[36] On the derivative of the H-function. Roumania Math. Pures. Appl. 15,189-191, 1970.

[37] Some integrals involving H-functions of generalized arguments. Math. Education 4, 32-38,1970.

[38] Some expansion formulae for the H-function. Labdev J. Sci. Tech. India, Part A, 8, 80-87, 1970.

[39] Expansion of the H-function involving generalized Legendre's associated function and H-function. Kyungpook Math. J. 10, 53-57, 1970.

[40] Some integrals involving Jacobi polynomials and H-function. Labdev J. Sci. Tech. Part A, 8, 145-149, 1970.

[41] An expansion for the H-function involving generalized Legendre's associated functions. Glasnik Mat. Ser. III, 5(25), 55-58, 1970.

[42] An expansion formula for the H-function involving associated Legendre function. J. Natur. Sci. Math. 10, No.1, 49-51, 1970.

[43] Some integrals involving associated Legendre function of first kind and the H-function. J. Natur. Sci. Math. 10, No. 1, 97-104,1970.

[44] Use of generalized Legendre associated function and the H-function in heat production in a cylinder. Kyungpook Math. J. 10, 107-113, 1970.

[45] Integration of products of generalized Legendre functions and the H-function with respect to parameters. Labdev J. Sci. Tech. 9A, 13-19, 1970.

[46] An expansion formula for the H-function involving generalized Legendre's associated functions. Portugal Math. 30, 173-180, 1971.

APPÉLL, P.

[47] Sur les Fonctions Hypergéometriques de Plusieurs Variables. Paris, Gauthier-Villars, 1925.

APPÉLL, P. and KAMPÉ DE FÉRIET, J.

[48] Fonctions Hypergéometriques et Hypersphériques; Polynomes d'Hermites. Paris, Gauthier-Villars, 1926.

ARORA, K.L. and KULSHRESHTHA, K.

[49] An infinite integral involving Meijer's G-function. Proc. Amer. Math. Soc. 26, 121-125, 1970.

BAILEY, W.N.

[50] A reducible case of the fourth type of Appèll's hypergeometric function of two variables. Quart. J. Math. Oxford Ser.(2) 4, 305-308, 1933.

[51] On the reducibility of Appéll's function F_4. Quart. J. Math. Oxford Ser. (2), 5, 291-292, 1934.

[52] Generalized Hypergeometric Series. Cambridge Tracts in Mathematics and Math. Physics. No. 32. Cambridge University Press, Cambridge and New York 1935.

[53] Some infinite integrals involving Bessel functions. Proc. London Math.Soc. 40, 37-48,1936.

[54] Some infinite integrals involving Bessel functions-II. J. London Math. Soc. 11, 16-20, 1936.

[55] On the sum of a terminating $_3F_2(1)$. Quart. J. Math. Oxford Ser.(2),4, 237-240, 1953.

[56] On two manuscripts of Bishop Barnes. Quart. J. Math. Oxford Ser.(2),10, 236-240, 1959.

BAJPAI, S.D.

[57] Some integrals involving MacRobert's E-function and Meijer's G-function. Prod. Nat. Acad. Sci. India, Sect. A. 36, 793-802. 1966.

[58] An integral involving Meijer's G-function and Jacobi polynomials J. Sci. Engrg. Res. 11, 113-115, 1967.

[59] Fourier series for Meijer's G-function. Gaz.Mat. (Lisboa) 28, No. 105-108, 40-42, 1967.

[60] Some Integrals involving Gauss' hypergeometric function and Meijer's G-function. Proc. Cambridge Philos. Soc. 63, 1049-1053, 1967.

[61] Integration of E-functions with respect to their parameters. Proc. Nat. Acad. Sci. India Sect. A, 37, 71-75, 1967.

[62] Summation of a series of products of two E-functions. (Arabic Summary). Bull. College Sci. (Baghdad) 10, 1967-68.

[63] Summation of some series of products of G-functions. Vijnana Parishad Anusandhan Patrika, 11, 31-38, 1968.

[64] Some expansion formulae for Meijer's G-function. Vijnana Parishad Anusandhan Patrika, 11, 177-191, 1968.

[65] Some expansion formulae for G-function involving Bessel functions. Proc. Indian Acad. Sci. Sect. A, 68, 285-290, 1968.

[66] On products of generalized hypergeometric functions. Gaz. Mat.(Lisboa) 29, No. 109-112, 15-21, 1968.

[67] Expansion theorems for generalized hypergeometric function-1. Gaz. Mat. (lisboa) 29, No. 109-112,22-35, 1968.

[68] Use of Gauss' hypergeometric function and Meijer's G-function in the production of heat in a cylinder. Proc. Cambridge Philos. Soc., 64, 1049-1054, 1968.

[69] Some results involving Meijer's G-function and exponential functions. Univ. Lisboa Revista Fac. Ci. A(2)12, 225-232, 1968-69.

[70] A finite integral involving a confluent hypergeometric function and Meijer's G-function. Portugal Math. 28, 55-61, 1969.

[71] An integral involving Fox's H-function and Whittaker functions. Proc. Cambridge Philos. Soc., 65, 709-712, 1969.

[72] On some results involving Fix's H-function and Jacobi polynomials. Proc. Cambridge Philos. Soc. 65, 697-701, 1969.

[73] Transformation of an infinite series of G-function. Proc. Cambridge Philos. Soc., 65, 467-469, 1969.

[74] Fourier series of generalized hypergeometric functions. Proc. Cambridge Philos. Soc., 65, 703-707, 1969.

[75] An expansion formula for Fox's H-function. Proc. Cambridge Philos. Soc., 65, 683-685, 1969.

[76] Special functions and forced symmetrical vibrations of a circular elastic plate. Proc. Cambridge Philos. Soc., 66, 349-353, 1969.

[77] An integral involving Fox's H-function and heat conduction. Math. Educ. 3, 1-4, 1969.

[78] An expansion formula for Meijer's G-function. Proc. Nat. Inst. Sci. India, Part A, 35, Suppl. 1, 90-94, 1969.

[79] Special functions and cooling of a heated cylinder. Labdev J. Sci. Tech. India, 7A, 59-61, 1969.

[80] An integral invilving Meijer's G-function and associated Legendre function. Ricerca (Napoli) (2)20, Settembre-Dicembre, 7-13, 1969.

[81] Associated Legendre functions and heat production in a cylinder. Proc. Nat. Inst. Sci. India Part A, 35, 366-374, 1969.

[82] An expansion formula for H-function involving Bessel function. Labdev J. Sci. Tech. Part A, 7, 18-20, 1969.

[83] Meijer's G-function and temperatures in a slab with faces at temperature zero. J. Sci. Engrg. Res. 13, 254-257, 1969.

[84] The potential about a spherical surface and Meijer's G-function. Vijnana Parishad Anusandhan Patrika. 12, 93-97, 1969.

[85] A contour integral involving G-function and Whittaker function. Indian J. Mech. Math. 7, 67-70, 1969.

[86] A new class of integral transform . Labdev J. Sci. Tech. 8A, 88-91, 1970.

[87] An expansion formula for Meijer's G-function involving Hermite polynomials Labdev J. Sci. Tech. Part A 8, 9-11, 1970.

[88] Some expansion formulae for Meijer's G-function involving Legendre functions. J. Mathematical and Physical Sci. 4, 173-175, 1970.

[89] A contour integral involving Legendre polynomial and Meijer's G-function. Proc. Indian Acad. Sci. Sect. A, 71, 209-214, 1970.

[90] Some expansion formulae for H-function involving exponential functions. Proc. Cambridge Philos. Soc., 67, 87-92, 1970.

[91] An integral involving Fox's H-function and its application. Univ. Lisboa, Revista Fac. Ci., II Ser. A,13, 109-110, 1969-70.

[92] On some results involving Fox's H-function and Bessel function. Proc. Indian Acad. Sci. Sect. A, 72, 42-46, 1970.

[93] Some results involving Fox's H-function. Portugal Math. 30, 45-52,1971.

BARNES, E.W.

[94] The asymptotic expansion of integral functions defined by generalized hypergeometric series. Proc. London Math. Soc. (2)5, 59-116, 1907.

[95] A new development of the theory of the hypergeometric functions. Proc. London Math. Soc. (2)6, 141-177, 1908.

[96] On functions defined by simple types of hypergeometric series. Trans. Cambridge Philos. Soc. 20, 253-279, 1908.

[97] A transformation of generalized hypergeometric series. Quart. J. Math. Oxford, 41, 136-140, 1910.

BHAGCHANDANI, L.K. and MEHRA, K.N.

[98] Some results involving generalized Meijer function and Jacobi polynomials. Univ. Nac. Tucuman Rev. Ser.A, 20, 167-174, 1970.

BHATT, R.C.

[99] A summation formula for Appéll's function F_2. Israel J. Math. 3, 87-88, 1965.

[100] Certain integrals involving the products of hypergeometric functions. Matematiche (Catania) 21, 6-10,1966.

BHISE, V.M.

[101] Inversion formulae for a generalized Laplace integral. Vikram-Quart. Res. J. Vikram Univ., India, 3,57-63, 1959.

[102] Certain rules and recurrence relations for Meijer-Laplace transform. Proc. Nat. Acad. Sci. India Sect. A,32, 389-404, 1962.

[103] On the derivative of Meijer's G-function and MacRobert's E-function. Proc. Nat. Acad. Sci. India Sect. A, 32, 349-354, 1962.

[104] Finite and infinite series of Meijer's G-function and the multiplication formula for G-function. J. Indian Math. Soc. (N.S)27, 9-17,1963.

[105] Some finite and infinite series of Meijer-Laplace transform. Math. Ann. 154, 267-272, 1964.

[106] Operators of fractional integration and a generalized Hankel transform. Collect. Math. 16, 201-209, 1964.

[107] Certain properties of Meijer-Laplace transform. Compositio Math. 18, 1-6, 1967.

BHONSLE, B.R.

[108] Some recurrence relations and series for the generalized Laplace transform. Proc. Glasgow Math. Assoc. 4, 119-121, 1960.

[109] A relation between Laplace and Hankel transforms. Proc. Glasgow Math. Assoc. 5, 114-115, 1962.

[110] On some results involving Jacobi polynomials. J. Indian Math. Soc. (N.S) 26, 187-190, 1962.

[111] Some series and recurrence relations for MacRobert's E-function. Proc. Glasgow Math. Assoc. 5, 116-117, 1962.

[112] A relation between Laplace and Hankel transform. Math. Japon 10,85-89,1965.

[113] Jacobi polynomials and heat production in a cylinder. Math. Japon. 11, No. 1, 83-90, 1966.

[114] Steady state heat flow in a shell enclosed between two prolate spheroids. Math. Japon. 12,(1), 1967.

BOCHNER, S.

[115] On Riemann's functional equation with multiple gamma factors. Ann. of Math. (2) 67, 29-41, 1958.

BOERSMA, J.

[116] On a function which is a special case of Meijer's G-function. Compositio Math. 15, 34-63, 1962.

BORA, S. L.

[117] An infinite integral involving product of hypergeometric functions. Vijnana Parishad Anusandhan Patrika, 11, 97-101, 1968.

[118] An infinite integral involving generalized function of two variables. Vijnana Parishad Anusandhan Patrika, 13 95-100, 1970.

BORA, S.L. and KALLA, S.L.

[119] Some results involving generalized function of two variables. Kyungpook Math. J. 10, 133-140, 1970.

BORA, S.L: KALLA, S.L. and SAXENA, R.K.

[120] On integral transforms. Univ. Nac. Tucuman Rev. Ser. A 20, 181-188, 1970.

BORA, S.L. and SAXENA, R.K.

[121] Integrals involving product of Bessel functions and generalized hypergeometric functions. Pub. Inst. Math. (Beograd) 11(25), 23-28,1971.

BOX, G.E.P.

[122] A general distribution theory for a class of likelihood criteria. Biometrika. 36, 317-346, 1949.

BRAAKSMA, B.L.J.

[123] Asymptotic expansions and analytic continuations for a class of Barnes integrals. Compositio Math. 15, 239-341, 1964.

BROMWICH, T.J. l'a.

[124] An asymptotic formula for the generalized hypergeometric series. Proc. London Math. Soc.(2),7, 101-106, 1909.

BURCHNALL, J.L.

[125] A relation between hypergeometric series. Quart. J. Math. Oxford Ser. 10, 145-150, 1932.

[126] The differential equations of Appéll's function F_4. Quart. J. Math. Oxford 10, 145-150, 1939.

[127] Differential equations associated with hypergeometric functions. Quart. J. Math. Oxford Ser. 13, 90-106, 1942.

BURCHNALL, J.L. and CHAUNDY, T.W.

[128] Expansions of Appéll's double hypergeometric functions. Quart. J. Math. Oxford Ser.11, 249-270, 1940.

[129] Expansion of Appéll's double hypergeometric functions-11. Quart. J. Math. Oxford Ser. 12, 112-128, 1941.

[130] The hypergeometric identities of Cayley and Orr and Bailey. Proc. London Math. Soc.(2) 50, 56-74, 1948.

BUSCHMAN, R.G.

[131] An inversion integral for a general Legendre transformation. SIAM Rev. 5, 232-233, 1963.

[132] Fractional integration. Math. Japon 9, 99-106, 1964.

[133] Integrals of hypergeometric functions. Math. Z. 89, 74-76,1965.

CARLITZ, L.

[134] Summation of some series of Bessel functions. Nederl. Akad. Wetensch.Proc. Ser. A. 65 = Indag. Math. 24, 47-54, 1962.

CARLITZ, L. and Al-SALAM, W.A.

[135] Some functions associated with Bessel functions. J. Math. Mech. 12,911-933, 1963.

CARSLAW, H.S. and JAEGER, J.C.

[136] Conduction of Heat in Solids. 2nd Ed. Oxford University Press, London, 1959.

CHATTERJEA, S.K.

[137] On a generalization of Laguerre polynomials. Rend. Sem.Math.Univ. Padova 34, 180-190, 1964.

CHAUNDY, T.W. (ALSO SEE BURCHNALL, J.L.)

[138] Expansions of hypergeometric functions. Quart. J. Math. Oxford Ser. 13, 159-171, 1942.

CHHABRA, S.P.

[139] Some series for Meijer's G-function. Proc. Nat. Acad. Sci. India Sect.A. 36, 575-584, 1966.

CHHABRA, S.P. and SINGH, F.

[140] An integral involving product of a G-function and a generalized hypergeometric function. Proc. Cambridge Philos. Soc. 65, 479-482, 1969.

CHOPRA, N.

[141] On certain infinite expansions of basic generalized MacRobert's E-function. Ganita 16, 43-50, 1965.

[142] Certain finite expansions of basic generalized MacRobert's E-function. Ganita, 16, 37-42, 1965.

[143] On certain recurrence formulae for the E_q-functions. Ganita, 16, 133-138, 1965.

CHURCHILL, R.V.

[144] Fourier Series and Boundary Value Problems McGraw-Hill, New York, 1942.

CONSTANTINE, A.G.

[145] Some non-central distribution problems in multivariate analysis. Ann. Math. Statist. 34, 1270-1285, 1963.

CONSUL, P.C.

[146] On some special G-functions. Acad. Roy. Belg. Bull. Cl. Sci. 52,1271-1274, 1966.

[147] On the exact distribution of the likelihood ratio criteria for testing linear hypotheses about regression coefficients. Ann. Math. Statist. 37, 1319-1330, 1966.

[148] On some inverse Mellin transforms. Acad. Roy Belg. Bull. Ci. Sci. (5) 52, 547-561, 1966.

[149] On the exact distribution of the likelihood ratio criteria for testing independence of sets of variates under the null hypothesis. Ann.Math. Statist. 38, 1160-1169, 1967.

[150] On the exact distribution of a criterion W for testing sphericity in a p-variate normal distribution. Ann. Math. Statist. 38,1170-1174, 1967.

[151] On the distribution of Votaw's likelihood ratio criterion L for testing the bipolarity of a covariance matrix. Math. Nachr. 36, 1-13, 1968.

[152] On the exact distribution of Votaw's criteria for testing compound symmetry of a covaraince matrix. Ann. Math. Statist. 40, 836-843, 1969.

COOKE, R.G.

[153] On the theory of Schlömilch series. Proc. London Math. Soc. 28,207-241, 1928.

DAHIYA, R.S.

[154] On double integrals involving Meijer's G-function. Kyungpook Math. J.11, 57-64, 1971.

DAHIYA, R.S. and SINGH, B.

[155] Fourier series of Meijer's G-function of higher order An. Sti.Univ. Al.l. Cuza.n.Ser. Sect.(1)17, 111-116, 1971.

DENIS, R.Y.

[156] Certain transformations of bilateral cognate trigonometrical series of hypergeometric type. Proc. Cambridge Philos. Soc. 64,421-424, 1968.

[157] A general expansion theorem for products of generalized hypergeometric series. Proc. Nat. Inst. Sci. India Part A, 35, 70-76, 1969.

DESHPANDE, V.L.

[158] Expansion theorems for the Kampé de Fériet function. Nederl. Akad. Wetensch. Proc. 74 = Indag.Math. 33,39-45, 1971.

DHAWAN, G.K.

[159] Series and expansion formulae for G-function of two variables. J. of M.A.C.T. 2, 88-94, 1969.

DIXON, A.L. and FERRAR, W.L.

[160] A class of discontinuous integrals. Quart. J. Math. Oxford Ser. 7, 81-96, 1936.

DOETSCH, G.

[161] Theorie und Anwendung der Laplce-Transformation. Dover, New York, 1943

[162] Einführung in Theorie und Anwendung der Laplace-Transformation. Birkhäuser-Verlag, Basel, 1958.

DZRBASJAN, V.A.

[163] On a theorem of Whipple, Z. Wyčysl. Mat. i.Mat. Fiz.4, 348-351, 1964.

EDELSTIEN, L.A.

[164] On the one-centre expansion of Meijer's G-function. Proc. Cambridge Philos. Soc. 60, 533-538, 1964.

[165] On the one-region two-centre expansions of a generalized Coulomb field in spherical polar and parabolic coordinates. Proc. Cambridge Philos. Soc. 60, 543-546, 1964.

ERDÉLYI, A.

[166] Inversion formulae for the Laplace transformation. Philos. Mag. (7) 34, 533-537, 1943.

[167] Note on an inversion formula for the Laplace transformation. J. London Math. Soc. 18, 72-77, 1943.

[168] On some functional transformation. Univ. Politec. Torino Rend. Sem.Mat. 10, 217-234, 1950-51.

[169] On a generalization of the Laplace transformation. Proc. Edinburgh Math. Soc. (2) 10, 53-55, 1954.

[170] Asymptotic Expansions. Dover, New York.

ERDÉLYI, A; MAGNUS, W; OBERHETTINGER, F and TRICOMI, F.G.

[171] Higher Transcendental Functions. Vol. I,II. MacGraw-Hill, New York, 1953.

[172] Tables of Integral Transforms. Vol. I,II. McGraw-Hill, New York, 1954.

[173] Higher Transcendental Functions. Vol. III. McGraw-Hill, New York, 1955.

ERDÉLYI, A. and WYMAN, M.

[174] The asymptotic evaluation of certain integrals. Arch. Rational Mech. Anal. 14, 217-260, 1963.

FASENMEYER, SISTER MARY CELINE.

[175] Some generalized hypergeometric polynomials. Bull. Amer. Math. Soc. 53, 806-812, 1947.

[176] A note on pure recurrence relations. Amer. Math. Monthly. 56, 14-17,1949.

FETTIS, H.E.

[177] Lommel-type integrals involving three Bessel functions. J. Math. Phys. 36, 88-95, 1957.

FETTIS, HENRY E. and SAYEED-UR-RAHMAN.

[178] Table errata: Higher Transcendental Functions. Vol. II. Math. Comp. 24, 239, 1970.

FIELDS, J.L.

[179] Rational approximations to generalized hypergeometric functions. Math. Comp. 19, 606-624, 1965.

[180] Asymptotic expansions of a class of hypergeometric polynomials with respect to the order III. J. Math. Anal. Appl. 593-601, 1965.

[181] A note on the asymptotic expansion of a ratio of gamma functions. Proc. Edinburgh Math. Soc. (2) 15, 43-45, 1966.

[182] A uniform treatment of Darboux's method. Arch. Rational Mech. Anal. 27, 289-305, 1968.

FIELDS, J.L. and LUKE, Y.L.

[183] Asymptotic expansion of a class of hypergeometric polynomials with respect to the order I,II. J. Math. Anal. Appl. 6, 394-403; 7,440-445,1963.

FIELDS, JERRY,L; LUKE, YUDELL, L and WIMP, JET

[184] Recursion formulae for generalized hypergeometric functions. J. Approximation Theory I, 137-166, 1968.

FIELDS, J.L. and WIMP, J.

[185] Expansions of hypergeometric functions in hypergeometric functions. Math. Comp. 15, 390-395, 1961.

[186] Basic series corresponding to a class of hypergeometric polynomials. Proc. Cambridge Philos. Soc. 59, 599-605, 1963.

FLETCHER, A; MILLER, J.C.P; ROSENHEAD, L. and COMRIE, L.J.

[187] An Index of Mathematical Tables Vol. I,II. Addison-Wesley, Reading, Massachusetts, 1962.

FOX, CHARLES.

[188] The expression of hypergeometric series in terms of similar series. Proc. London Math. Soc. 26(2), 201-210, 1927.

[189] The asymptotic expansion of generalized hypergeometric functions. Proc. London Math. Soc. (2) 27, 389-400, 1928.

[190] The G and H-functions as symmetrical Fourier kernels. Trans. Amer. Math. Soc. 98, 395-429, 1961.

[191] Integral transforms based upon fractional integration. Proc. Cambridge Philos. Soc. 59, 63-71, 1963.

[192] A formal solution of certain dual integral equations. Trans. Amer. Math. Soc. 119, 389-398, 1965.

[193] An inversion formula for a kernel $K_\nu(x)$. Proc.Cambridge Philos.Soc. 61, 457-467, 1965.

[194] Solving integral equations by L and L^{-1} operators. Proc. Amer. Math.Soc. 29, 299-306, 1971.

FRÖMAN, N. and FRÔMAN, P.O.

[195] JWKB Approximation. North-Holland Publ., Amsterdam, 1965.

GOKHROO, D.C.

[196] Some complex inversion formulae for the generalized stieltjes transform. Ann. Soc. Sci. Bruxelles Ser. 1, 79, 107-112, 1965.

[197] A definite integral involving Meijer's G-function. Proc. Nat. Acad.Sci. India, 36,Sect.A, 841-842, 1966.

[198] Infinite integrals involving Meijer's G-function. Portugal Math. 26(2), 169-174, 1967.

[199] The Laplace transform of the product of Meijer's G-functions. Univ.Nach. Tucuman Rev. Ser.A 20, 59-62, 1970.

GOLAS, P.C.

[200] Integration with respect to parameters. Vijnana Parishad Anusandhan Patrika, 11, 71-76, 1968.

[201] On a generalized Stieltjes transform. Proc. Nat.Acad.Sci. India Sect.A, 39, 42-48, 1969.

GOLDSTEIN, S.

[202] Operational representation of Whittaker's confluent hypergeometric functions and Weber's parabolic cylinder functions. Proc. London Math. Soc. (2), 34, 103-125, 1932.

GOURSAT, E.

[203] Sur l'equation differentielle lineaire qui admet pour integrale la serie hypergeometrique. Ann. de l'ecole normale, (2)10, Suppl. 3-142, 1881.

GOYAL, A.N.

[204] Some infinite series of H-functions. 1. Math. Student 37, 179-183,1969.

GOYAL, A.N. and GOYAL, G.K.

[205] On the derivatives of the H-function. Proc. Nat. Acad. Sci. India Sect. A. 37, 56-59, 1967.

[206] Expansion theorems of H-function. Vijnana Parishad Anusan and Patrika. 10, 205-217, 1967.

GOYAL, G.K.

[207] Some relations between Hankel transform and Meijer's Bessel function transform. Proc. Nat. Acad. Sci. Sect. A, 36, 9-15, 1966.

[208] A finite integral involving H-function. Proc. Nat. Acad. Sci. Sect. A. 39, 201-203, 1969.

GOYAL, R.P.

[209] Convergence of Meijer-Laplace transform. Ganita 17, 57-67, 1966.

GRAY, A. and MATHEWS, G.B.

[210] A Treatise on Bessel Functions and Their Applications to Physics. 2nd Ed. Prepared by A. Gray and T.M. MacRobert, Dover, 1966.

GUPTA, K.C. (ALSO SEE SRIVASTAVA, A).

[211] Some theorems concerning Meijer transform. Collect Math. 16, 33-44, 1964.

[212] On the inverse Meijer transform of the G-function. Collect Math. 16, 45-54, 1964.

[213] Certain integral representations for the Meijer transform. Proc. Nat. Acad. Sci. India Sect. A, 34, 541-548, 1964.

[214] A theorem concerning Meijer and Varma transforms. Proc. Nat. Acad. Sci. India Sect. A. 34, 163-168, 1964.

[215] Some theorems on Laplace Meijer and Varma transform. Proc. Nat. Acad. Sci. India Sect. A, 34, 185-194, 1964.

[216] On the H-function. Ann. Soc. Sci. Bruxelles Ser. 1, 79, 97-106, 1965.

[217] On generalized Hankel and Meijer's Bessel function transforms. Proc. Nat. Acad. Sci. India, 35 A, 105-112, 1965.

[218] On the integration of G-functions with respect to parameters. Proc. Nat. Acad. Sci. India Sect. A. 36, 193-198, 1966.

[219] Integrals involving H-functions. Proc. Nat. Acad. Sci. India, Sect. A, 36, 504-509, 1966.

[220] On Fourier kernels. Vijnana Parishad Anusandhan Patrika. 13, 85-94, 1970.

GUPTA, K.C. and JAIN, U.C.

[221] The H-function - II. Proc. Nat Acad. Sci. India Sec.A,36, 594-609, 1966.

[222] The H-function - III. Proc. Nat. Acad. Sci. India Sect. A,38,189-192,1968.

[223] On the derivative of the H-function. Proc. Nat. Acad. Sci. India Sect.A,38 189-192, 1968.

[224] The H-function - IV. Vijnana Parishad Anusandhan Patrika. 12, 25-30,1969.

GUPTA, K.C. and MITTAL, P.K.

[225] The H-function transform. J. Austral. Math. Soc. 11, 142-148, 1970.

[226] The H-function transform-II, J. Austral. Math. Soc. 12,444-450, 1971.

GUPTA, K.C. and MITTAL, S.S.

[227] On Gauss' hypergeometric transform-II. Vijnana Parishad Anusandhan Patrika. 12, 133-137, 1969.

GUPTA, K.C. and OLKHA, G.S.

[228] Integrals involving products of generalized hypergeometric functions and Fox's H-function. Univ. Nac. Tucuman Rev. Ser.A 19, 205-212, 1969.

GUPTA, K.C. and SAXENA, R.K.

[229] Certain properties of generalized Stieltjes transform involving Meijer's G-function. Proc. Nat. Inst. Sci. India Sect. A 30, 707-714, 1964.

[230] On Laplace transform. Riv. Mat. Univ. Parma. Italy 5, 159-164, 1964.

GUPTA, K.C. and SRIVASTAVA, A.

[231] On certain recurrence relations. Math. Nachr. 46, 13-23, 1970.

[232] On certain recurrence relations-II. Math. Nachr. 49, 187-197, 1971.

[233] On integrals involving generalized hypergeometric functions. Indian J. Pure Appl. Math. 2, 495-500, 1971.

GUPTA, P.N. and RATHIE, P.N.

[234] Some results involving generalized hypergeometric functions. Riv. Mat. Univ. Parma, (2) 9,91-96, 1968.

GUPTA, S.C.

[235] Integrals involving products of G-functions. Proc. Nat. Acad. Sci. India Sect. A. 39(2), 193-200, 1969.

[236] Reduction of G-function of two variables. Vijnana Parishad Anusandhan Patrika 12, 51-59, 1969.

[237] Theorems on Hankel and G-function transform -I. Vijnana Parishad Anusandhan Patrika. 13, 113-128. 1970.

HERZ, C.S.

[238] Bessel functions of matrix argument. Ann. of Math. 61, 474-523, 1955.

JAIN, N.C.

[239] An expansion formula for Meijer's G-function. Labdev J. Sci. Tech. Part A 8, 7-8, 1970.

JAIN, P.C. and SHARMA, B.L.

[240] An expansion for generalized function of two variables. Univ. Nac. Tucuman Rev. Ser. A. 18, 7-15, 1968.

[241] Some new expansions of the generalized function of two variables. Univ. Nac. Tucuman Rev. Ser. A. 18, 25-33, 1968.

JAIN, R.N.

[242] Some infinite series of G-functions. Math. Japon 10, 101-105, 1965.

[243] A finite series of G-functions. Math. Japon. 11, 129-131, 1966.

[244] A generalized hypergeometric polynomial. Ann. Polon. Math. 19,177-184, 1967.

[245] General series involving H-functions. Proc. Cambridge Philos. Soc. 65, 461-465, 1969.

JAIN, U.C.

[246] On a chain of genealized Laplace transform involving the H-function. Riv. Mat. Univ. Parma (2), 8, 125-130, 1967.

[247] Certain recurrence relations for the H-function. Proc. Nat. Inst. Sci. India Part A, 33, 19-24, 1967.

[248] On chains for Meijer's Bessel function transform invilving the H-function. Vijnana Parishad Anusandhan Patrika. 10, 91-97, 1967.

[249] On an integral involving the H-function. J. Austral. Math. Soc. 8, 373-376, 1968.

JAMES, A.T.

[250] Normal multivariate analysis and the orthogonal group. Ann. Math. Statist. 25, 40-75, 1954.

[251] Zonal polynomials of the real positive definite symmetric matrices. Ann. of Math. 74, 456-469, 1961.

KALIA, R.N.

[252] A new class of integral transforms. Glasnik Mat. 5 (25),269-276, 1970.

KALLA, S.L.

[253] Infinite integrals involving product of Bessel and hypergeometric functions. Vijnana Parishad Anusandhan Patrika 9, 165-170, 1966.

[254] Finite integrals involving generalized hypergeometric functions in several variables. Vijnana Parishad Anusandhan Patrika. 10, 99-106,1967.

[255] Some infinite integrals involving generalized hypergeometric functions ψ_2 and F_c. Proc. Nat. Acad. Sci. India Sect. A 37,195-200,1967.

[256] Finite integrals involving product of Bessel functions. Proc. Nat. Acad. Sci. India Sect. A 38, 229-233, 1968.

[257] Integral operators involving Fox's H-function. Acta Mexicana Ci. Tecn.3, 117-122, 1969.

[258] Some theorems on fractional integration-II. Proc. Nat. Acad. Sci. India Sect. A, 39, 49-56, 1969.

[259] Infinite integrals involving Fox's H-function and confluent hypergeometric functions. Proc. Nat. Acad. Sci. India 39 A, 3-6, 1969.

[260] Some finite integral involving product of Bessel functions. Vijnana Parishad Anusandhan Patrika 13, 101-105, 1970.

KALLA, S.K. and KUSHWAHA, R.S.

[261] Production of heat in an infinite cylinder. Acta Mexicana Cien. Tecn. 4, 89-93, 1970.

kalla, S.L. and MUNOT, P.C.

[262] An expansion formula for the generalized Fox function of two variables Repub. Venezuela Bol. Acad. Ci. Fis. Mat.Natur 30, No.86,87-93, 1970.

KALLA, S.L. and SAXENA, R.K.

[263] Integral operators involving hypergeometric functions. Math. Z. 108, 231-234, 1969.

KAPILEVIČ, M.B.

[264] A certain class of hypergeometric functions of Horn (Russian). Differencial'nye Uravneniya 4, 1465-1483, 1968.

KAPOOR, V.K.

[265] On a generalized Stieltjes transform. Proc. Cambridge Philos. Soc. 64, 407-412, 1968.

KAPOOR, V.K. and GUPTA, S.K.

[266] Fourier series for H-function. Indian J. Pure Appl. Mat . 1, No.4, 433-437, 1970.

KAPOOR, V.K. and MASOOD, S.

[267] On a generalized L-H transform. Proc. Cambridge Philos, Soc. 64,399-406, 1968.

KAUFMAN, H; MATHAI, A.M. and SAXENA, R.K.

[268] Distributions of random variables with random parameters. South Afr. Statist. J. 3, 1-7, 1969.

KESARWANI, R.N. (= ROOP NARAIN = NARAIN R)

[269] On an integral transform involving G-functions. SIAM J. Appl.Math. 20, 93-98, 1971.

KHATRI, C.G. and PILLAI, K.C.S.

[270] Some results on non-central multivariate Beta distribution and moments of traces of two matrices. Ann. Math. Statist. 36, 1511-1520, 1965.

[271] On the non-central distributions of two criteria in multivariate analysis of variance. Ann. Math. Statist. 39, 215-216, 1968.

KHATRI, C.G. and SRIVASTAVA, M.S.

[272] On exact non-null distributions of likelihood ratio criterion for sphericity test and equality of two covariance matrices. Sankhya Ser.A. 33, 201-206, 1971.

KNOTTNERUS, U.J.

[273] Approximation Formulae for Generalized Hypergeometric Functions for Large Values of the Parameters with Applications to Expansion Theorems for the Function $G^{m,n}_{p,q}(z)$, J.B. Wolters, Groningen, 1960.

KOBER, H.

[274] On fractional integrals and derivatives. Quart. J. Math. Oxford Ser. 11, 193-211, 1940.

KUIPERS, L. and MEULENBELD, B.

[275] Une généralisation d'une formule de Hobson. Nieuw Arch. Wisk. (3) 12, 99-101, 1964.

[276] Developpement de la fonction $Q^{m,n}_k(z)$ en série de fonctions de Legendre associées. Nieuw. Arch. Wisk (3), 14,114-118, 1966.

KUSH, S.N.

[277] Certain formulae involving a q-analogue of MacRobert's E-function. Ann. Soc. Sci. Bruxelles Ser. 1, 77, 134-139, 1963.

KUZNECOV, D.S.

[278] Special Functions (in Russian). Izdat. Vysŝ Š̄kola, Moscow, 1965.

LAWRYNOWICZ, J.

[279] On expansions of Meijer's functions-1 : The object of the paper and auxialiary results. Ann. Polon Math. 17, 245-257, 1966.

[280] On expansion of Meijer's functions-II: The method of the exponential factor. Ann. Polon. Math. 18, 43-52, 1966.

[281] On expansions of Meijer's functions-III: A problem of the changed parameters and particular cases. Ann. Polon. Math. 18, 147-161, 1966.

[282] Remarks on the preceeding paper of P. Anandani. Ann. Polon. Math. 21, 120-123, 1969.

LEBEDEV, N.N

[283] Special Functions and Their Applications. (translation from Russian) Prentice-Hall, New Jersey, 1965.

LEVELT, A.H.M.

[284] On a formula of C.S. Meijer. Proc. Nederl. Akad. Wetensch. Ser.A 63, 102-105, 1960.

LUKE, Y.L. (ALSO SEE FIELDS, J.L; WIMP, J)

[285] Integrals of Bessel Functions. McGraw-Hill, New York, 1962.

[286] On the approximate inversion of some Laplace transforms. 4th U.S. Congr. Appl. Mechan. Berkeley, California, 1962, pp.269-276. Amer. Soc.Mech. Engrs. New York, 1962.

[287] Recursion formulas for polynomials in rational approximations to generalized hypergeometric functions. Blanch Anniversary Volume pp. 161-167, Aerospace Research Lab., U.S. Air Force, Washington, D.C., 1967.

LUKE, YUDELL, L. and COLEMAN, R.L.

[289] Expansion of hypergeometric functions in series of other hypergeometric functions. Math. Comp. 15, 233-237, 1961.

LUKE, YUDELL, L. and WIMP, JET

[290] Expansions of a generalized hypergeometric function over a semi-infinite ray. Math. Comp. 17, 395-403. 1963.

MACROBERT, T.M. (MACROBERT, THOMAS MURRAY)

[291] Induction proofs of the relations between certain asymptotic expansions and corresponding generalized hypergeometric series. Proc. Royal Soc. Edinburgh 58, 1-13, 1937-1938.

[292] A proof by induction of the analytic continuations of certain generalized hypergeometric functions. Philos. Magazine (7)25, 848-851, 1938.

[293] Solution in multiple series of a type of generalized hypergeometric equation. Proc. Royal Soc. Edinburgh 59, 49-54, 1939.

[294] Expressions for generalized hypergeometric functions in multiple series. Proc. Royal Soc. Edinburgh 59, 141-144, 1939.

[295] Some formulae for the E-function. Philos. Mag (7) 31, 254-260, 1941.

[296] Some integrals involving E-functions and confluent hypergeometric functions. Quart. J. Math. Oxford Ser. 13, 65-68, 1942.

[297] On an identity involving E-functions. Philos. Mag. (7), 39, 466-471, 1948.

[298] Proofs of some formulae for the hypergeometric functions and the E-functions. Philos.Mag.(7) 34, 422-426, 1943.

[299] An integral involving an E-function and associated Legendre function of the first kind. Proc. Glasgow Math. Assoc. 1, 111-114, 1953.

[300] Integral of an E-function expressed as a sum of two E-functions. Proc. Glasgow Math. Assoc. 1, 118, 1953.

[301] Some integrals involving E-functions. Proc. Glasgow Math. Assoc. 1, 190-191, 1953.

[302] Integrals involving a modified Bessel function of the second kind and an E-function. Proc. Glasgow Math. Assoc. 2, 93-96, 1954.

[303] Integrals involving E-functions and associated Legendre functions. Proc. Glasgow Math. Assoc. 2, 127-128,1955.

[304] Integrals allied ty Airy's integrals. Proc. Glasgow Math. Assoc. 3,91-93, 1957.

[305] Infinite series of E-functions. Proc. Glasgow Math. Assoc. 4, 26-28, 1958.

[306] Integrals invovling E-functions. Math. Z. 69, 234-236, 1958.

[307] Integrals involving hypergeometric functions and E-functions. Proc. Glasgow Math. Assoc. 3, 196-198, 1958.

[308] Integrals of products of E-functions. Math. Ann. 137, 412-416, 1959.

[309] Infinite series of E-functions. Math. Z. 71, 143-145, 1959.

[310] Integration of E-functions with respect to their parameters. Proc. Glasgow Math. Assoc. 4, 84-87, 1959.

[311] The multiplication formula for the gamma function and E-function series. Math. Ann. 139, 133-139, 1959.

[312] Multiplication formulae for the E-functions regarded as functions of their parameters. Pacific J. Math. 9, 759-761, 1959.

[313] Applications of the multiplication formula for the gamma function to E-function series. Proc. Glasgow Math. Assoc. 4, 114-118, 1960.

[314] Recurrence formulae for the E-functions. Math. Z. 73, 254-255, 1960.

[315] Expression for an E-function as a finite series of E-functions. Math.Ann. 140, 414-416, 1960.

[316] Evaluation of an E-function when two of the upper parameters differ by an integer. Proc. Glasgow Math. Assoc. 5, 30-34, 1961.

[317] Transformation of series of E-functions. Pacific J. Math. 11, 309-311, 1961.

[318] Fourier series for E-functions. Math.Z. 75, 79-82, 1960-61.

[319] Beta function formulae and integrals involving E-functions. Math. Ann. 142, 450-452. 1960-61.

[320] Integrals involving Gegenbauer function and E-functions. Math. Ann. 144, 299-301, 1961.

[321] Barnes integral as a sum of E-function. Math. Ann. 147, 240-243, 1962.

[322] Functions of a Complex Variable. 5th Ed. MacMillan, London, New York,1962.

[323] Evaluation of an E-function when three of its upper parameters differ by integral values. Pacific J. Math. 12, 999-1002, 1962.

[324] Integrals involving E-functions. Proc. Glasgow Math. Assoc. 6, 31-33,1963.

MACROBERT, T.M. and RAGAB, F.M.

[325] E-function series whose sums are constants. Math. Z. 78, 231-234, 1962.

MAGNUS, W; OBERHETTINGER, F. and SONI, R.P.

[326] Formulas and Theorems for the Special Functions of Mathematical Physics. (52), Springer-Verlag, New York, 1966.

MAHESHWARI, M.L.

[327] On Meijer's G-function. Math. Vesnik 6(21),283-287, 1969.

MALOO, H.B.

[328] Integrals involving Bessel functions and G-function. Vijnana Parishad Anusandhan Patrika 8, 37-42, 1965.

[329] Integrals involving products of Bessel function and Meijer's G-function Monatsh. Math. 70, 127-133, 1966.

[330] Integrals involving products of Bessel function and Meijer's G-function. Monatsh. Math. 70, 350-356, 1966.

[331] Integral involving products of Bessel function. Univ. Nac.Tucuman Rev. Se. A, 17, 37-46, 1967.

[332] An integral involving products of Bessel function and Meijer's G-function -III. Vijnana Parishad Anusandhan Patrika, 10, 25-36, 1967.

MANOCHA, H.L.

[333] A formula involving hypergeometric functions. Ricerca (Napoli) No. 1, 28-30,1968.

MANOCHA, H.L. and SHARMA, B.L.

[334] Summation of infinite series. J. Austral. Math. Soc. 6 (4), 470-476, 1966.

MATHAI, A.M. (ALSO SEE KAUFMAN, H)

[335] Applications of Generalized Special Functions in Statistics. Monograph. Indian Statistical Institute and McGill University, 1970.

[336] The exact distribution of a criterion for testing the hypothesis that several populations are identical J. Indian Statisticall Assoc.8,1-17,1970. 1970.

[337] The exact distribution of Bartlett's criterion for testing equality of covariance matrices, Publ. L'ISUP, Paris, 19, 1-15, 1970.

[338] Statistical theory of distribution and Meijer's G-function. Metron, 28, 122-146, 1970.

[339] A representation of an H-function suitable for practical applications. Indian Statistical Institute, Calcutta, Technical Report, Math. Statist. 29-70.

[340] Hard limiting of several sinusoidal signals. (Communicated for publication).

[341] An expansion of Meijer's G-function in the logarithmic case with applications. Math. Nachr. 48, 129-139, 1971.

[342] On the distribution of the likelihood ratio criterion for testing linear hypotheses on regression coefficients. Ann. Inst. Statist. Math. 23, 181-197, 1971.

[343] An expansion of Meijer's G-function and the distribution of product of independent beta variates. S. Afr. Statistic. J. 5,71-90, 1971.

[344] The exact non-null distributions of a collection of multivariate test statistics. Publ. L'ISUP, Paris, 20, No. 1 (to appear), 1971.

[345] The exact distributions of three criteria assocaited with Wilks' concept of generalized variance. Sankhya Ser. A., 34, 161-170, 1972.

[346] The exact non-central distribution of the generalized variance. Ann. Inst. Statist . Math., 24,53- 65, 1972.

[347] The exact distribution of a criterion for testing that the covariance matrix is diagonal. Trab. Estadistica 28,111-124, 1972.

[348] The exact distribution of a criterion for testing the equality of diagonal elements given that the covariance matrix is diagonal. Trab. Estadistica. (to appear).

[349] A few remarks on the exact distributions of likelihood ratio criteria-1 Ann. Inst. Statist. Math. 24, 1972.

[350] A review of the different methods of obtaining the exact distributions of multivariate test criteria. Sankhya Ser. A. (to appear).

MATHAI, A.M. and RATHIE, P.N.

[351] An expansion of Meijer's G-function and its application to statistical distributions. Acad. Roy. Belg. Ci. Sci.(5) 56, 1073-1084, 1970.

[352] The exact distribution of Votaw's criterion. Ann. Inst. Statist. Math. 22, 89-116, 1970.

[353] The exact distribution for the sphericity test. J. Statist. Res.(Dacca), 4, 140-159, 1970.

[354] Exact distribution of Wilks' criterion. Ann. Math.Statist. 42,1010-1019, 1971.

[355] The exact distribution of Wilks' generalized variance in the non-central linear case. Sankhya Ser. A, 33, 45-60, 1971.

[356] The problem of testing independence. Statistica, 31, 673-688, 1971.

MATHAI, A.M. and SAXENA, R.K.

[357] On a generalized hypergeometric distribution. Metrika,11, 127-132, 1966.

[358] Distribution of a product and the structural setup of densities. Ann. Math. Statist. 40, 1439-1448, 1969.

[359] Applications of special functions in the characterization of probability distributions. S. Afr. Statist. J. 3, 27-34, 1969.

[360] Extensions of Euler's integrals through statistical techniques. Math. Nachr. 51, 1-10, 1971.

[361] On linear combinations of stochastic variables. Metrika (to appear).

[362] Multiple integral representation of a G-function through statistical techniques. Ricerca (Napoli) (to appear).

[363] Meijer's G-function with a matrix argument. Acta Mexicana Ci. Tecn. 5,

[364] A generalized probability distribution. Univ. Nac. Tucuman Rev. Ser. A, 21, 193-202, 1972.

[365] Expansions of Meijer's G-function of two variables when the upper parameters differ by integers. Kyunpook Math. J. 12,61-68, 1972.

[366] Generalized Hypergeometric Functions with Applications in Statistics and Physical Sciences Monograph. Department of Mathematics, McGill University, July, 1972.

MATHUR, S.N.

[367] Integrals involving H-function. Univ. Nac. Tucuman Rev. Ser.A, 20, 145-148, 1970.

MCLACHLAN, N.W.

[368] Bessel Functions for Engineers. 2nd Ed. Oxford, 1961.

MEIJER, C.S.

[369] Über Whittakersche bezw. Besselsche funktionen and deren produkte. Nieuw. Arch. Wisk. (2), 18, 10-39, 1934.

[370] Einige integraldarstellungen fur produkte von Whittakerschen funktionen. Quart. J. Math. Oxford, 6, 241-248, 1935.

[371] Neue integraldarstellungen aus der theorie der Whittakerschen und Hankelschen funktionen. Math. Ann. 112, 439-489, 1936.

[372] Über produkte von Legendreschen funktionen. Nederl. Adad. Wetensch. Proc. 42, 930-937, 1939.

[373] Über Besselsche, Lommelsche und Whittakersche funktionen. (Zweit. Mitteilung). Nederl. Adad. Wetensch. Proc. 42, 938-947, 1939.

[374] Über Besselsche Struvesche und Lemmelsche funktionen-I,II. Nederl. Akad. Wetensch. Proc. 43, 189-210; 366-378, 1940.

[375] Multiplikationstheoreme fur die funktion $G^{m,n}_{p,q}(z)$. Nederl. Akad. Wetensch. Proc. 44, 1062-1070, 1941.

[376] Neue integraldarstellungen fur Whittakersche funktionen. 1-V. Nederl. Akad. Wetensch. Proc. 44, 81-92; 186-194; 298-307;442-451; 590-598,1941.

[377] Integraldarstellungen für Whittakersche funktionen und ihre produkte. I-II. Nederl. Akad. Wetensch. Proc. 44, 435-441; 559-605, 1941.

[378] On the G-function, I-VIII. Nederl. Akad. Wetensch. Proc. 49, 227-237; 344-356; 457-469; 632-641; 765-772; 936-943; 1063-1072; 1165-1175 = Indag. Math. 8, 124-134; 213-225; 312-324; 391-400; 468-475; 595-602; 661-670; 713-723, 1946.

[379] Expansion theorems for the G-function. I-XI. Nederl. Akad. Wetensch. Proc. Ser. A, 55 = Indag. Math. 14, 369-379, 483-487, 1962; Proc. Ser.A, 56 = Indag. Math. 15, 43-49, 187-193, 349-357, 1953; Proc. Ser. A, 57 = Indag. Math. 16, 77-82, 83-91, 273-279, 1954; Proc. Ser. A,58 = Indag. Math. 17, 243-251, 309-314, 1955; Proc. Ser. A 59 = Indag. Math. 17, 70-82, 1956.

[380] Expansions of generalized hypergeometric functions. Simon Stevin 31, 117-139, 1957.

MELLIN, H.J.

[381] Abrip einer einheitlichen theorie der Gamma und der hypergeometrischen funktionen. Math. Ann. 68, 305-337, 1910.

MILLER, K.S.

[382] Multidimensional Gaussian Distribution. Wiley, New York, 1964.

MITRINOVIC, D.S. und DJOKOVIC, D.Ž.

[383] Specijalne Funkcije. Gradjevinska Knjiga, Belgrade, 1964.

MUKHERJEE, S.N.

[384] On some results involving Gegenbauer polynomials und associated Legendre polynomials. J. Sci. Res. Banaras Hindu Univ. 14, No. 1, 145-150, 1963-64.

NAGEL, GENGT; OLSOON, PER and WEISSGLAS, PETER.

[385] Certain recurrence relations for Appéll functions useful in the calculation of matrix elements in an angular momentum representation. Ark. Fys 23, 137-143, 1963.

NAIR, V.C. and SAMAR, M.S.

[386] An integral involving the product of three H-functions. Math. Nachr. 49, 101-105, 1971.

NARAIN, R. (= KESERWANI, R.N. = ROOP NARAIN)

[387] The G-functions as unsymmetrical Fourier kernels -I. Proc. Amer. Math. Soc. 13, 950-959, 1962.

[388] The G-functions as unsymmetrical Fourier kernels-II. Proc. Amer. Math. Soc. 14, 18-28, 1963.

[389] The G-functions as unsymmetrical Fourier kernels-III. Proc. Amer. Math. Soc. 14, 271-277, 1963.

[390] A pair of unsymmetrical Fourier kernels. Trans. Amer. Math. Soc. 115, 356-369, 1965.

[391] The G-functions as kernels in chain transforms. Portugal. Math. 24, 39-45, 1965.

[392] Fourier series for Meijer's G-functions. Compositio Math. 17, 149-151, 1965.

[393] Fractional integration and Hankel transform. Univ. Politec. Torino Rend. Sem. Mat. 26, 87-91, 1966-67.

[394] Fractional integration and functions self-reciprocal in Hankel transform. Portugal Math. 26, 473-478, 1967.

[395] Fractional integration and certain dual integral equations. Math. Z. 98, 83-88, 1967.

[396] G-functions as self-reciprocal in an integral transform. Compositio Math. 18, 181-187, 1967.

[397] Correction to: Fractional integration and certain dual integral equations Math. Z. 107, 82, 1968.

NIBLETT, J.D.

[389] Some hypergeometric identities. Pacific J. Math. 2. 219-225, 1952.

NIELSEN, N.

[399] Handbuch der Theorie der Gammafunktionen. Band I; Theorie des Integrallogarithmus und Verwandter Transzendenten. Band-II. Chelsia, New York,1965.

NIGAM, HIRDAI,N.

[400] A note on Fox's H-function. Ganita, 20, No. 2, 47-52, 1969.

NORLUND, N.E.

[401] Vorlesungen Über Differenzenrechnung. Chelsea, New York, 1954.

[402] Hypergeometric functions. Acta. Math. 94, 289-349, 1955.

OBERHETTINGER, F. (ALSO SEE MAGNUS, W).

[403] Tabellen Zur Fourier Transformation. Springer, Berlin, 1957.

[404] Tables of Laplace and Mellin Transforms. Berlin, Gottingen, Heidelberg, New York, Springer, 1972.

OBERHETTINGER, F. and HIGGINS, T.P.

[405] Tables of Lebedev, Mehler and Generalized Mehler Transforms. Math. Note No. 246, Boeing Sci. Lab., Seattle, 1961.

OBERHETTINGER, F. and MAGNUS, W.

[406] Anwendung der Elliptischen Funktionen in Physik and Technik. Springer, Berlin, 1949.

OLKHA, G.S. (ALSO SEE GUPTA, K.C)

[407] Some finite expansions for the H-function. Indian J. Pure Appl. Math. 1, No. 3, 425-429, 1970.

OLSSON,P.O.M.

[408] A hypergeometric function of two variables of importance in perturbation theory -I. Ark. Fys. 30, 187-191, 1965.

[409] A hypergeometric function of two variables of importance in perturbation theory -II. Ark. Fys. 29, 459-465, 1965.

PARASHAR, B.P.

[410] Fourier series for H-functions. Proc. Cambridge Philos. Soc. 63, 1083-1085, 1967.

[411] Some theorems on a generalized Laplace transform and results involving H-function of Fox. Riv. Mat. Univ. Parma (2) 8, 375-384, 1967.

[412] Domain and range of functional integration operators. Math. Japon 12, 141-145, 1967.

PARIHAR, C.L.

[413] On integrals involving hypergeometric functions. Proc. Indian Acad. Sci. Sect. A. 65, 291-297, 1967.

[414] Fourier series for Meijer's G-function and MacRobert's E-function. Proc. Nat. Inst. Sci. India Part A, 35, 135-139, 1969.

PATHAK, R.S.

[415] Definite integrals involving G-functions. Ganita, 17, 96-98, 1966.

[416] Definite integrals involving generalized hypergeometric functions. Proc. Nat. Acad. Sci. India Sect. A 36, 849-852, 1966.

[417] Some results involving G- und H- functions. Bull. Calcutta Math. Soc. 62, 97-106, 1970.

PATHAN, M.A.

[418] A theorem on Laplace transform. Proc. Nat. Acad. Sci. India Sect. A, 37, 124-130, 1967.

[419] Certain recurrence relations. Proc. Cambridge Philos. Soc. 64, 1045-1048, 1968.

[420] Self reciprocal functions. Proc. Nat. Acad. Sci. India Sect. A(2), 39 140-144, 1969.

PETERS, A.S.

[421] Certain integral equations and Sonine's integrals. IMM-NYU 285, Institute of Mathematical Sciences, New York University, 1961.

PILLAI, K.C.S; AL-ANI, S. and JOURIS, G.M. (ALSO SEE KHATRI, C.G)

[422] On the distributions of the roots of a covariance matrix and Wilks's criterion for tests of three hypotheses. Ann. Math. Statist.40,2033-2040, 1969.

PILLAI, K.C.S. and JOURIS, G.M.

[423] Some distribution problems in multivariate complex Gaussian case. Ann. Math. Statist. 42,517-525, 1971.

PILLAI, K.C.S. and NAGARSENKER, B.N.

[424] On the distribution of the sphericity test criterion in classical and complex normal populations having unknown covariance matrices. Ann. Math. Statist. 42, 764-767, 1971.

POCHHAMMER, L.

[425] Über die differentialgleichung der allgemeineren hypergeometrischen Reihe mit zwei endlichen singulären punkten. Journ. für die reine und angewandte Math. 102, 76-159, 1888.

RAGAB, FOUAD M. (ALSO SEE MACROBERT, T.M)

[426] An integral involving the product of a Bessle function and an E-function. Proc. Glasgow Math. Assoc. 1, 8-9, 1952.

[427] Generalizations of some intengrals involving Bessl functions and E-functions. Proc. Glasgow Math. Assoc. 1, 72-75, 1952.

[428] Generalization of an integral due to Hardy. Proc. Glasgow Math. Assoc. 1, 115-117-1953.

[429] An integral involving a modified Bessel function of the second kind and an E-function. Proc. Glasgow Math. Assoc. 1, 119-120, 1953.

[430] Integrals involving E-functions. Proc. Glasgow Math. Assoc. 1,129-136, 1953.

[431] Integrals of E-functions expressed in terms of E-functions. Proc. Glasgow Math. Assoc. 1, 192-195, 1953.

[432] Recurrence formulae for the E-functions. Proc. Math. Phys. Soc. Egypt. 4, No. 4, 127-136, 1952-53.

[433] A linear relation between- E-functions. Proc. Glasgow, Math. Assoc. 1, 185-186,1953.

[434] Linear relations between E-functions and Bessel functions. Acta Math. 92, 1-11, 1954.

[435] Integrals involving E-functions and Bessel functions. Nederl. Akad. Wetensch. Proc. Ser. A. 57 = Indag. Math. 16, 414-423, 1954.

[436] Integrals involving E-functions and Bessel functions of the second kind. Proc. Glasgow Math. Assoc. 2, 52-56, 1954.

[437] Further integrals involving E-functions. Proc. Glasgow Math. Assoc, 2, 77-84, 1954.

[438] An integral involving a product of two modified Bessle functions of the second kind. Proc. Glasgow Math. Assoc. 2, 85-88, 1954.

[439] New integrals involving Bessel functions. Math. Z. 61, 386-390, 1955.

[440] Some formulae for the product of two Whittaker functions. Nederl. Akad. Wetensch. Proc. Ser. A. 58 = Indag. Math. 17, 430-434,1955.

[441] Some formulae for the product of three modified Bessel functions of the second kind. Nederl. Akad. Wetensch. Proc. Ser. A 58 = Indag. Math. 17, 621-626, 1955.

[442] A product of two E-functions expressed as a sum of two E-functions. Proc. Glasgow Math. Assoc. 2, 124-126, 1955.

[443] New integrals involving Bessel functions. Acta. Math. 95, 1-8, 1956.

[444] Integrals involving products of Bessel functions. Proc. Glasgow Math. Assoc. 2, 180-182, 1956.

[445] On the product of two confluent hypergoemetric functions. Arch. Math. 8, 180-183,1957.

[446] Integrals involving products of E-functions, Bessel functions and generalized hypergeometric functions. Boll. Un. Mat. Ital. (3) 12,535-551,1957.

[447] Expansions for products of two Whittaker functions. Div. Electromag. Res. Inst. Math. Sci. New York. Univ. Res. Rep. No. BR-23,i + 13pp, 1957.

[448] Some formulae for the products of E-functions and Whittaker functions. Div. Electromag. Res. Inst. Math. Sci. New York, Univ. Res. Rep. No.BR-24, i + 8 pp, 1957.

[449] The inverse Laplace transform of an exponential function. Div. Electromag. Res. Inst. Math. Sci. New York, Univ. Res. Rep. No. EM-107, pp.16, 1957.

[450] On the product of two confluent hypergeometric functions. Monatsh. Math. 61, 312-317, 1957.

[451] Some formulae for the product of hypergeometric functions. Proc. Cambridge Philos. Soc. 53, 106-110, 1957.

[452] Integration of E-functions with regard to their parameters. Proc. Glasgow Math. Assoc. 3, 94-98, 1957.

[453] The inverse Laplace transform of an exponential function. Comm. Pure. Appl. Math. 11, 115-127, 1958.

[454] On an identity involving Bessel polynomials. Illinois J. Math. 2, 236-239, 1958.

[455] Some formulae for the product of Bessel and Legendre functions. Math. Z. 68, 338-339,1958.

[456] Integration of E-functions and related functions with respect to their parameters. Nederl. Akad. Wetensch. Proc. Ser. A. 61 = Indag. Math. 20, 335-339, 1958.

[457] An expansion involving confluent hypergeometric functions. Nieuw. Arch. Wisk.(3) 6,52-54,1958.

[458] Transcendental addition theorems for the hypergeometric function of Gauss. Pacific J. Math. 8, 141-145, 1958.

[459] On the product of Legendre and Bessel functions. Proc. Amer. Math. Soc. 9, 26-31, 1958.

[460] A formula similar to Barnes' lemma. Ann. Polon Math. 5, 149-152, 1958.

[461] Expansion of an E-function in a series of products of E-functions. Proc. Glasgow Math. Assoc. 3, 194-195, 1958.

[462] Series of products of Bessel polynomials. Canda. J.Math. 11, 156-160,1959

[463] An integral involving the associated Legendre functions of the first kind. Michigan Math.J. 6, 97-99, 1959.

[464] Series of products of two generalized hypergeometric functions of the type ${}_1F_2$ and of the Bessel function $[S_v(x)]^2$. Bull. Math. Soc. Sci. Math. Phys. R.P. Roumaine. (N.S). 4(52) No.3-4, 93-102, 1960.

[465] Two series of products of two transcendental hypergeometric functions of Gauss. Bull. Math. Soc. Sci. Math. Phys. R.P. Roumaine (N.S) 4(52), No. 3-4, 89-92, 1960.

[466] Integrals involving products of Bessel functions. (English summary). Ann. Mat. Pura. Appl. (4)56, 301-311, 1961.

[467] Series associated with an identity of Ramanujan. Nieuw. Arch. Wisk.(3) 9, 113-116, 1961.

[468] The Orr hypergeometric identity and associated series. Riv. Mat. Univ. Parma. (2)2, 37-45, 1961.

[469] Integrals involving products of Bessel functions. Math. Z. 80, 177-183, 1962.

[470] Expansions of generalized hypergeometric functions in series of products of generalized Whittaker functions. Proc. Cambridge Philos. Soc. 58, 239-243, 1962.

[471] Summation of a series of products of E-functions. Proc. Glasgow Math. Assoc. 5, 118-120, 1962.

[472] Neue integraldarstellungen aus der theorie der MacRobertschen und Besselschen funktionen. Math. Z. 79, 147-157, 1962.

[473] Reihen von produkten MacRobertscher E-functionen. Math. Z. 78,222-230, 1962.

[474] Infinite series of E-functions. Math. Ann. 148, 94-98, 1962.

[475] The inverse Laplace transform of the product of two Whittaker functions. Proc. Cambridge Philos. Soc. 58, 580-582, 1962.

[476] The inverse Laplace transform of the modified Bessel function $K_{mn}(a^{1/2m}p^{1/2m})$. J. London Math. Soc. 37, 391-402, 1962.

[477] The Laplace transform of the modified Bessel function $K(t^{\pm m}x)$ where $m = 1,2,3,\ldots,n$. Proc. Edinburgh Math. Soc. (2)13, 325-329, 1962-63.

[478] Some formulae similar to Barnes' lemma. Ann. Polon. Math. 12, 265-272, 1962-63.

[479] Expansions of Kampé de Fériet's double hypergeometric function of higher order. J. Reine Angew. Math. 212, 113-119, 1963.

[480] The inverse Laplace transform of the product of two modified Bessel functions $K_{n\nu}(a^{1/2n}p^{1/2n})\,K_{n\mu}(a^{1/2n}p^{1/2n})$ where $n = 1,2,3,\ldots$, Ann. Polon. Math. 14, 77-83, 1963.

[481] Reihen fur MacRobertsche und Whittakersche funktionen. (Russian summary) Czechoslovak Math. J. 13(88), 284-289, 1963.

[482] Integrals involving products of modified Bessel functions of the second kind. Proc. Glasgow Math. Assoc. 6, 70-74, 1963.

[483] Reihen von produkten MacRobertscher E-funktionen. Bull.Math.Soc. Sci.Math. Phys. R.P. Roumaine. (N.S) 6(54), 107-115,1964.

[484] Expansions into double Kampé de Fériet hypergeometric functions of higher orders. (Russian) Dokl. Akad. Nauk. SSSR, 519-521, 1964.

[485] The inverse Laplace transform of the product $\exp(-a^{1/n}p^{1/n}/2)K_{n\nu}(a^{1/n}p^{1/n}/2)$, $n = 1,2,3,\ldots$, Boll. Univ. Mat. Ital. (3)19, 26-30, 1964.

[486] E-transforms. J.Res. Nat. Bur. Standards, Sect. B, 71B, 23-37, 1967.

[487] E-transforms-II. J. Res. Nat. Bur. Standards, Sect. B, 71B, 77-89, 1967.

[488] Infinite series of Kampé de Fériet's double hypergeometric function of higher order. Rend. Circ. Mat. Palermo (2)16, 225-232, 1967.

RAGAB, F.M. and SIMARY, M.A.

[489] Integration of E-functions and related series. Monatsh. Math. 70, 65-83, 1966.

[490] Integrals involving E-functions. Proc. Glasgow Math. Assoc. 7,174-177, 1966.

RAGAB, F.M. and HAMZA, A.M.

[491] Integrals involving E-functions and Kampe de Fériet's functions of higher order. Ann. Mat. Pura. Appl. (4), 87,11-24,1970.

RAINVILLE, E.D.

[492] Special Functions. The MacMillan Co. New York, 1965.

RAO, C.R.

[493] Linear Statistical Inference and its Applications. Wiley, New York,1965.

RATHIE, C.B.

[494] Some results involving hypergeometric and E-functions. Proc. Glasgow Math. Assoc.(2), 132-138,1955.

[495] A few infinite integrals involving E-functions. Proc. Glasgow Math. Assoc. 2,170-172. 1956.

[496] A theorem in operational Calculus and some integrals involving Legendre, Bessel and E-functions. Proc. Glasgow Math. Assoc. 2,173-179, 1956.

[497] Integrals involving E-functions. Proc. Glasgow Math. Assoc. 4,186-187, 1960.

RATHIE, P.N. (ALSO SEE GUPTA, P.N; MATHAI, A.M.)

[498] Theorems concerning Bessel transforms. Proc. Nat. Acad. Sci. India Sec.A 34, 501-506, 1964.

[499] A theorem on Laplace and Meijer transforms. Vijnana Parishad Anusandhan Patrika 8, 63-69, 1965.

[500] Integrals involving Appell's function F_4. Proc. Nederl. Akad. Wetensch. Ser. A. 68 = Inad. Math. 27, 113-118,1965.

[501] The inverse Laplace transform of the product of Bessel and Whittaker functions. J. London Math. Soc. 40, 356-369, 1965.

[502] Some results for generalized hypergeometric series. Nieuw. Arch. Wisk. 14, 261-267, 1966.

[503] Some infinite series for F_c, F_4, G-functions. Collect Math. 18, 195-205. 1966-67.

[504] Some finite integrals involving F_4 and H-functions. Proc. Cambridge Philos. Soc. 63, 1071-1081, 1967.

[505] Some finite and infinite series for F_c,F_4, ψ_2 and G-functions. Math. Nachr. 35, 125-136, 1967.

[506] Some results for generalized hypergeometric series. Portugal Math. 26, 175-184, 1967.

[507] Infinite integrals involving product of Bessel and Meijer's G-functions. Proc. Nat. Acad. Sci. India Sect. A. 38, 256-258, 1968.

[508] Some series involving hypergeometric functions. Vijnana Parishad Anusandhan Patrika, 11, 65-69,1968.

[509] On definite integrals involving special functions. Ricerca (Napoli) (2) 20, Maggio-agosto, 13-18, 1969.

RATHIE, P.N. and GUPTA, O.P.

[510] Some infinite integrals involving ψ_2,F_c and H-functions. Vijnana Parishad Anusandhan Patrika, 11, 11-29, 1968.

RATTI, J.S.

[511] On an infinite series of G-functions. Studies and Essays (Presented to Yu-Why Chen on his 60th birthday, April, 1970), pp.269-272. Math. Res. Center.Nat.Taiwan Univ. Taipei, 1970.

ROBERTS, G.E. and KAUFMAN, H.

[512] Tables of Laplace Transforms. W.B. Saunders Co, Philadelphia, London,1966.

SAKSENA, K.M.

[513] An inversion theory for the Laplace integral. Nieuw Arch. Wisk.(3) 15, 218-224,1967.

SARAN, S.

[514] A definite integral involving the G-function. Nieuw Arch. Wisk(3) 13, 223-229,1965.

SAXENA, R.K. (ALSO SEE AGARWAL, I; BORA, S. L; KAUFMAN,H; MATHAI, A.M)

[515] A theorem on Meijer transform. Proc. Nat. Inst. Sci. India Part A, 25, 166-170, 1959.

[516] A study of the generalized Stieltjes transform. Proc. Nat. Inst. Sci. India Part A 25, 340-355, 1959.

[517] Some theorems on generalized Laplace transform -I. Proc. Nat. Inst. Sci. India Part A 26, 400-413, 1960.

[518] Some integrals involving E-functions. Proc. Glasgow Math. Assoc. 4, 178-185, 1960.

[519] An integral involving G-function. Proc. Nat. Inst. Sci. India Part A, 26, 661-664, 1960.

[520] Some theorems in operational calculus and infinite integrals involving Bessel function and G-functions. Proc. Nat. Inst. Sci. India, Part A 27, 38-61,1961.

[521] Some theorems on generalized Laplace transform -II. Riv. Mat. Univ.Parma, (2), 287-299, 1961 .

[522] A definite integral involving associated Legendre function of the first kind. Proc. Cambridge Philos. Soc. 57, 281-283, 1961.

[523] Integrals involving Legendre functions. Math. Ann. 147, 154-157, 1962.

[524] Some infinite integrals involving E-functions. Proc. Glasgow Math. Assoc. 5, 183-187, 1962.

[525] Definite integrals involving G-functions. Proc. Cambridge Philos. Soc. 58, 489-491, 1962.

[526] Some formulae for the G-function. Proc. Cambridge Philos. Soc. 59,347-350, 1963.

[527] Some formulae for the G-function-II. Collect Math. 15, 273-283, 1963.

[528] A relation between generalized Laplace and Hankel transforms. Math. Z. 81, 414-415, 1963.

[529] Integrals involving Bessel functions and Whittaker functions. Proc. Cambridge Philos. Soc. 60, 174-176, 1964.

[530] Integrals involving Legendre functions -II. Math. Ann. 154,181-184,1964.

[531] Integrals involving G-functions. Ann. Soc. Sci. Bruxelles Ser. 1, 8,157-162, 1964.

[532] Integrals involving products of Bessel functions. Proc. Glasgow Math. Assoc. 6, 130-132, 1964.

[533] Some theorems on Laplace transforms. Proc. Nat. Inst. Sci. India, Sect.A, 30, 230-234, 1964.

[534] On some results involving Jacobi polynomials. J. Indian Math.Soc. (N.S) 28, 197-202, 1964.

[535] Certain properties of Varma transform involving Whittaker functions. Collect. Math. 16, 193-200, 1964.

[536] Some formulae for the associated Legendre function of the first kind. Proc. Nat. Acad. Sci. India, Sect. A, 19-24, 1965.

[537] Some theorems on generalized Laplace transform -III. Riv. Math. Univ. Parma, (2), 6, 135-146, 1965.

[538] Integrals involving products of Bessel functions -II. Monatsh. Math. 70, 161-163, 1966.

[539] An inversion formula for Varma transform. Proc. Cambridge Philos. Soc. 62, 467-471, 1966.

[540] An inversion formula for a kernel involving a Mellin-Barnes type integral Proc. Amer. Math. Soc. 17, 771-779, 1966.

[541] An integral involving products of G-functions. Proc. Nat. Acad. Sci. India Sect. A, 36, 47-48, 1966.

[542] On the formal solution of certain dual integral equations involving H-functions. Proc. Cambridge Philos. Soc. 63, 171-178, 1967.

[543] On the formal solution of dual integral equations. Proc. Amer. Math. Soc. 18, 1-8, 1967.

[544] On fractional integration operators. Math. Z. 96, 299-291, 1967 .

[545] A new generating function for the Jacobi polynomials. Proc. Cambridge Philos. Soc. 66, 345-347, 1969.

[546] Integration involving Kampe de Fériet function and Gauss' hypergeometric function. Ricerca (Napoli)2, 21-27, 1970.

[547] Definite integrals involving Fox's H-functions. Acta Mexicana Cien.Tech. 5, 6-11, 1971.

[548] Integrals involving products of H-functions. Univ. Nac. Tucuman Rev. Ser. 21, (to appear).

SAXENA, R.K. and BHAGCHANDANI, L.K.

[549] An expansion formula for hypergeometric functions of two variables. Univ. Nac. Tucuman Rev. Ser. 19, 201-204, 1969.

SAXENA, R.K. and GUPTA, K.C.

[550] On Laplace transform. Riv. Mat. Univ. Parma, (2), 5, 159-164, 1964.

[551] Certain properties of generalized Stieltjes transform involving Meijer's G-function. Proc. Nat. Inst. Sci. India Part A, 30, 230-234, 1964.

SAXENA, R.K. and KUSHWAHA, R.S.

[552] Applications of Jacobi polynomials to non-linear oscillations. Proc. Nat. Acad. Sci. India Sect. A, 40, 65-72, 1970.

[553] Applications of Jacobi polynomials to non-linear differential equation associated with confluent hypergeometric function. Proc. Nat. Acad. Sci. India Sect. A, 40, 281-288, 1970.

SAXENA, R. K. and MATHUR, S.N.

[554] A finite series for H-function. Univ. Nac. Tucuman Rev.Ser. 21,(to appear)

[555] Some formulae for the G-function of two variables. Portugal. Math. 32, (to appear).

SAXENA, R.K. (of INDORE)

[556] On self reciprocal functions involving two complex variables. Ganita, 15, 19-30, 1964.

[557] Certain integral equations and self reciprocal functions. Proc. Nat. Acad. Sci. India Sect. A. 37, 179-186, 1967.

[558] Definite integrals involving self-reciprocal functions. Proc. Nat. Inst. Sci. India, Part A, 34, 326-336, 1968.

SCHÄFKE, F.W.

[559] Einfuhrung in die Theorie der Spezielle Funktionen. Springer, Berlin,1963.

SEVARIA, K.S.

[560] An integral involving product of Legendre function and Meijer's G-function. Vijnana Parishad Anusandhan Patrika, 10, 227-229, 1967.

SHAH, MANILAL.

[561] Expansion formula for a generalized hypergeometric polynomial in series of Hermite polynomial. Proc. Indian Acad. Sci. Sect. A, 68, 324-328,1968.

[562] Some results on Fourier series for H-functions. J. Natur. Sci. Math. 9, No. 1, 121-131, 1969.

[563] On some results involving generalized hypergeometric polynomials. Proc. Cambridge Philos. Soc. 87-92, 1969.

[564] Some results on generalized hypergeometric polynomials. Proc. Cambridge Philos. Soc. 66, 95-104, 1969.

[565] Some infinite integrals involving Whittaker functions and generalized hypergeometric polynomials with their applications. Proc. Cambridge Philos. Soc. 65, 483-488, 1969.

[566] Some results on the H-functions involving the generalized Lagurre polynomial. Proc. Cambridge Philos. Soc. 65, 713-720, 1969.

[567] On some results of the H-functions invovling Hermite polynomials J. Natur. Sci. Math. 9, 223-233, 1969.

[568] On some results involving H-functions and associated Legendre functions. Proc. Nat. Acad. Sci. India Sect. A, 39, 503-507, 1969.

[569] On some results involving generalized hypergeometric and Gegenbauer (Ultraspherical) polynomials. Proc. Nat. Acad. Sci. India Sect. A, 39, 493-502, 1969.

[570] On applications of Mellin's and Laplace's inversion formulae to H-functions. Labdev J. Sci. Tech. Part A, 7, 10-17, 1969.

[571] Some results on Fourier series for H-functions. J. Natur. Sci. Math. 9, 121-131, 1969.

[572] Fourier series for generalized Meijer functions. An. St. Univ. Al. I Cuza. Lasi. n. Ser. Sect. 16, 293-313, 1970.

SHARMA, B.L. (SHARMA, BHAGIRATH LAL: ALSO SEE ABIODUN, R.F.A)

[573] Some definite integrals involving Legendre and G-functions. Vikram-Quart. Res. J. Vikram Univ. 8 No. 1 and 3, 25-29, 1964.

[574] Integrals involving G-functions. Collect. Math. 16, 3-13, 1964.

[575] On the generalized function of two variables -I. Ann. Soc. Sci. Bruxelles Ser. 1, 79, 26-40, 1965.

[576] An integral involving G-function. Quart. Res. J. Vikram Univ. 10, 79-83, 1966.

[577] Some formulae for Jacobi polynomials. Proc. Cambridge Philos. Soc.62, 459-462, 1966.

[578] Integrals associated with generalized function of two variables. Mathematica (Cluj), 361-374, 1967.

[579] Some theorems in operational calculus. Mathematica (Cluj) 9 (32), 2, 375-381, 1967.

[580] A formula for G-function. Ricerca (Napoli), No. 1, 3-7, 1967.

[581] Integrals involving generalized function of two variables -II, Proc. Nat. Acad. Sci. India Sect. A 37, 137-148, 1967.

[582] Infinite integrals involving Legendre and G-functions. Arch. Math. (Basel) 18, 293-298, 1967.

[583] Some formulae for generalized function of two variables. Math. Vesnik 5 (20),43-52, 1968.

[584] An integral involving products of G-function and generalized function of two variables. Univ. Nac. Tucuman Rev. Ser. A, 18, 17-23, 1968.

[585] A new expansion formula for hypergeometric functions of two variables. Proc. Cambridge Philos. Soc. 64, 413-416, 1968.

[586] An integral involving generalized function of two variables. Ricerca (Napoli) (2) 20, gennaio-aprile, 1969.

[587] Some formulae for G-function. Rev. Un. Mat. Argentina 24, 159-166,1968-69.

[588] A new expansion formula for G-function. An. Sti. Univ. Al. L Cuza, Iasi. Sect. la. Mat. (N.S) 15, 43-46, 1969.

[589] Some formulae for G-function. Ricerca (Napoli) (2) 21,gennaio-aprile, 8-18, 1970.

[590] Sum of a series involving Leguerre polynomials and generalized functions of two variables. An. Sti. Univ. Al. I. Cuza.Iasi. n. Ser.Sect. (1) 17, 117-122, 1971.

SHARMA, B.L. and JINDIA, R.K.

[591] Some definite integrals involving Legendre and generalized function of two variables. Univ. Nac. Tucuman Rev. Ser. A. 17, 67-78, 1967.

SHARMA, K.C.

[592] A theorem on Meijer transform and infinite integrals involving G-function and Bessel functions. Proc. Nat. Inst. Sci. India Part A, 24,113-120,1958.

[593] Infinite integrals involving E-function and Bessel function. Proc. Nat. Inst. Sci. India Part A, 25, 161-165, 1959.

[594] Infinite integrals involving E-function and Bessel function. Proc. Nat. Inst. Sci. India Part A 25, 337-339, 1959.

[595] Theorems relating Hankel and Meijer's Bessel transforms. Proc. Glasgow Math. Assoc. 6, 107-112, 1963.

[596] Integrals involving products of G-function and Gauss' hypergeometric function. Proc. Cambridge Philos. Soc. 60, 539-542, 1964.

[597] Theorems on Meijer's Bessel function transform. Proc. Nat. Inst. Sci. India Part A 30, 360-366, 1964.

[598] A few inversion formulae for Meijer transform. Proc. Nat. Inst. Sci. India, Part A 30, 736-742, 1964.

[599] An integral involving G-function. Proc. Nat. Inst. Sci. India Part A 30, 597-601, 1964.

[600] A theorem on Meijer's Bessel function transform. Proc. Nat. Acad. Sci. India Sect. A, 35, 153-160, 1965.

[601] On an integral transform. Math. Z. 89, 94-97, 1965.

SHARMA, M.C.

[602] A theorem on generalized Hankel transform. Proc. Nat. Inst. Sci. India Part A 29, 114-118, 1963.

SHARMA, O.P.

[603] Some finite and infinite integrals involving H-function and Gauss' hypergeometric functions. Collect. Math. 17, 197-209, 1965.

[604] On self-reciprocal functions. Proc. Nat. Acad. Sci. India Sect. A, 36, 719-733, 1966.

[605] Certain infinite and finite integrals involving H-function and confluent hypergeometric functions. Proc. Nat. Acad. Sci. India Sect. A, 36, 1023-1032, 1966.

[606] On the Hankel transformations of H-functions. J. Math. Sci. 3,17-26,1968.

[607] On H-function and heat production in a cylinder. Proc. Nat. Acad. Sci. India Sect. A, 39, 355-360, 1969.

[608] On generalized Hankel and K-transforms. Math. Student, 37, 109-116,1969.

[609] On generalized Hankel and Meijer-Laplace transforms. Math. Student, 37, 159-168, 1969.

SINGH, F.

[610] An integral involving product of G-function, generalized hypergeometric function and Jacobi polynomial. J. Sci. Engrg. 12, 155-160, 1968.

[611] Application of the E-operator to evaluate an infinite integral. Proc. Cambridge Philos. Soc. 65, 725-730, 1969.

SINGH, R.

[612] On some results involving H-function of Fox. Proc. Nat. Acad. Sci. India Sect. A, 38, 240-250, 1968.

[613] Tow theorems on H-function of Fox. Proc. Nat. Acad. Sci. India Sect. A 38, 155-160,1968.

[614] On Fox's H-transform in two variables. Proc. Nat.Acad. Sci.India Sect. A, 39, 149-160, 1969.

SINGH, R.P.

[615] A note on double transformations of certain hypergeometric functions. Proc. Edinburgh Math. Soc. (2) 14, 221-227, 1964-65.

SINGH, S.P.

[616] On certain properties of generalized Hankel transforms. J. Indian Math. Soc. (N.S) 26, 35-52, 1962.

SINGHAL, J.P.

[617] Integration of certain products involving a generalized Meijer function. Proc. Nat. Acad. Sci. India Sect A, 36, 976- 986, 1966.

[618] An integral involving Kampé de Fériet's function. Vijnana Parishad Anusandhan Patrika 10, 37-42,1967.

SKIBIŃSKI, P.

[619] Some expansion theorems for the H-function. Ann. Polon. Math. 23,125-138, 1970.

SLATER, LUCY, J.

[620] An integral of hypergeometric type. Proc. Cambridge Philos. Soc. 48, 578-582, 1952.

[621] Integrals representing general hypergeometric transformations. Quart. J. Math. Oxford Ser. (2)3, 206-216, 1952.

[622] Two double hypergeometric integrals. Quart. J. Math. Oxford Ser. (2)4, 127-131, 1953.

[623] Integrals for asymptotic expansions of hypergeometric functions. Proc. Amer. Math. Soc. 6, 226-231, 1955.

[624] The integration of hypergeometric functions. Proc. Cambridge Philos. Soc. 51, 288- 296, 1955.

[625] Hypergeometric Mellin transforms. Proc. Cambridge Philos. Soc. 51, 577-589, 1955.

[626] Confluent Hypergeometric Functions. Cambridge University Press, 1960.

[627] Generalized Hypergeometric Series . Cambridge University Press, 1961.

SMITH, F.C.

[628] Relations among the fundamental solutions of the generalized hypergeometric equation when $p = q + 1$. I : Non-logarithmic cases. Bull. Amer. Math. Soc. 44, 429-433, 1938.

SMITH-WHITE, W.B. and BUCHWALD, V.T.

[629] A generalization of $z!$ J. Austral. Math. Soc. 4, 327-341, 1964.

SNEDDON, IAN N(ALSO SEE ERDÉLYI, A)

[630] Special Functions of Mathematical Physics and Chemistry. Oliver and Boyd, London, 1956.

[631] A relation between Hankel transforms with applications to boundary value problems in potential theory. J. Math. Mech. 14, 33-40, 1965.

[632] Mixed Boundary Value Problems in Potential Theory. North Holland Publishing Co., 1966.

SRIVASTAVA, ARUNA and GUPTA, K.C.

[633] On certain recurrence relations. Math. Nachr. 46, 13-23, 1969.

[634] On certain recurrence relations-II, Math. Nachr. 49, 187-197, 1971.

SRIVASTAVA, H.M.

[635] On integrals associated with certain hypergeometric functions of three variables. Proc. Nat Acad. Sci. India Sect. A, 34, 309-316, 1964.

[636] Some expansions of generalized Whittaker functions. Proc. Cambridge Philos. Soc. 61, 895-896, 1965.

[637] On the sums of certain generalized hypergeometric series. Proc. Nat. Acad. Sci. India Sect. A, 35, 37-42, 1965.

[638] On transformations of certain hypergeometric functions of three variables. Publ. Math. Debrecen 12, 65-74, 1965.

[639] A hypergeometric transformation associated with the Appell function F_4. Proc. Cambridge Philos. Soc. 62, 765-767, 1966.

[640] Some expansions in products of hypergeometric functions. Proc. Cambridge Philos. Soc. 62, 245-247, 1966.

[641] The integration of generalized hypergeometric functions. Proc. Cambridge Philos. Soc. 62, 761-764, 1966.

[642] On a summation formula for the Appell function F_2. Proc. Cambridge Philos. Soc. 63, 1087-1089, 1967.

[643] Generalized Neumann expansions involving hypergeometric functions. Proc. Cambridge Philos. Soc. 63, 425-429, 1967.

[644] Finite summation formulas associated with a class of generalized hypergeometric polynomials. J. Math. Anal. Appl. 23, 453-458, 1968.

[645] On a generalized integral transform. Math. Z. 108, 197-201, 1969.

[646] Infinite series of certain products involving Appell's double hypergeometric functions. (Serbo-Creation Summary). Glasnik Mat. Ser. III. 4(24) 67-73, 1969.

[647] An infinite summation formula associated with Appell's function F_2. Proc. Cambridge Philos. Soc. 65, 679-682, 1969.

[648] On a generalized integral transform-II. Math. Z. 121, 263-272, 1971.

SRIVASTAVA, H.M. and DAOUST, MARTHA C.

[649] Certain generalized Neumann expansions associated with the Kampé de Fériet function. Nederl. Akad. Wetensch. Proc. Ser. A 72 = Indag.Math. 31, 449-457, 1969.

SRIVASTAVA, H.M. and JOSHI, C.M.

[650] Certain double Whittaker transforms of generalized hypergeometric functions. Yokohama Math. J. 15, 17-32, 1967.

[651] Certain integrals involving a generalized Meijer function. Glasnik Mat. Ser. III. 3(23), 183-191, 1968.

[652] Integration of certain products associated with a generalized Meijer function. Proc. Cambridge Philos. Soc. 65, 471-477, 1969.

SRIVASTAVA, H.M. and SINGHAL, J.P.

[653] Double Meijer transformations of certain hypergeometric functions. Proc. Cambridge Philos. Soc. 64, 425-430, 1968.

[654] Certain integrals involving Meijer's G-function of two variables. Proc. Nat. Inst. Sci. India Part A, 35, 64-69, 1969.

SRIVASTAVA, KRISHNA JI.

[655] Certain integral representation of MacRobert's E-function. Ganita, 8, 51-60, 1957.

SRIVASTAVA, K.N.

[656] Inversion integrals involving Jacobi polynomials. Proc. Amer. Math. Soc. 15, 635-638, 1964.

[657] On some integral transforms involving Jacobi polynomials. Ann. Polon. Math. 16, 195-199, 1965.

SRIVASTAVA, R.P.

[658] A pair of dual integral equations involving Bessel functions of the first kind and the second kind. Proc. Edinburgh Math. Soc. (2) 14,149-158, 1964-1965.

SRIVASTAVA, T.N. and SINGH, Y.P.

[659] On Maitland's generalized Bessel function. Canad. Math. Bull. II, 739-741, 1968.

SUGIURA, N. and FUJIKOSHI, Y.

[660] Asymptotic expansions of the non-negative distribution of the likelihood ratio criteria for multivariate linear hypothesis and independence. Ann. Math. Statist. 40, 942-952,1969.

SUNDARARAJAN, P.K.

[661] Some integrals invovling E-function of MacRobert and G-function of Meijer. Proc. Nat. Acad. Sci. India Sect. A 34, 97-104, 1964.

[662] Some finite and infinite integrals involving G-function. Proc. Nat. Acad. Sci. India Sect. A, 36, 435-440, 1966.

[663] On the derivative of a G-function whose argument is a power of the variable. Compositio Math. 17, 286-290, 1966.

SWAROOP, RAJENDRA.

[664] On a generalization of the Laplace and the Stiltjes transformations. Ann. Soc. Sci. Bruxelles Ser. 1, 78, 105-112, 1964.

[665] A study of Varma transform. Collect. Math. 16, 15-32, 1964.

[666] A general expansion involving Meijer's G-function. Ann. Soc. Sci. Bruxelles Ser. 1, 79, 47-57, 1965.

SZEGŎ, G.

[667] Orthogonal Polynomials. (Colloq. Publ. Vol. 23). Amer. Math. Soc. Providence, Rhode Island, 1959.

TITCHMARSH, E.C.

[668] Introduction to the Theory of Fourier integrals. Oxford University Press, London, New York, 1948.

THOMAE, J.

[669] Über die hoheren hypergeometrischen Reihen, insbesondere uber die Reihe.. Math. Ann. 2, 427-444, 1870.

TOTOV, GEORGI.

[670] A hypergeometric function of n-variables (Bulgarian, Russian Summary). Godišnik Viss. Tech. Včebn. Zavad. Mat. I, Kn. 2, 37-77, 1965.

TRANTER, C.J.

[671] A further note on dual integral equations and applications to the diffraction of electromagnetic waves. Quart. J. Mech. Appl. Math. 7, 318, 1954.

[672] Integral Transforms in Mathematical Physics. 2nd Ed. Methuen and Co., London, John Wiley, New York, 1956.

TRICOMI, F.G.

[673] Funzioni Ipergeometriche Confluenti. Ed. Cremonese, Rome, 1954.

[674] Expansion of the hypergeometric function in series of confluent ones and application to the Jacobi polynomials. Comment. Math. Helv. 25, 196-204, 1951.

TRIPATHI, C.S.

[675] On Wimp's G-transform. Ganita. 18, 35-48, 1967.

VAN DER CORPUT, J.G.

[676] On Kummer's solution of the hypergeometric differential equation. Nederl. Akad. Wetensch. Proc. 39, 1056-1059, 1936.

VAN TUYL, A.H.

[677] The evaluation of some definite integrals involving Bessel functions which occur in hydrodynamics and elasticity. Math. Comp. 18, 412-432, 1964.

VARMA, R.C.

[678] On some integrals involving Jacobi polynomials. Proc. Nat. Acad. Sci. India Sect. A, 36, 465-468, 1966.

[679] Generalization of the results due to Wrinch for orthogonal polynomials. J. Indian Math. Soc. (N.S) 31, 141-148,1967-68.

VARMA, R.S.

[680] On a generalization of Laplace integral. Proc. Nat. Acad. Sci. India, 20, Sec. A, 209-216, 1951.

VARMA, V.K.

[681] On some infinite integrals invovling the E-function of MacRobert and operational images. Bull. Calcutta Math. Soc. 53, 185-192, 1961.

[682] On another representation of an H-function. Proc. Nat. Acad. Sci. India, Sect. A (2) 33, 275-278, 1963.

[683] On a multiple integral representation of a kernel of Fox. Proc. Cambridge Philos. Soc. 61, 469-474, 1965.

[684] On a new kernel and its relation with H-function of Fox. Proc. Nat. Acad. Sci. India Sect. A (2), 36, 389-394, 1966.

[685] On sums of finite series involving modified Bessel functions. J. Indian Math. Soc. (N.S) 30, 173-178, 1966.

VERMA, ARUN.

[686] General expansions involving E-functions. Math. Z. 83, 29-36, 1964.

[687] Series involving products of two E-functions. Proc. Glasgow Math. Assoc. 6, 172-176, 1964.

[688] Integration of bilateral hypergeometric series with respect to their parameters. J. London Math. Soc. 39, 673-684, 1964.

[689] Integration of E-functions with respect to their parameters. Bull. Math. Soc. Sci. Math. R.S. Roumanie 9 (57) 343-349, 1965.

[690] A class of expansions of G-functions and the Laplace transform. Math. Comp. 19, 664-666,1965.

[691] A note on the evaluation of certain integrals involving G-functions. Ganita 16, 51-54, 1965.

[692] Certain expansions involving Meijer's G-functions. Ganita, 17,118-126,1966.

[693] A note on an expansion of hypergeometric functions of two variables. Math. Comp. 20, 413-417, 1966.

[694] Expansions involving hypergeometric functions of two variables. Math. Comp. 20, 590-596, 1966.

[695] Certain expansions of the basic hypergeometric functions. Math. Comp. 20, 151-157, 1966.

[696] A note on expansions involving Meijer's G-functions. Math. Comp. 21,107-112, 1967.

[697] A note on the summation of the generalized hypergeometric functions. Math. Comp. 21, 232-236, 1967.

[698] On integration of Meijer's G-functions. Bull. Calcutta Math. Soc. 59, 67-72, 1967.

[699] Integration of E-function with respect to their parameters-II, Bull. Math. Soc. Sci. Math. R.S. Roumanie II (59), 205-217, 1967-68.

[700] A transformation connecting products of generalized basic hypergeometric functions. Canad. Math. Bull. II, 241-248, 1968.

VERMA, C.B.L.

[701] On chains of integral transform and E-function of MacRobert. Proc. Nat. Acad. Sci. India Sect. A (2) 33, 285-294, 1963.

[702] Some operational images of infinite series involving MacRobert's E-function. Proc. Nat. Acad. Sci. India Sect. A 34, 147-156, 1964.

[703] G-function and self reciprocity. Proc. Nat. Acad. Sci. India Sect. A. 34, 139-146, 1964.

[704] On H-function of Fox. Proc. Nat. Acad. Sci. India Sect. A(3), 36, 637-642, 1966.

VERMA, R.U.

[705] Integrals involving Meijer's G-functions. Ganita 16, 65-68, 1965.

[706] Certain integrals involving the G-function of two variables. Ganita 17, 43-50, 1966.

[707] On some integrals involving Meijer's G-function of two variables. Proc. Nat. Inst. Sci. India Part A, 32, 509-515, 1966.

[708] A new generalization of Laplace transform of two variables. Ganita 18, 8-12, 1967.

[709] On an integral transform of two variables. Nieuw. Arch. Wisk. (3)15, 64-68, 1967.

[710] On some infinite series of the G-function of two variables. Mat. Vesnik 4(19), 265-271, 1967.

[711] On a generalized transform of one variable. Ganita 19, 73-80, 1968.

[712] A generalized integral transform of two complex variables. Proc. Nat. Acad. Sci. India Sect. A 39, 265-267, 1969.

[713] Addition theorem on G-function of two variables. Math. Vesnik 7 (22), 165-168, 1970.

[714] Expansion formula for the G-function of two variables. An. Sti. Univ. Al. I. Cuza Iasi. n.Ser. Sect. la 16, 289-291, 1970.

[715] On the H-function of two variables -II. An. Sti.Univ. Al.I Cuza, Iasi, 17, 103-109, 1971.

VYAS, R.C. and MATHUR, S.N.

[716] Integrals involving products of modified Bessel functions of the second kind. Math. Nachr. 40, 225-227, 1969.

WATSON, G.N. (ALSO SEE WHITTAKER, E.T.)

[717] A Treatise on the Theory of Bessel Functions. Cambridge University Press, London, New York, 1945.

WHITTAKER, E.T. and WATSON, G.N.

[718] A Course of Modern Analysis. Cambridge University Press, London, New York, 1927.

WILKS, S.S.

[719] Certain generalizations in the analysis of variance. Biometrika 24, 471-494, 1932.

WIMP, J. (ALSO SEE FIELDS, J.L; LUKE, Y.L)

[720] Polynomial expansions of Bessel functions of some associated functions. Math. Comp. 16, 446-458, 1962.

[721] A class of integral transforms. Proc. Edinburgh Math. Soc. (2) 14, 33-40, 1964-65.

[722] On the zeros of a confluent hypergeometric function. Proc. Amer. Math. Soc. 16, 281-283, 1965.

[723] The asymptotic representation of a class of G-functions of large parameters. Math. Comp. 21, 639-646, 1967.

[724] Recursion formula for hypergeometric functions. Math. Comp. 22, 363-373, 1968.

[725] On a certain class of Laguerre series. Ist. Veneto Sci. Lett. Arti. Atti. Ci. Sci. Mat. Natur. 127, 141-148, 1968-69.

WIMP, JET and LUKE, Y.L.

[726] Expansion formulas for generalized hypergeometric functions. Rend. Circ. Mat. Palermo (2) 11, 351-366, 1962.

[727] Jacobi polynomial expansions of a generalized hypergeometric function over a semi-infinite ray. Math. Comp. 17, 395-404, 1963.

WINKLER, E.

[728] Über die hypergeometrische Differentialgleichung der ordnung mit zwei endlichen Singulären Punkten. Dissertation, München, 1931.

WRIGHT, E.M.

[729] The asymptotic expansion of the generalized hypergeometric function. J. London Math. Soc. 10, 286-293, 1935.

[730] The asymptotic expansion of integral functions defined by Taylor series. Phil. Trans. Royal. Soc. London (A) 238, 423-451, 1940.

YOUNG, A.T.

[731] Table errata: Tables of Integral Transforms. Math. Comp. 24, 239, 1970.

APPENDIX

ORDERS OF THE SPECIAL FUNCTIONS FOR LARGE AND SMALL VALUES OF THE ARGUMENT

In this section A,B and C stand for some numerical constants different in each case and ϵ is an arbitrary small positive number.

Functions	Order for small values	Order for large values
$\mathrm{bei}(x)$	x^2	$x^{1/2}\exp(x/2^{1/2})\sin(x/2^{1/2}-\pi/8)$
$\mathrm{ber}(x)$	1	$x^{-1/2}\exp(x/2^{1/2})\cos(x/2^{1/2}-\pi/8)$
$\mathrm{ber}_v^2\,(2(2x)^{1/2}) + \mathrm{bei}_v^2\,(2(2x)^{1/2})$	x^v	$x^{-1/2}\exp(4x^{1/2})$
$\mathrm{Ci}(x)$	$x^{-\epsilon}$	$x^{-1}\sin x$
$D_v(x)$	$A + Bx$	$x^v \exp(-x^2/4)$
$E(p;a_r : q;b_s : x)$	x^δ $\delta = \min R(a_i)$ $(i = 1,\ldots,p)$	1
${}_0F_3\,(-;\, a,b,c;\, -4x^4)$	1	$x^{3/2-a-b-c}\exp(4x)$
${}_1F_1\,(a;c;-x)$	1	x^{-a}
${}_1F_1(a;c;\,x)$	1	$x^{a-c}\exp(x)$
${}_1F_2\,(a;c,d;\,x)$	1	$A\,x^{1/4+(a-c-d)/2} + B\,x^{-a}$
${}_1F_2\,(a;b,c;\,-x^2/4)$	1	$A\,x^{a-b-c+\frac{1}{2}}\cos(x+B) + C\,x^{-2a}$

Functions	Order for small values	Order for large values
${}_2F_1\,(a,b:c:\ -x)$	1	$Ax^{-a} + Bx^{-b}$
${}_2F_2\,(a,b;\ c,d;\ -x)$	1	$Ax^{-a} + Bx^{-b}$
${}_2F_3\,(a,b:\ c,d,e;\ -x)$	1	$Ax^{-a} + Bx^{-b} + Cx^{(a+b-c-d-e+\frac{1}{2})/2}$
$G^{m,0}_{p,q}\left(x\middle\vert\begin{matrix}a_p\\ b_q\end{matrix}\right)$	$\lvert x\rvert^{\alpha}$ where, $p \leq q$; $\alpha = \min R(b_h)$; $h = 1,\ldots,m$	$\exp[(p-q)x^{(1/(q-p))}]\ x^{\beta}$; where, $\beta=(1/(q-p))[\ \frac{p-q+1}{2} + \sum_{i=1}^{q} b_i - \sum_{i=1}^{p} a_i]$ $\lvert\arg x\rvert < (q-p+\delta)\,\pi$, ($\delta = \frac{1}{2}$ if $q = p+1$; $\delta = 1$, if $q \geq p+2$.
$G^{m,n}_{p,q}\left(x\middle\vert\begin{matrix}a_p\\ b_q\end{matrix}\right)$	$\lvert x\rvert^{\alpha}$, $p \leq q$; $\alpha = \min R(b_h)$ $(h = 1,\ldots,m)$	$\lvert x\rvert^{\beta}$, where $p < q$ and $n \geq 1$; $\beta = \max R(a_j) - 1,\ j = 1,\ldots,n$; $m+n > \frac{1}{2}\,p + \frac{1}{2}\,q$ and $\lvert\arg x\rvert < (m+n-\frac{1}{2}\,p - \frac{1}{2}\,q)\pi$.
$H^{m,n}_{p,q}\left(x\middle\vert\begin{matrix}(a_p,A_p)\\ (b_q,B_q)\end{matrix}\right)$	$\lvert x\rvert^{\alpha}$ where, $\alpha = \min R(b_h/B_h)$ $h = 1,\ldots,m$ and $\delta = \sum_{1}^{q} (B_j) - \sum_{1}^{p} A_j \geq 0.$	$\lvert x\rvert^{\beta}$, where, $\beta = \max R\ (\frac{a_j-1}{A_j})$ $j = 1,\ldots,n$; $\delta > 0$ $\lvert\arg x\rvert < 1/2\ \pi\sigma$ and $\sigma = \sum_{1}^{n} A_i - \sum_{n+1}^{p} A_i + \sum_{1}^{m} B_j - \sum_{m+1}^{q} B_j > 0.$

Functions	Order for small values	Order for large values
$H_v(x) - Y_v(x)$	$Ax^v + Bx^{-v}$	x^{v-1}
$I_v(x)$	x^v	$x^{-1/2}(A\exp(x)+B\exp(-x))$
$I_v(x)-L_v(x)$	x^v	x^{v-1}
$I_v(2x)\ J_v(2x)$	x^{2v}	$x^{-1}\exp(2x)\cos(2x+A)$
$J_v(x)$	x^v	$x^{-1/2}\cos(x+A)$
$J_{m,n}(x)$	x^{m+n}	$x^{-1}\exp(x/2)\cos(\frac{\sqrt{3}}{2}x+A)$
$J_\mu^\lambda(x)$	1	$x^{-k(\mu+\frac{1}{2})}\exp\{\frac{(\lambda x)^k(\cos k\pi)}{\lambda k}\}$ where, $k = \frac{1}{(1+\lambda)}$
$K_v(x)$	$Ax^v + Bx^{-v}$	$x^{-1/2}\exp(-x)$
$\log x$	$x^{-\epsilon}$	x^{ϵ}
$L_v(x) - Y_v(x)$	$Ax^v + Bx^{-v}$	x^{v-1}
$M_{k,m}(x)$	$x^{m+\frac{1}{2}}$	$Ax^{-k}\exp(x/2)+Bx^k\exp(-x/2)$
$S_2(b_1,b_2,b_3,b_4:x)$	$\|x\|^\alpha$, $\alpha = 2\min R(b_j)+1$ $(j=1,2)$	$x^\eta\cos(4x^{1/2}+A)$ where $\eta = \frac{(2\sum_{i=1}^{4} b_i + 1)}{4}$
$S_4(b_1,b_2,b_3,b_4:x)$	$\|x\|^\alpha$ $\alpha = 2\min R(b_j) + 1$ $(j = 1,2)$	$x^\eta\exp(-4x^{1/2})$, where η is defined above
$S_n(b_1,b_2,b_3,b_4:x)$ $(n = 1,2,3,4)$	$\|x\|^\alpha$ $\alpha = 2\min R(b_j)+1$ $(j = 1,2,3,4)$	$x^\eta\exp\{-4x^{1/2}\exp\frac{(4-n)}{4}i\pi\}$ where η is defined above

Functions	Order of small values	Order of large values
$Si(x)$	1	$x^{-1} \cos x$
$W_{k,m}(x)$	$Ax^{m+\frac{1}{2}} + Bx^{m-\frac{1}{2}}$	$x^k \exp(-x/2)$
$Y_v(x)$	$Ax^v + Bx^{-v}$	$x^{-\frac{1}{2}} \sin(x + C)$

Functions	Order at 1	Order at -1
$P_v^{\mu}(x)$	$(1-x)^{-\frac{\mu}{2}}$ $\mu \neq 1,2,3,\ldots$	$(1+x)^{-\frac{\mu}{2}}$ for $R(\mu) > 0.$ $(1+x)^{\frac{\mu}{2}}$ for $R(\mu) < 0$.
$Q_v^{\mu}(x)$	$(1-x)^{-\frac{\mu}{2}}$ for $R(\mu) > 0$ $(1-x)^{\frac{\mu}{2}}$ for $R(\mu) < 0$	$(1+x)^{-\frac{\mu}{2}}$ for $R(\mu) > 0$ $(1+x)^{\frac{\mu}{2}}$ for $R(\mu) < 0$.

Functions	Order for large values
$P_v^{\mu}(x)$	$Ax^v + Bx^{-v-1}$
$Q_v^{\mu}(x)$	x^{-v-1}

INDEX OF SYMBOLS

Miscellaneous Notations.

AUTHOR INDEX

SUBJECT INDEX